AF327637

CISM COURSES AND LECTURES

INTERNATIONAL CENTRE FOR MECHANICAL SCIENCES

COURSES AND LECTURES - No. 359

PROTECTION OF THE ARCHITECTURAL HERITAGE AGAINST EARTHQUAKES

EDITED BY

V. PETRINI
POLYTECHNIC OF MILAN

M. SAVE
POLYTECHNICAL UNIVERSITY OF MONS

SpringerWienNewYork

Le spese di stampa di questo volume sono in parte coperte da
contributi del Consiglio Nazionale delle Ricerche.

This volume contains 189 illustrations

In order to make this volume available as economically and as
rapidly as possible the authors' typescripts have been
reproduced in their original forms. This method unfortunately
has its typographical limitations but it is hoped that they in no
way distract the reader.

ISBN 3-211-82805-2 Springer-Verlag Wien New York

PREFACE

This book devoted to the protection of our architectural heritage from earthquakes has its distant origin in a series of lectures held one week in December 1985 at the "European University Center for Cultural Heritage" in Ravello, Italy, followed by a second, similar series in early 1987, and then by various smaller seminars. These meetings, organized by archeologists and seismologists, were of wide scope, bringing together specialists representing many various fields: archeology and seismology, of course, but also history, sociology, urbanism, architecture, and engineering. From the discussions during the meetings, it appeared that the vulnerability of cultural artifacts to earthquakes had to be considered the problem of a "system", composed not only of the constructions involved (with their morphological, historical, and structural aspects), but also of the community using these buildings, or, more generally, these cultural artifacts (with its culture, methods of construction and maintenance, capacity of reaction to earthquakes, etc.). Examining the problem in this context, it proved possible to define some of the elements that may influence the vulnerability of the "system" and consequently a program of operation, namely:

- *the possibility of representing the constructions concerned by satisfactory structural models;*
- *the knowledge of the materials and techniques used in the constructions, and an evaluation of their quality;*
- *the knowledge of the steps in the construction process, and, more generally, of the history of the construction;*
- *the present ownership of the construction (State, private institution or persons);*
- *the necessity to consider the problem as that of a "system" was not known beforehand and did not command the reunion of that variety of specialists; on the contrary it came out of the discussion between these various specialists;*
- *the "feasible programme" was studied, but we only achieved the listing of the elements capable of influencing such a program;*
- *the nature of resources available for the operation;*
- *the present use of the construction.*

It could then be understood that the "local specificity" of the protection problem (encountered on numerous occasions in the course of the meetings) might result from differences in the elements of the "systems". For example, architectural artifacts can be divided into three groups, each of which present distinctive characteristics:

A. constructions of archeological value, where
- *modeling is simple;*
- *materials and construction techniques are easily identified and of high economic value;*
- *history of the construction steps is easily identified or retraced;*
- *ownership is public;*
- *management of the operation is public;*
- *resources available for the operation are sufficient, or nearly sufficient;*
- *present use is not the original (normal) function, tourist use having often to be taken into account;*

B. monuments (and groups of monuments), where
- *modeling is sometimes difficult, but almost always possible;*
- *materials and construction techniques can almost always be identified, and their value exceeds the average at the time of construction;*
- *construction steps are rather well known, or retraceable;*
- *ownership is usually public, or semi-public;*
- *management of the operation is public, or under direct public control;*
- *financial resources for the operation are hardly ever sufficient;*
- *the constructions are nearly always still in use, sometimes with a function different from their original one;*

C. current constructions of general architectural interest, more often with the following features:
- *modeling is possible only in particular cases;*
- *materials and construction techniques cannot always be identified, and are not always adequate;*
- *history of the construction process is nearly always unknown, or difficult to rediscover;*
- *ownership is usually private;*
- *management of the operation is usually private, with very little, and always indirect, public control*
- *resources for the operation are sufficient only in particular cases, or when*

the value of the construction is enhanced (for example, in historic centres, or tourist sites);
- use is nearly always as originally conceived.

The conclusion reached from the analysis summarized here was that archeological remains should be treated with special emphasis on the reversibility and "legibility" of the operation, while categories B and C could be considered together. In all instances, two salient aspects are the impact of "local" influences and the importance of a multidisciplinary approach.

This very general study of the conservation and rehabilitation of architectural artifacts in seismic regions enabled participants to define various lines of long-term research. But some of us, both engineers and architects, were already convinced in 1987 that efforts should also be made, within the broad framework defined, to provide some immediate help to those persons in charge of protecting our architectural heritage in their everyday operations. This conviction led to a two-week course in 1989 at the Faculties of Architecture of Florence and Athens, repeated at these institutions and also at the Faculty Polytechnique of Mons, Belgium in 1991. The experience gained and the response from the numerous participants were such that the group of lecturers decided, at a special meeting in Tremezzo in May 1992, to formalize the course by publishing a text.

This volume is the first attempt to produce such a book. As indicated by the title, the contributions focus on architectural artifacts, more than on archeological remains. They have been conceived and written for both architects and engineers, since both categories are likely to take part in the protection and restoration of architectural artifacts in general and in seisimic regions specifically. Architects, as a rule, are not over familiar with either structural dynamics (or solid mechanics in general) or seismology. Their knowledge (and practice) of the protection and rehabilitation of old buildings may also vary greatly from one person to another, because these problems are not regularly taught in all schools of architecture. For their part, engineers are, as a rule, completely unfamiliar with the protection and rehabilitation of architectural artifacts, and tend to confine their expertise to structures built in conventional modern materials (metals and reinforced and pre-stressed concrete), within the framework of the classical assumptions of homogeneity, isotropy, and linear elasticity, assumptions all three that are, obviously, not satisfied in the majority of old buildings.

These considerations inform the nature of the contributions, which begin with general indications on how renovation and rehabilitation are viewed today, idependently of the origin of the destruction (earthquake or other). Then the core of the text is presented, first examining:

1) structural dynamics in relation to historic buildings;

2) the knowledge that can be extracted from the study of historic buildings subjected to earthquake, in the specific case of Greece;

3) an example of the possible modeling of stone buildings; and

4) tentative guidelines for operations, in the light of the preceding chapters.

Some recent experimental results on the mechanical behaviour of traditional building materials are then given, before a critical discussion, in the second part, of examples of operations in both Greece and Italy.

The conclusion restates and summarizes the main features, and suggests some orientations for further progress.

As can be deduced, this book is somewhat different from the texts usually produced in an International Centre for Mechanical Sciences. Alhough we firmly believe that, despite all other aspects of the problem, the response and resistance of a given construction at a given time to a given earthquake is nothing but a problem of solid mechanics, we must always remember that this problem occurs, as is shown here, in conditions of uncertainty and inaccuracy, while the process of protection or rehabilitation is itself subject to strong architectural and economic constraints.

The understanding and, insofar as possible, definition of these constraints, uncertainties, and inaccuracies are of utmost importance in achieving good solutions.

V. Petrini
M. Save

CONTENTS

PROTECTION AND REHABILITATION OF THE ARCHITECTURAL HERITAGE

H. Wilquin

Polytechnical University of Mons, Mons, Belgium

ABSTRACT:

How are we to define a method for approaching restoration and rehabilitation of buildings? Reexamining the way in which until now we have considered the stratification of history and the way of dealing with old buildings in the light of practices since the Antiquity to the 19th century is too relativistic an approach. A review of the various theories since the writings of Aloïs RIEGL at the turn of the century and up to the articles of the Charter of Venice set down by The International Council of Monuments and Sites (ICOMOS) in 1964 sheds light on thought and practice relative to both restoration and rehabilitation. The latter distinction hinges on the value of the building which is the object of our consideration although, of course, a sharp dividing line cannot be drawn between the two. Restoration is best applied to intentional monuments (temples, churches, castles, etc.) for which the greatest respect must be maintained whereas rehabilitation is directed more to non intentional monument; moreover, rehabilitation tolerates greater freedom and scope owing to its lesser importance as an art and historical document.

1. STRATIFICATION OF HISTORY

In his "Journal of travels in Italy", Montaigne writes "*Sur les brisures mêmes des vieux bâtiments, comme la fortune les a logés, en se dissipant, ils ont planté le pied de leurs palais nouveaux, comme sur de gros lopins de rochers, fermes et assurés*" [1].

Indeed, awareness of a stratification of History, the re-using of materials and the conversion of existing buildings have profoundly influenced the evolution of architecture from the 15th century onwards. This process is more the outcome of a state of affairs than of theoretical reflection. Thus, in the middle ages, the abandoning of ancient monuments and the re-using of materials is characteristic of urban life; the Diocletian palace becomes the town of Split, the arenas of Nîmes and Arles are turned into fortified villages, the theatre of Marcellus in Rome is turned into a residence of flats, Hagia Sophia in Istanbul a mosque; on the other hand, a church is squeezed In the great mosque of Cordoba, an entire village in Turkey is built with the stones of ancient cities. Moreover, some buildings have multiple lives such as the mausoleum of Augustus in Rome which was turned into a fortress, then a garden, the arenas and finally a concert hall.

By the Renaissance, other rehabilitations appear, e.g. Roman temples converted into churches by Alberti, Diocletian spas are turned into churches by Michelangelo, the palazzo della Ragione is recast by Palladio who reinforces and increases the scale of the building by new peripheral construction.

The end of the 19th century sees the appearance, in France, of the pseudo-purism of E. Viollet-Le-Duc who, at that time, in his "Dictionnaire raisonné de l'architecture"[2] defines restoration as follows: "*To restore a building is not merely to maintain, repair or renew it, it is to give it back a completely new shape which may never have existed at any time*". We should note the paradoxical opposition between giving back and having never existed. Of course, E.Viollet-Le-Duc must be seen in the context of his time where the limited knowledge and the lack of technical means for investigation and archaeological surveying are obvious. We are then in the heart of this seething period where romantic atmosphere of the century of Chateaubriand and Hugo envelops everything under its black wing (this was the time of the bogus ruins and of the Burchgraves, etc.). Other parallel attitudes such as those of Ruskin in Great Britain should be considered in this connection.

Viollet-Le-Duc converts in a trivial way, adding a roof here, an arrow there, modifying Notre Dame de Paris, Vézelay, Saint-Guilhem-Le-Désert or ramparts of Carcassonne.

He rewrites History, failing to distinguish his personal contributions from the original part handed down by the vicissitudes of time. A sublime and inspired artist he may have been, but as an archaeologist how he clouded the issue!

Until then, this supposed unity of style was never in season. As F.Choay points out: "*From antiquity to the 15th century, only a few broad, not very significant, conceptual types can be uncovered as well as a few protective measures, but these are exceptional, occasional, not systematic and mainly confined to Rome.*" [3]

Indeed, as Viollet-Le-Duc points out, before going on to reject this idea: "*If in a 12th century building, a broken arch had to be replaced, a capital of the 13th,*

14th or 15th centuries was laid in its place; if a piece was missing for a long frieze of crockets of the 13th century, adornment in the taste of the time was inlaid" [2].

2. THE CHARTER OF VENICE (1964)

At this stage of our survey, three main periods must be singled out : the writings of Aloïs Riegl in 1903 In Vienna, a conference of Athens for archaeologists in 1931 and the Charter of Venice of the ICOMOS in 1964.

In 1903, in Austria, Alois Riegl speaks of the modern attitude towards our heritage of monuments. He distinguishes three values the Altwert (oldness: building is old), the Denkmalswert (the value of recollection something happened in this building), the Kunsthistorischeswert (the historical and artistic value of the building: this building contains remarkable and emblematic elements for art And for History).

Riegl emphasizes the fact that the average man which he calls "modern", is not interested in scholarly information which say be decoded in the details of an ornament, in the peculiar quality of a structure or in the arrangement of colonnade but he points out, indeed unfortunately, that the population consider oldness as the all-purpose value which holds all historical values together.[4]

The Conference of Athens on restoration for archaeologists in 1931 recommends the use of modern materials. A recommendation is expressly made for a marked difference between new or adjusted elements, for an absence of decoration in the new parts and for simplicity of geometry and technology. This approach ratifies, to a large extent, the criteria worked out at this time, by the architects of the modernist school. The sensitivity of the time is characterized by the contrast between old and new. Thus, Köller's treatises on psychology (1929) as well as Koffka(1935) show a systematic organisation of conceptual principles in which the notions of ground-form and contrast are fundamental. In the explanation of perception and meaning, he says *"No sensible object exists except in relation to a certain background "* [5].

At the same period of time, the teachings of Kandinski, Alberg and Moholy-Nagy and even Paul Klee at the beginning of the Bauhaus use the same psychological categories in the training of draughtsmen and designers; they teach the phenomenon of meaning in every field of visual art is produced by means of juxtapositions, interrelations, contrasts of scale, texture or heterogeneous materials.

Following the writings of A. Riegl and the recommendations of the Conference of Athens In 1931, the present attitude to restoration stems from the Charter of Venice (1964), issued by the International Council of Monuments and Sites, a charter which was confirmed by the Declaration of Amsterdam in 1975.

The Charter of Venice stipulates:
- in its article 9
" The process of restoration is a highly specialized operation. Its aim is to preserve and reveal the aesthetic and historical value of the monument and is

based on respect for original material and authentic documents. It must stop at the point where conjecture begins, and in this case moreover any extra work which is indispensable must be distinct from the architectural composition and must bear a contemporary stamp. The restoration in any case must be preceeded and followed by an archeological and historical study of the monument."
- in its article 11
" The valid contributions of all periods to the building of a monument must be respected, since unity of style is not the aim of a restoration. When a building includes the superimposed work of different periods, the revealing of the underlying state can only be justified in exceptional circumstances and when is removed is of little interest and the material which is brought to light is of great historical, archeological, or aesthetic value, and its state of preservation good enough to justify the action. Evaluation of the importance of the elements involved and the decision as to what may be destroyed cannot rest solely on the individual in charge of the work"
- in its article 12
" Replacements of missing parts must integrate harmouniously with the whole, but at the same time must be distinguishable from the original so that restoration does not falsify the artistic or historical evidence"
     This is a stumbling block to be avoided, for the teaching of archaeological museography, which, although it presents well identified parts, unfortunately leaves them dislocated in their space.
- in its article 13
" Additions cannot allowed except in so far as they do not detract from the interesting parts of the building, its historical setting, the balance of its composition and its relation with its surroundings"
- in its article 15
" Ruins must be maintained and measures necessary for the permanent conservation and protection of architectural features and of objects discovered must be undertaken. Furthermore, every means must be taken to facilitate the understanding of the monument and to reveal it without ever distorting its meaning. All reconstruction work should however be ruled out a priori. Only anastylosis, that is to say, the reassembling of existing but dismembered parts can be permitted. The material used for integration should always be recognizable and its use should be the least that will ensure the conservation of a monument and the reinstatement of its form"
- in its article 16
" In all works of preservation and restoration there should always be precise documentation in the form of analytical and critical reports, illustrated with drawings and photographs. every stage of the work is clearing, consolidation, rearrangement, and integration as well as technical and formal features identified during the course of the work should be included. This record should be placed in the archives of a public institution and made available to research workers. It is recommended that the report should be published"

But what about the Charter of Venice today? [6] This text was inspired by European scientists and architects and specially dedicated to European way of

thinking and doing. But is the Charter always relevant, for example, for the Japanese way of protection of the architectural heritage? (to destroy the former temple and to rebuild it new like it was before with the same ways of construction and doing that every twenty or thirty years. For them, the most important thing is to preserve and to transmit up the knowledge of building.)
Is the Charter applicable to earth constructions in some aeras in Africa?
What do we think about the way the Polishmen rebuild the town of Warsaw after the secund world war, not always following scientific documents as it is recommended in articles 11, 12, 13 and 15 (but the Polish people had in mind its chief city and wanted to get it back).[7]

3. THE DISTINCTION BETWEEN MONUMENT AND ORDINARY BUILDING

A distinction needs to be drawn, though not of course a hard-and-fast one, between monuments and their restoration and ordinary buildings and their rehabilitation (although a monument, if it loses its original function, must not only be restored but must undergo a subtle rehabilitation if it is to provide a structure suitable to a new function on which the building's survival depends).

Intentional monuments, according to A. Riegl's terminology (churches, temples, ...), have different degrees of Kunsthistorischeswert. The most scientifically sophisticated means (hence also the most costly) are then widely used. In cases where the most insistent doubt reigns, the enlightened opinion of the greatest number of scientists will help to resolve the most delicate problems. Moreover - and this goes for all buildings, irrespective of their importance - not only must diagnosis and protocol regarding the work to be undertaken be as scientific as possible, but, for intentional monuments and non-intentional buildings, it is advisable to enlighten analysis and action with the opinion of manual practitioners, from craftsmen to mere builders and carpenters. The work undertaken will endeavour to be discreet and efficient, out of respect for the important and pregnant testimony to our civilisation which these monuments are. Ordinary buildings, on the other hand, possess a certain Alteswert but sometimes no real Kunsthistorischeswert. Their importance usually comes from their Denkmalswert since they bear witness to some past activity, to a decisive event in the collective memory, or they represent an outstanding type (archetype) within a style of building characteristic of a certain context. Often the available financial means are much less substantial, coming as they to from private and local sources. This doesn't mean, however, that the scientific procedure used will be a careless one, but obviously it will have to be cruder. Greater scope and freedom may be taken in view of the lesser importance of such an artistic and historical document.

4. EXAMPLES

Let us speak about two intentional monuments, a non-intentional building and a unique work about the re-edification of villages.

The superb example of Castelvecchio of Verona dealt with by Carlo Scarpa shows in a pertinent and quite superlative manner that the balance between this didactic and academic attitude and the expression of an open and vibrant sensitivity is the mark of a unique talent. C.Scarpa here resorts to a subtle neoplastic language, which is not immediately discernable; these historicist figures, reinterpreted analogically in the historical authenticity of construction, thus transcends the rather closed and introverted building as it exists by a dialectical game with spatial continuity on the inside by means of visual axis, openings and complex overall geometries. He accumulates analogical images of architectural works from the early Middle Ages to various periods of the most recent European experiences (from Vienna to Glasgow). The analogical approach here is not based on a visible synchronism of interdependent forms, but on the association made by the observer through time. Relationships are established between the real and the imaginary historical building on the one hand, and the elements of new design, inspired by the connotative capacity of the past languages being used, which indeed serve to create a dependent cohesion in the cacophony of different historical parts.

Another opposed example can be found in the way following during the restoration of the collegiate church of St Gertrude at Nivelles (Belgium).

This Romanesque abbey church dates from the 11th and the 12th centuries. Through the ages, it has undergone sizeable changes to both the exterior (in the 17th century) and the interior (in the 18th century). The project of restoring it to its supposed original Romanesque state came into being in the middle of the 19th century, saw a start of the work at the beginning of the 20th century (work on the eastern choir), was envisaged in its entirely when the monument was classified in 1931 and put into operation in two phases, following the damages caused by the bombardments of 1940. From 1948 to 1959, work was undertaken on the naves, transept, and eastern choir above the crypt; in september 1971, work began on the forepart which is now completely achieved.[8]

The real problem began when it was decided to rebuild, from scratch, the apse at the western end destroyed several centuries ago. Moreover, once this already questionable conjectural restoration had carried out, the architects grew anxious for the tower to be brought into harmony with the new feature it would be overlooking. The tower itself had had so much reinforcements and so many additions of every sort that it was extremely hazardous to pick any original homogeneous "core" which could be the basis of a "reinstatement" specially as opinion as the latter was not unanimous. Was the tower to be square imagined it have been following the historic and scientific knowledges about the Romanesque period, or polygonal as desired by S.Brigode, one of the three architects in charge of the restoration?

The "original" church was, before destruction by bombing in 1940, with a gothic tower but without any apsis which had disappeared during centuries.

Worse still, yet further ideas were by now being championed: removal of the existing bell-tower and building a saddle-back roof over the Romanesque body; retention of the tower with a new gothic spire on top; or building of a modern tower and spire on a design selected by international competition.[9]

Finally, after twenty years of wavering, on the lighting of the specialized barristers (international well-known professors and researchers), the population of the town was asked to choice by vote between the various suggestions- with the exception of the last (the contemporean version) alleging the lack of money to organize an international competition of architecture!

So, may be the most incompetent part of the inhabitants had to settle by giving a casting vote because the specialists, putting themselves in a impregnable position, didn't make up their mind.

A little of 58% opted for the Romanesque polygonal tower (the most technical and historical unlikely). For S.Brigode, according to Viollet-Le-Duc, the matter was not one of preserving an intricate combination of styles belonging to successive periods and embodied in features of most unequal values, but reconstructing what must certainly have been in the mind of any architect working in the days of Romanesque [9] but if to make that, a large quantity of concrete is necessary like they used at Nivelles , we can ask, according to L.H. De Koninck [10], "*if the solution doesn't appear to be an architectural imitation, specially as far as the new parts of the suggestion are concerned because historically the archeological conformity is questionable due to lack of conclusive documents. (The Romanesque normally avoid interpenetration of roofing and this penetration of a tower is not satisfactory because it destroys any structural value, which in turn affects the whole architecture)*".

The third example is an application of the idea of the rehabilitation on a non-intentional building, a little café, in the centre of the old town of Mons ,(Belgium). The original building, a classic one of the secund half of the 18th century made of stones and bricks, had been partially destroyed for the ground-floor and the street façade. This destruction, the first one probably during the end of the 19th century, had let an iron beam and the marks that some parts had completely changed of designs and uses (for example, one window became the secundary door of the right side). After a lot of researches, it was obvious that no old sketches, no old writings, no complete traces on the building itself (which can help us to find out the" original" façade) didn't exist any more. So, according to the new shape of that café (along a longitudinal axis) and following the Charter of Venice, I decided to design a contemporean façade for the missing and completely unknown parts. Four arches were discovered beyond the wood sign-board, so they were kept and restored (the basement and the thresholds of the façade had beencompletely destroyed).

The main architectural idea was also to get a wide glass opening which could completely open during summertime. So, according to the structural behaviour, the new load bearing system, in place of the iron beam, is made of two new columns. To distinguish them clearly and also to "write" them in a contemporean way (art. 9 of the Charter of Venice), they are made of painted steel. to play with the perspective effects, the wide glass-window placed one meter beyond the undercarriage was specially designed with U profiles and steel tubes to avoid large dimensions (the U profiles have the great advantage to get shadows which grow thin the dimensions of the window undercarriage). The basements of the new columns are thiner also to enlight the connection with the new stone threshold, made of stone but in a today plate engraving (art. 12 of the Charter of Venice).

The two columns are pierced through three holes to find a local transparency and to show out the thickness and the density.

If we look now to the original part of the façade, we can see different levels of architectural composition. the main line is made of the expression of the load bearing system in a stone grid (of course decorated in the mood of the 18th century).A secundary line is compouned by the windows and the scrolls, a third one is designed by the upper fixed parts of the windows.

Following that way of composing, the perforated sheet-irons are placed between the upper parts of the columns and also used for the secundary door (it is a way to find together a transparency and a border, a virtual wall). To find a new third level of composing and also an human scale, two hanging circular tubes (also supporting the new sign-board) are used like "eyelids" of the windows holes. Finally, the lightning system for the façade is established by the two metallic candlesticks soldered to the columns. The choice of the new metallic façade has been taken of the idea of an harmony with light differences.

The wonderful example of the re-edification of villages in Friuli (Italy) after two earthquakes in 1976 (17th may and 15th september) will end cleverly that list.

The main luck in the unhappiness of the Friuli was to dispose of, a.o., a "Centro Regionale per la Catalogazione del Patrimonio Culturale e Ambientale del Friuli-Venezia Giulia" since 1973 [11], [12], [13], [14] and to have enough time between the two earthquakes to protect, to photography, to measure, to sketch (a.o. by photogrammetry,..),...

So, they largely used the anastylosis (art. 15 of the Charter of Venice) to rebuild the destroyed villages. according to the Charter of Venice, the missing and new parts were replaced by similar pieces of materials but such in a way they are distinguishable (shapes, cuttings,...). Of course, they don't forget to stabilize each building with a asismic structure; a.o., the unique re-edification of the church of Venzone keeps also the parts of the walls already on the way to collapse but still standing like a "frozen" mecanism during an earthquake, reinforces the new structure with concrete walls and uses largely anastylosis like it is described supra.

5. THE RELATIONSHIP BETWEEN FORM AND FUNCTION

It is essential to breath new life into an old building if it is not to wither and die out. The usual architectural procedure on a new piece of land is that function and method of construction together create the form, yet are themselves partly determined in the process. But what are we do to with the form when the function for which it was designed has vanished ? If we use the allegory of the egg whose shell is explained by the functions of conception and incubation, what must be done when the shell is empty ? And more still, what must be done when part of this: shell has disappeared with having left any trace, and weakening what remains of the protective surrounding ?

Rehabilitation is successful only if there is a judicious appropriateness between a new function and the existing form.

It is always advisable to wonder what function an existing building might accommodate, and to do so in view of this intrinsic nature. As Claude Soucy says:

"The meeting of an old shell with new needs and means will give birth unique object which is not a mere juxtaposition, but at once a constructive and an architectural synthesis."

6. THE CONTEXT

For a given building, and since a feature belongs to, influences and is influenced by its physical and metaphysical environment, it is important at the outset to go from the general to the particular, and especially to determine the characteristics (fragmented, volumetric, typologies, rhythms, sequences, structures and building systems,...) of surrounding buildings, their evolution through History as well as that of the symbols and signs related to them.

The socioeconomic context will In no way be neglected, for often it has a considerable, if not a decisive, bearing on the functional and technical choices.

For the building itself, as E. Viollet-Le-Duc (among others) wrote, and quite rightly:"*It is thus essential, before any restoration work, to ascertain the exactage and character of each part, to compile a kind of report, based on the most certain documents, in the form either of written notes or sketches*" [2]

The procedure will therefore consist in analysing minutely that which already exists, referring either to the building itself (attempting to discover the original code or an overlaid code, whilst recording the non-system of deviations In relation of this code), or to descriptive writings, drawings and photographs of earlier states of the building.

7. METAPHOR AND METONYMY

To simplify a bit crudely, two types of approach are possible

- the metaphoric approach transferring meaning by analogical substitution

- the metonymic approach : designating a concept by another concept to which the first is related.

Let us rather listen to U.Eco:

"When aesthetic holds that it is possible to wake out the totality of the work even if it is mutilated, ruined and disintegrated by time, this is only because, from

the which stands out on the still perceptible layers, one can deduce the code which shaped the missing parts and guess them.

In fact, the art of restoration is based this possibility of deducing from existing parts a message for those to be reconstructed. This operation ought to be impossible since one is reconstructing parts which the artist invented by going beyond the systems of norms and expectations that were valid in his time. But this restorer ... is precisely the person who rediscovers the laws which govern a work, its idiolect, the structural diagram which governs all its parts." [15]

Today's user of past forms learns to distort them so as to read messages which no longer belong to them with the help of his own or erroneous criteria, but also to discover the right criteria. His cultural knowledge serves to motivate him to recover philosophical codes while his flexibility in recovering them often provides him with relevant insights.

"If, in the past, the normal growth and "obsolescence" of communication systems proceeded according to his sinusoidal curve (which is why Dante was definitively inaccessible to the rationalist reader of the 18th century), our age proceeds along a continual spiral so that every discovery is an increase" [15].

8. INTERVENTIONS

"From the beginning of recorded time, our age is unique in the attitude it has taken to the past. In Asia (...) when a temple or a palace underwent the degradations of time, another one was built beside it. The Romans reconstructed but did not restore, and the proof of this is that Latin has no word corresponding to our word restoration according to the meaning given to it today. Instaurare, reficiere, renovare do not mean restore but rebuild" [2].

When the work of analysis has been done, conclusions must be drawn and the methods of restoration or rehabilitation defined. Jean-Pierre Paquet [16], quoted by Georges Duval [17], expressed them in three categories:
- healing;
- substitution;
- recentring of loads.

The first method, so-called healing, is in itself a definition. It requires, as for a piece of fabric, replacing a material no longer able to fulfil its function by another sound one, which, as far as it is possible, should be identical to it, and, - in this following the Charter of Venice, - discernable (for example, replacing the damaged stones by sound stones, but of a slightly different hew and shaping; or consolidation of the framework and wooden floors with the help of resins, the latter merging harmonious but always remaining perceptible to specialists, ...).

The second method of restoration, so-called substitution, is that where by a weakening part is replaced by another part, this time of a different nature (for example, the framework of the cathedral of Reims, in France, repaired by Deneux after the first World War by prefabricated parts which can be dismantled).

The third method, recentring of loads, is used when the damage to be stopped stems from alterations in the building equilibrium.

The Conference of Athens for archaeologists in 1931 recommended using modern materials, and expressly recommended that there be a pronounced difference for new or altered parts, and a lack of decoration. This predominance of contrast is clearly less widely used nowadays although it is sometimes found, other things being equal, in more conceptual operations.

The fundamental aesthetic principle now used, and the most wide spread one, is the analogical procedure, which consists of repetitions and differences in respect of the former basis on which it is founded.

"As an aesthetic operation, the intervention is the imaginative arbitrary and free proposal by which one seeks not only recognise the significant structures of the existing historical material but also to use them as analogical marks of the new construction.

Comparison as difference and similarly, from within the only possible system, that particular system defined by the existing object, is the foundation of any analogy. On this analogy is constructed every possible and unpredictable meaning" [18].

9. IN FINE

The procedure of restoration or rehabilitation of an old building will start by revealing the structure of this signifier, will then reveal the structure of the signified, of the diachronic of these parameters, before reinforcing certain elements recognised as important (historical, artistic, of collective memory ...), and then orchestrating a dialectical interplay of new analogous elements based on what exists, on the traces of what once existed, or even on plausible fiction (what might have happened).

"It is no longer a question of looking for a prefabricated symbolism, archetypes which no longer have any meaning and which relate to a memory which has changed. What we need to look for is a multiplicity of languages, which, however, is not eclecticism. A multiplicity of languages means accepting something which is not in itself settled, but open, and which accepts non-corresponding correspondences" [19].

A cross-bred contextualisation would be the most appropriate procedure. By contextualisation, we mean assessment of the context and of its wealth of possibilities, the positioning of elements within it, whether in fusion, in contrast, or in crystallization (for example, in a mediocre and inconsistent context, a strong element may become the central organizing referent to which the rest is gauged, measured and positioned).

By cross breeding, we mean combinations consisting of unions but also of differences, of sometimes violent clashes and diversified poles in the image of modern societies. Old and new then interact to create the place through their interplay.

As Mallarmé writes: "What takes place is the place".

9. REFERENCES

[1]   Montaigne: Le voyage en Italie.
[2]   Viollet-Le-Duc, E.: Dictionnaire raisonné de l'architecture française du XIème au XVIème siècle - T.VIII, (1854-1868)
[3]   Choay, F.: Le sens de la ville, sémiologie et urbanisme, Le Seuil Ed., Paris.
[4]   Riegl, A.: Le culte moderne des monuments, son essence et sa genèse, Seuil Ed., Paris (1904).
[5]   Guillaume, P.: Psychologie de la forme, Champs Flammarion Ed., Paris (1979).
[6]   ICOMOS 1990:Actes du colloque, Lausanne 6-11 octobre 1990, The Swiss Committee ICOMOS Ed. (1990).
[7]   Ciborowski, A.:Warsaw, a city destroyed and rebuild, Interpress Publishers, Warsaw (1969).
[8]   Ladrière, G./Donnay-Rocmans, C.: Les restaurations de la collégiale Sainte Gertrude à Nivelles, Monumentum, vol.XX, XXI, XXII,ICOMOS Ed. (1982), 97-116.
[9]   Martiny, V.G.: Vox Populi, Vox Dei-A propos de la restauration de l'avant-corps de la collégiale Sainte Gertrude de Nivelles, Monumentum, vol. XX, XXI, XXII, ICOMOS Ed. (1982), 117-123.
[10]   Coll.: L.H. De Koninck, Archives d'Architecture Moderne Ed., Bruxelles (1980),357-361.
[11]   Doglioni, F.-Moretti,A.-Petrini,V.:Le chiese e il terremoto, Edizioni Lint, Trieste (1994).
[12]   Bellina, A.:L'anastilosi nella ricostruzione del Friuli, Bollettino dell'associazione "Amici di Venzone", anno XV (1986).
[13]   De Lucca, S.:Fotogrammetria e recupero nei centri storici terremotati del Friuli, Gemona,Venzone,Artegna, Bollettino dell'associazone "Amici di Venzone", Anno XVI-XVII (1987-1988).
[14]   Coll.:Per il restauro della chiesa dei ss. Giacomo e Anna, Bollettino dell'associazione "Amici di Venzone",Anno X (1981).
[15]   Eco, U.: La structure absente, Mercure de France Ed., Paris (1984).
[16]   Paquet, J-P.: Structures des monuments anciens et leur consolidation, Annales de l'Institut technique du bâtiment et des travaux publics, 122, quatrième année (1958) et les monuments historiques de la France (1955), vol. II, 81-87.
[17]   Duval, F.: Restauration et réutilisation des monuments anciens, techniques contemporaines, Mardaga* Ed., Liège (1990), 16.
[18]   De Sola Morales, I.: From contrast to analogy, Interpretation of the past, Lotus international 46, Electa Ed., Milano (1985).
[19]   De Carlo, G.C.: Débat sur la place, La piazza e la citta. 50, rue de Varenne, Mondadori Ed., Paris (1985).

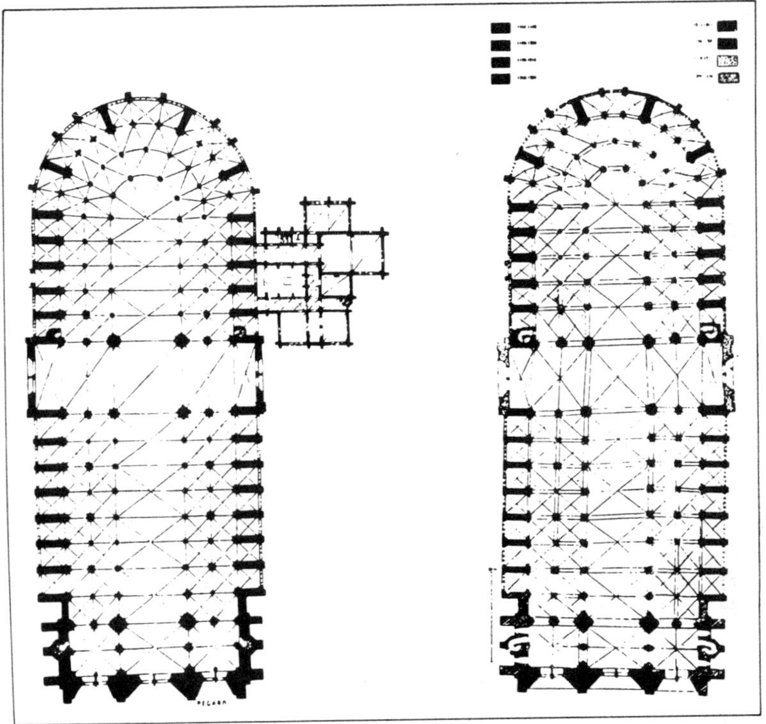

Fig. 1, comparison between a sketch of Notre-Dame of Paris by E.Viollet-Le-Duc and the true one.

Fig. 2, Castelvecchio of Verona- Arch. C.Scarpa.

Fig.3, Sainte Gertrude of Nivelles, different proposals for the front tower.

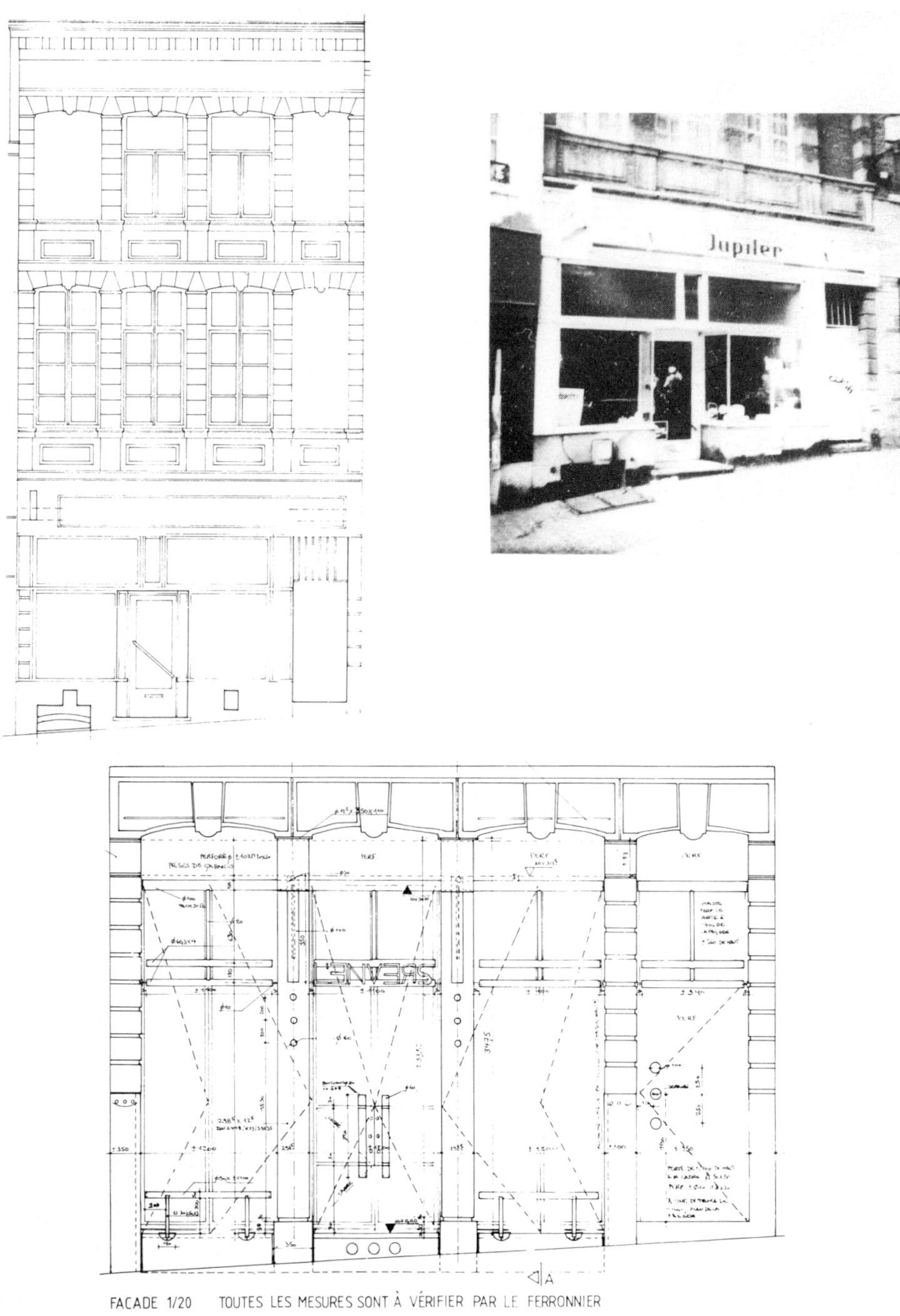

Fig.4,5 and 6, Café "L'Envers" at Mons-Arch. H.Wiquin-Façade as it was before and two proposals.

Fig. 7,8,9 and 10, Café "L'Envers", photographs and details.

Fig. 11 and 12, the church of Venzone-Arch. F.Doglioni.

INTRODUCTION TO STRUCTURAL DYNAMICS

C.A. Syrmakezis
National Technical University of Athens, Athens, Greece

ABSTRACT:

This document deals with modal analysis under dynamic loading. It is assumed that all masses are concentrated in mass particles and a lumped mass system is used for the simulation of the structure. The solutions are given using the modal analysis method. The free vibration response of a system, the dynamic loadings and the forced vibration of one degree of freedom systems (one mass particle systems), are examined. The same points for multi-degree of freedom systems (multi-mass particle systems).

1. INTRODUCTION

Dynamic actions, applied several times on a historical structure during its lifetime, are of several types: explosions, wind loads, traffic vibrations etc.

The most important dynamic action is of course earthquake. Special attention has always to be paid on the earthquake action, as for the case of a historical building, and especially for the common case of a masonry historical building built in an area of high seismicity, the problems to be answered by the engineer become more complex.

For the analysis of a structure under dynamic loading, actions applied on a "static" way is not enough: inertial forces of the masses, due to the fast variation of the dynamic actions, has necessarily to be taken into account. This physical situation leads to more complex mathematical tools for the solution of the problem.

This text includes introductory notions, equations and solutions for the structural dynamics problem, and is addressed to persons contacting the problem for the first time. Hopefully, it can help them for further discussion on technical features of earthquake engineering.

Due to space limitations, the text refers to the general theoretical aspects of the problem only. The transfer of the theoretical results on the study of the real historical structures response are going to be discussed elsewhere.

2. FREE VIBRATIONS

2.1 General Definitions:

Some general basic principles related with the dynamic response of mass particle systems are the following:

2.1.1 Simulation of structures

A mass particle is defined as dimensionless material point of mass m.

A given moving mass m, can be considered as a mass particle of mass m if all the points of the mass are moving in one direction, with common displacement.

A structure with one mass moving in one direction (the direction of its total loading), can be considered as an one degree of freedom system.

A structure with a system of several masses, moving each in one direction, can be simulated as a multi-degree of freedom system.

2.1.2 Principle of conservation of momentum for a mass particle

According to the principle of the conservation of momentum, the variation of momentum J of a particle of mass m, equals, at every moment of time t, to the total force P(t) acting on the mass m (fig.1).

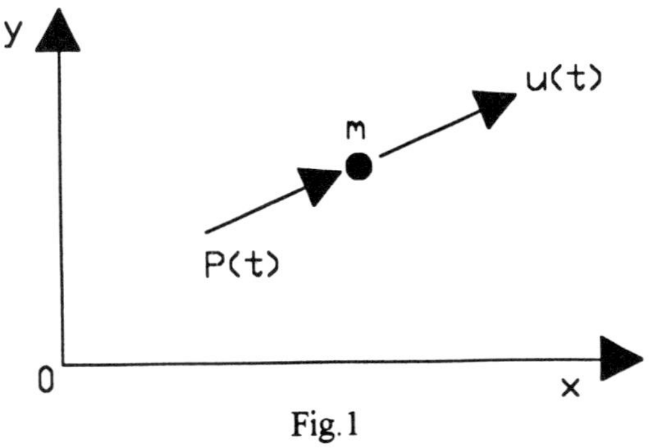

Fig.1

$$P(t) = \frac{dJ}{dt} = \frac{d}{dt}\left[m\frac{du(t)}{dt}\right] = \frac{d}{dt}[m\dot{u}(t)]$$

u(t)=displacement of the particle,

$\dot{u}(t)$ =velocity of the particle.

If mass m remains constant with time, then:

$$P(t) = m\frac{d\dot{u}(t)}{dt} = m\ddot{u}(t)$$

2.1.3 Principle of conservation of momentum for a system of particles (fig.2)

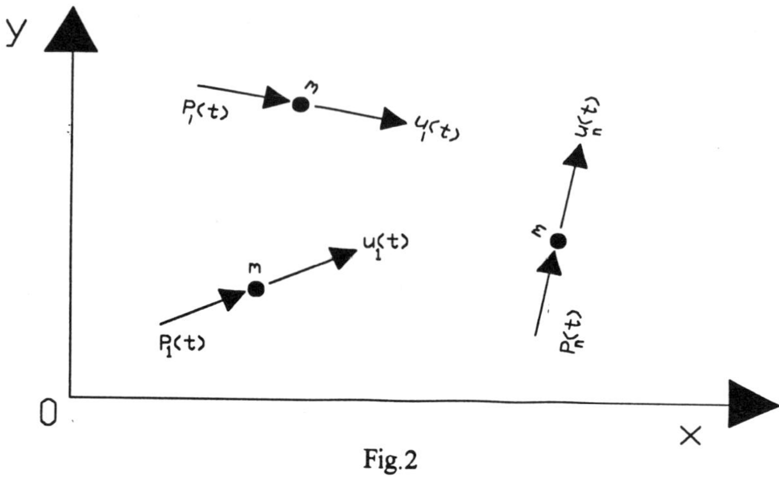

Fig.2

For constant masses:

$$\begin{bmatrix} P_1(t) \\ P_i(t) \\ P_n(t) \end{bmatrix} = \begin{bmatrix} m_1 & & \\ & m_i & \\ & & m_n \end{bmatrix} \begin{bmatrix} u_1(t) \\ u_i(t) \\ u_n(t) \end{bmatrix}$$

or in a matrix form:

$$[P(t)] = [m]\left[\ddot{u}(t)\right]$$

2.1.4 D'Alembert principle

During the motion of a particle there is a dynamic equilibrium between external and inertial forces, acting on the particle.

$$\boxed{P(t) + \left[-m\ddot{u}(t)\right] = 0}$$

Integration of this differential equation leads to the description of motion of the particle.

For a system of n mass particles, the equivalent equation is:

$$[P(t)] + \{-[m][\ddot{u}(t)]\} = 0$$

3. SINGLE DEGREE OF FREEDOM SYSTEM

3.1 Equation of motion

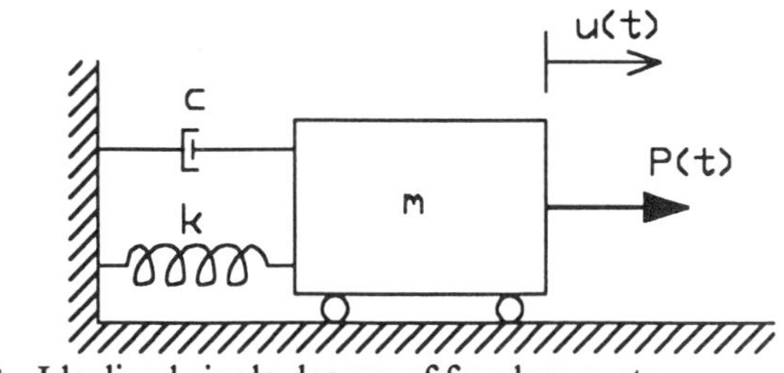

Fig.3   Idealized single degree of freedom system:
basic configuration

Assumption: The moving body is considered as a mass particle (fig 3).

Equation of motion:

$$\boxed{m\ddot{u} + c\dot{u} + ku = p(t)}$$

m=mass of the system
c=damping coefficient
k= spring stiffness

3.2 Derivation of the equation of motion

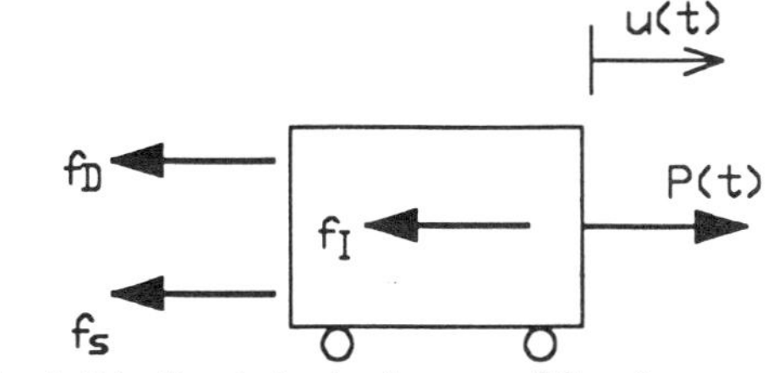

Fig.4 Idealised single degree of freedom system:
forces in equilibrium

Equilibrium equation:

$$f_I + f_D + f_S = P(t)$$

where:

$$f_I = m\ddot{u}(t), \quad f_D = c\dot{u}(t), \quad f_s = ku(t)$$

m=mass of the system
c=damping coefficient
k= spring stiffness

Equation of motion:

$$m\ddot{u} + c\dot{u} + ku = p(t)$$

4. TYPES OF FREE VIBRATION:

4.1 Undamped systems

4.1.1 Equation of motion

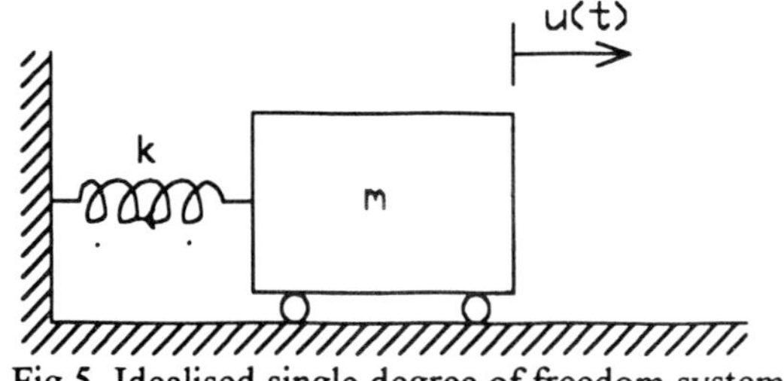

Fig.5 Idealised single degree of freedom system:
basic components

Initial conditions: displacement $u(0)$ and/or velocity $\dot{u}(0)$ (fig.5).
Equation of motion:

$$m\ddot{u} + ku = 0$$

Angular natural frequency: $\omega = \sqrt{\dfrac{k}{m}}$

$$\text{Natural period: } T = \frac{2\pi}{\omega} = 2\pi\sqrt{\frac{m}{k}}$$

4.1.2 Solution of the equation

$$u(t) = \frac{\dot{u}(0)}{\omega}\sin\omega t + u(0)\cos\omega t$$

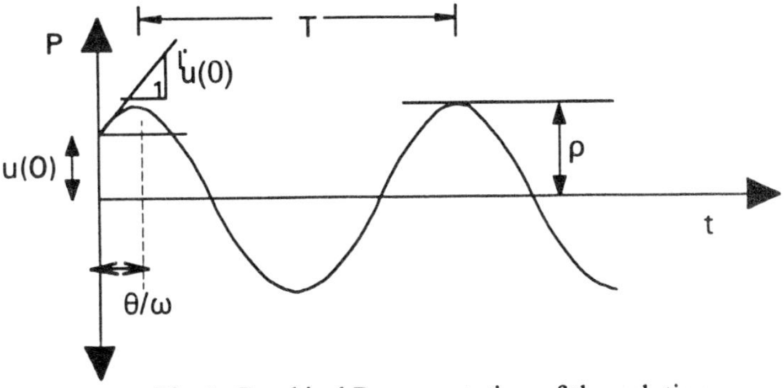

Fig.6 Graphical Representation of the solution.

4.1.3 Real Example

One story-two column frame, with infinite stiffness girder, mass m_0, stiffness k_0 (fig7):

$$k_0 = 2\frac{12EI}{h^3}$$

E=modulus of elasticity
I=moment of inertia

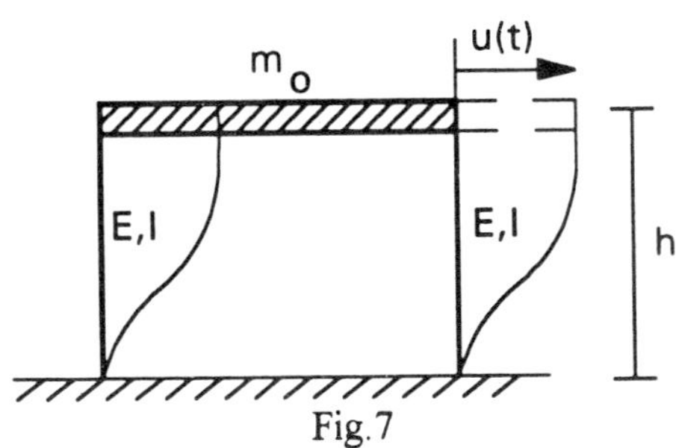

Fig.7

$$T_0 = 2\pi \sqrt{\frac{m_0}{k_0}}$$

4.1.4 Physical interpretation

Modification of initial mass m_0 and stiffness k_0 affect directly the initial natural period T_0:

$$\frac{T}{T_0} = \frac{\sqrt{k/k_0}}{\sqrt{m/m_0}}$$

<u>Examples</u>:

	k/k_0	m/m_0	T/T_0
1	1	1	1.000
2	2	1	1.414
3	3	1	1.732
4	1	2	0.707
5	1	3	0.577

4.1.5 Graphical representation

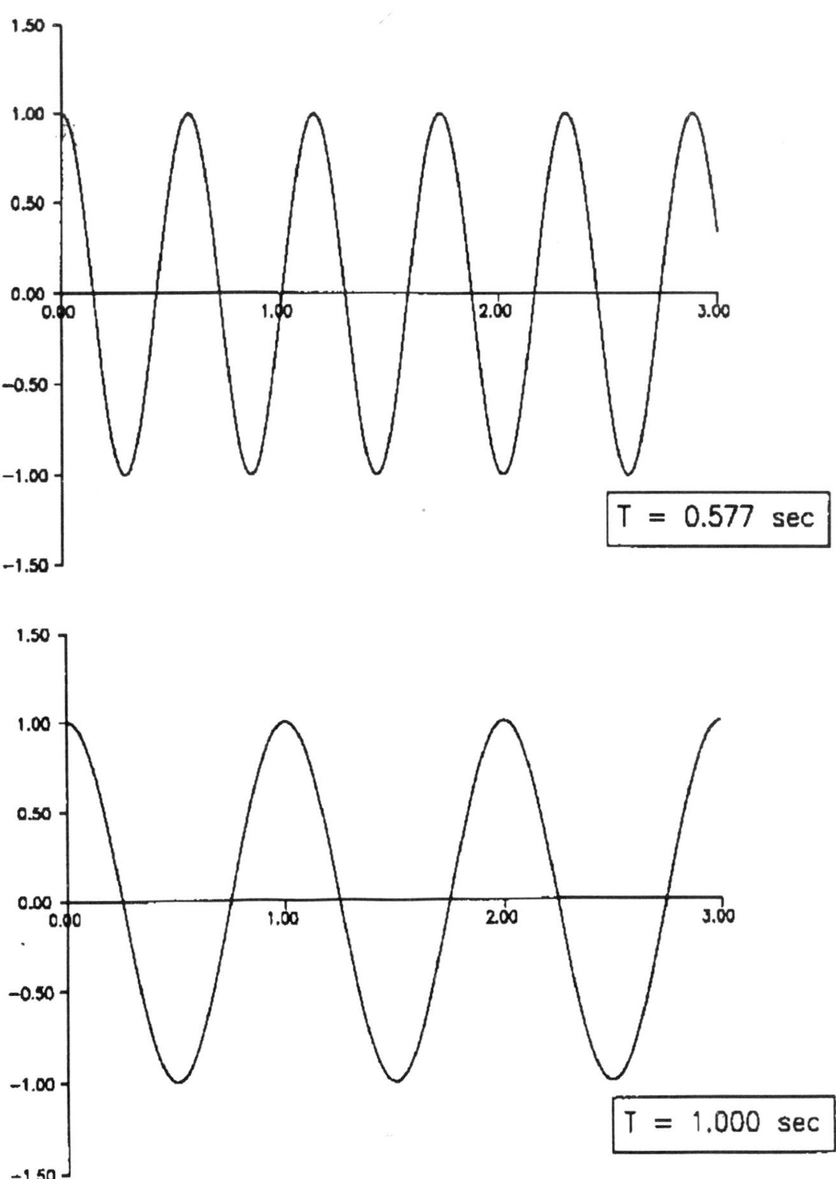

Fig.8

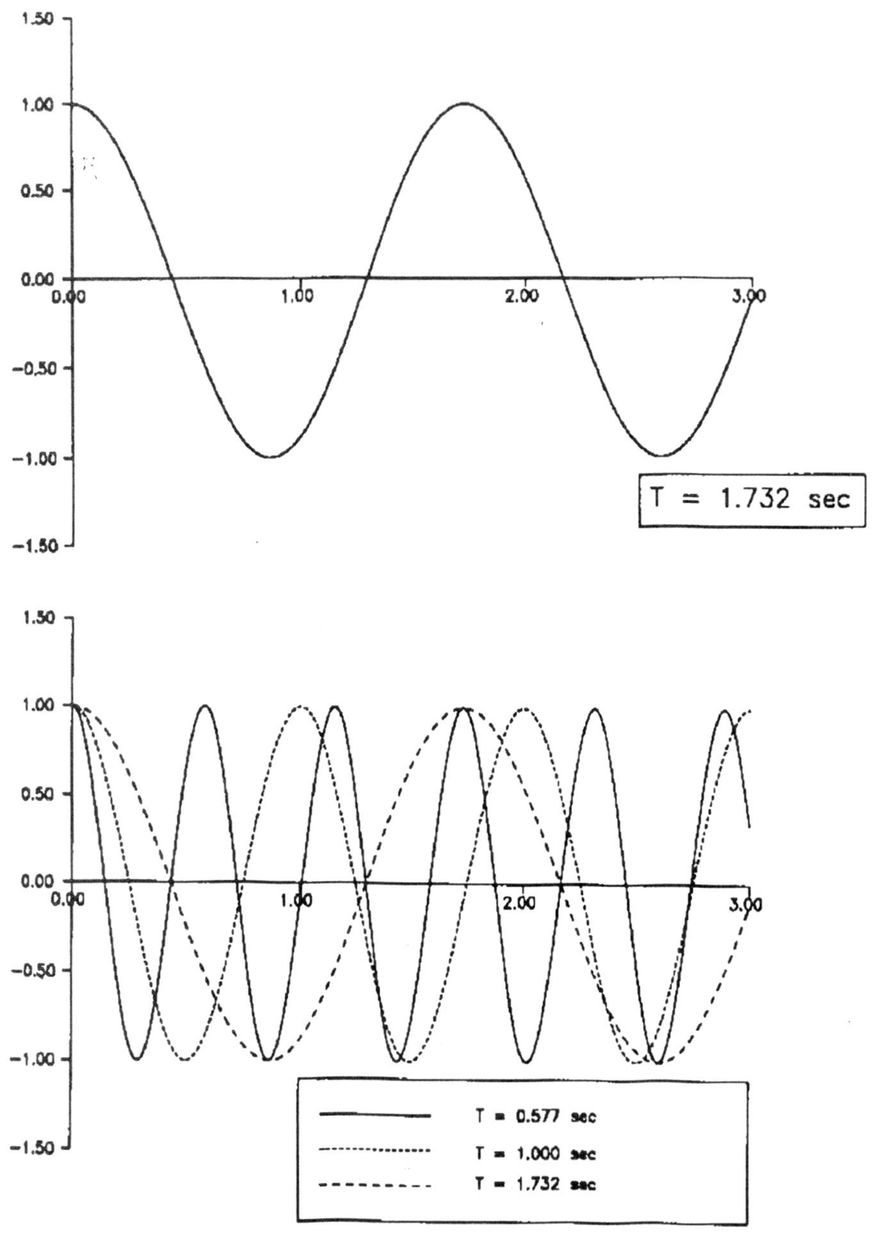

Fig.9

4.2 Underdamped systems ($\xi < 1$)

4.2.1 New Definitions

Damping coefficient $c = 2\xi m\omega$

ξ = damping ratio

Damped angular natural frequency: $\omega_D = \omega\sqrt{1-\xi^2}$

Damped angular natural period: $T_D = T / \sqrt{1-\xi^2}$

Indicative values of ξ	
Reinforced concrete	5 - 8%
Steel structures	2 - 4%
Masonry sructures	6 - 15%
Wood structures	7 - 9%

4.2.2 Equation of motion

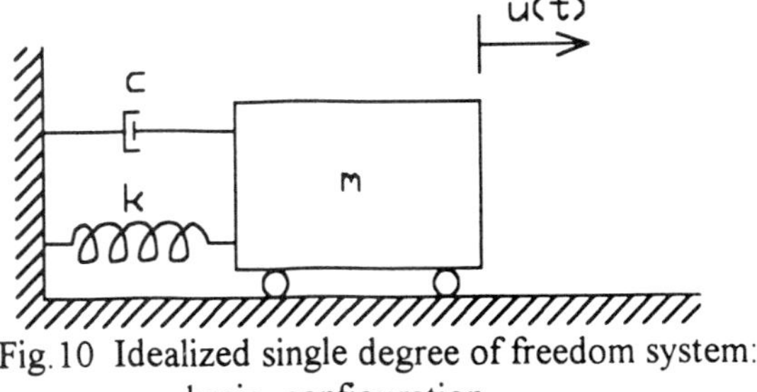

Fig.10 Idealized single degree of freedom system:
basic configuration

Assumption: The moving body (fig.10) is considered as a mass particle.

Equation of motion:

$$m\ddot{u} + c\dot{u} + ku = 0$$

m=mass of the system

c=damping coefficient
k= spring stiffness

4.2.3 Solution of the equation

$$u(t) = e^{-\xi\omega t}\left[\frac{\dot{u}(0) + u(0)\xi\omega}{\omega_D}\sin\omega_D t + u(0)\cos\omega_D t\right]$$

or:

$$u(t) = \rho e^{-\xi\omega t}\cos(\omega_D t - \theta)$$

Phase angle: $\theta = \tan^{-1}\dfrac{\dot{u}(0)}{\omega u(0)}$

Amplitude: $\rho = \sqrt{[u(0)]^2 + \left[\dfrac{\dot{u}(0)}{\omega}\right]^2}$

4.2.4 Experimental Determination of Damping Ratio

In order to determine the damping ratio ξ the following steps are applied:

1) Displacement u_0 of the mass from its equilibrium position.

2) Releasing

3) Recording free vibrations of the mass (fig.11)

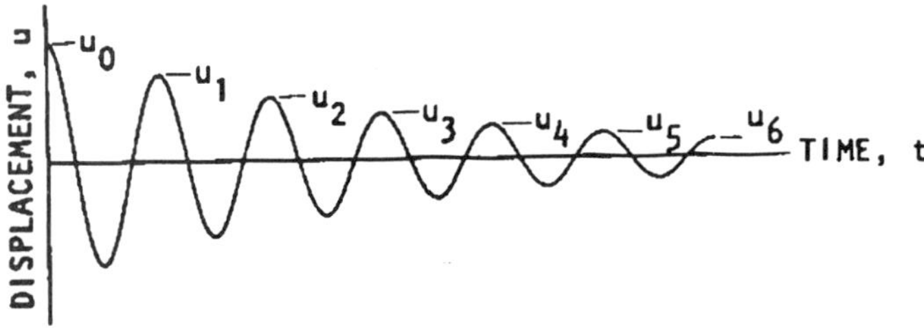

Fig.11

4) Measurement of time required for j complete cycles of vibration (i.e. j-times the period T)

5) Measurement of u_i (at cycle i) and u_{i+j} (after j-cycles)

6) Compute $\delta = \ln\dfrac{u_i}{u_{i+j}}$

7) Compute $\xi = \dfrac{\delta}{\sqrt{4j^2\pi^2 + \delta^2}}$

4.3 Critically damped systems $(\xi = 1)$

4.3.1 Equation of motion

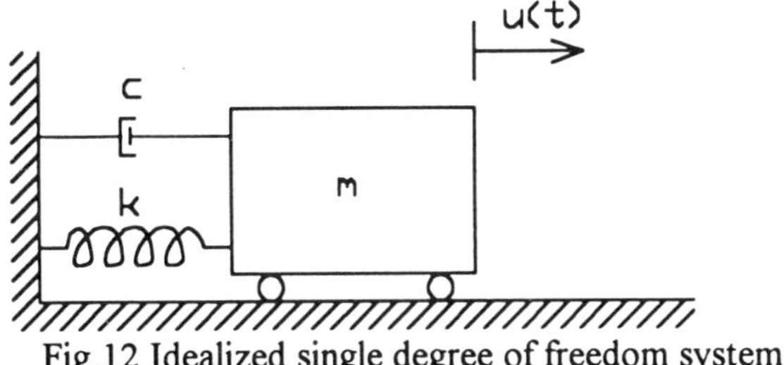

Fig. 12 Idealized single degree of freedom system:
basic configuration

Assumption: The moving body (fig. 12) is considered as a mass particle.

Equation of motion:

$$m\ddot{u} + c\dot{u} + ku = 0$$

m=mass of the system
c=damping coefficient
k= spring stiffness

4.3.2 Solution of the equation

$$u(t) = [u(0)(1+\omega t) + \dot{u}(0)t]e^{-\omega t}$$

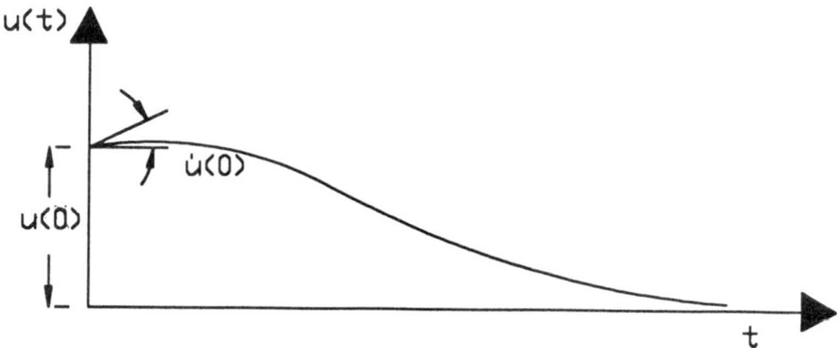

Fig.13 Free vibration with critical damping

In this case no oscillation occurs, the mass returns to zero in accordance with the exponential decay.

4.4 Overdamped systems $(\xi > 1)$

4.4.1 Equation of motion

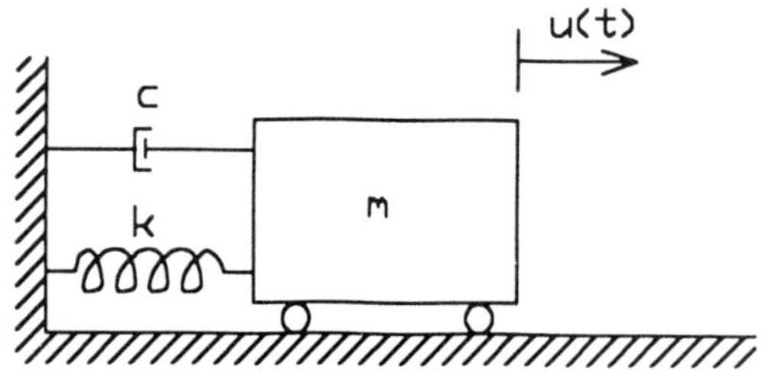

Fig.14 Idealized single degree of freedom system:
basic configuration

Assumption: The moving body (fig.14) is considered as a mass particle.

Equation of motion:

$$m\ddot{u} + c\dot{u} + ku = 0$$

m=mass of the system
c=damping coefficient
k= spring stiffness

4.4.2 Solution of the equation

$$u(t) = e^{-\xi\omega t}(A\sinh\hat{\omega}t + B\cosh\hat{\omega}t)$$

where $\quad \hat{\omega} = \omega\sqrt{\xi^2 - 1}$

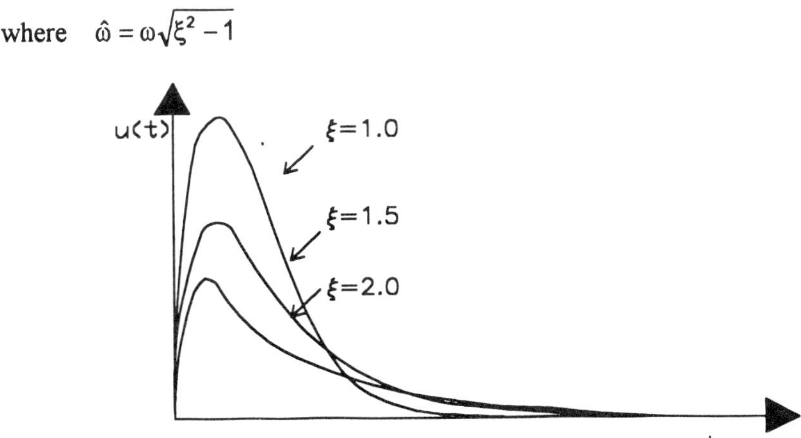

Fig.15 Response of overdamped systems

Like for the critically damped case $(\xi = 1)$ no oscillation occurs but the amplitude decays more slowly (fig.15).

5. MULTI-DEGREE OF FREEDOM SYSTEMS:

5.1 Definitions

Mass matrix:
$$[m] = \begin{bmatrix} m_1 & & & \\ & & m_i & \\ & & & \cdot & \\ & & & & m_n \end{bmatrix}$$

m_i the mass corresponding to the degree of freedom i.

Stiffness matrix:
$$[k] = \begin{bmatrix} k_{11} & k_{12} & \cdot & \cdot & k_{1n} \\ k_{21} & k_{22} & \cdot & \cdot & k_{2n} \\ \cdot & \cdot & \cdot & \cdot & \cdot \\ \cdot & \cdot & \cdot & \cdot & \cdot \\ k_{n1} & k_{n2} & \cdot & \cdot & k_{nn} \end{bmatrix}$$

k_{ij}=The force corresponding to the degree of freedom i due to a unit displacement of the degree of freedom j and all the other degrees of freedom constrained (i,j=1,2,3,...n).

$$\text{Damping matrix } [c] = \begin{bmatrix} c_{11} & c_{12} & \cdot & \cdot & \cdot & c_{1N} \\ c_{21} & c_{22} & \cdot & \cdot & \cdot & c_{2N} \\ \cdot & \cdot & \cdot & \cdot & \cdot & \cdot \\ \cdot & \cdot & \cdot & \cdot & \cdot & \cdot \\ \cdot & \cdot & \cdot & \cdot & \cdot & \cdot \\ c_{N1} & c_{N2} & \cdot & \cdot & \cdot & c_{NN} \end{bmatrix}$$

c_{ij}=force corresponding to degree of freedom i due to a unit velocity of degree j

$$\text{Displacement matrix:} \quad [u(t)] = \begin{bmatrix} u_1(t) \\ u_i(t) \\ \cdot \\ u_n(t) \end{bmatrix}$$

5.2 Equation of motion

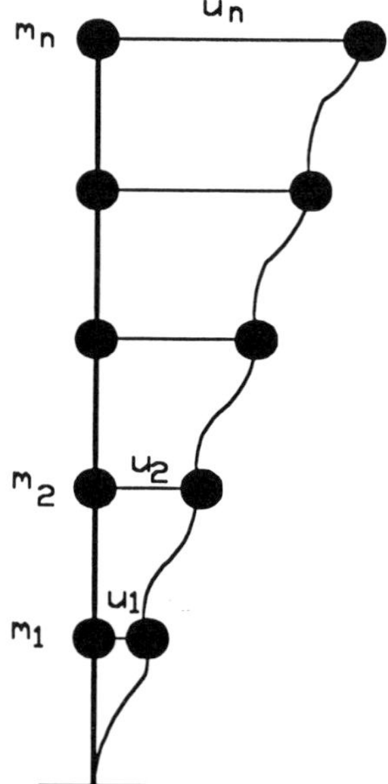

Fig.16 Each moving mass is considered as an one mass particle.

The motion is governed by ordinary differential equations, as many as the number of the degrees of freedom of the system.

Equation of motion:

$$[m][\ddot{u}] + [c][\dot{u}] + [k][u] = [0]$$

5.3 Derivation of the equation

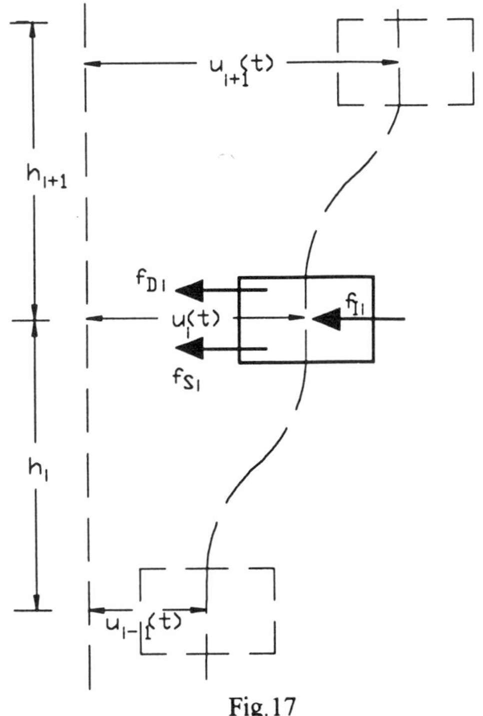

Fig. 17

Equilibrium equation:

$$f_{Ii} + f_{Di} + f_{Si} = 0$$

where:

$$f_{Ii} = m_i \ddot{u}_i(t)$$
$$f_{Di} = c_i \dot{u}_i(t)$$
$$f_{si} = k_i\big(u_i(t) - u_{i-1}(t)\big) - k_{i+1}\big(u_{i+1}(t) - u_i(t)\big)$$

$$k_i = \frac{12E_i I_i}{h_i^{\ 3}} \ , \quad k_{i+1} = \frac{12E_{i+1} I_{i+1}}{h_{i+1}^{\ 3}} \ ,$$

Equation of motion:

$$[m][\ddot{u}] + [c][\dot{u}] + [k][u] = [0]$$

5.4 Undamped systems

5.4.1 Equation of motion

Initial conditions: $u_i(0)$, $\dot{u}_i(0)$ (i=1,2,....N)

Equation of motion:

$$[m][\ddot{u}] + [k][u] = [0]$$

Angular natural frequencies of vibration: ω_i i=1,2,....n

Natural periods of vibration $T_i = \dfrac{2\pi}{\omega_i}$,   i=1,2,....n, is the time required for one complete cycle of the harmonic motion following one of the n natural modes of vibration.

Natural mode of Vibration is the characteristic deflected shape such that if the structure displaced and released, it will vibrate in harmonic motion.

The equation giving each natural circular frequency ω_i, and the corresponding eigen-vector $[\beta_i]$ of the vibration is the following:

$$[k - \omega_i^{\ 2} m][\beta] = 0$$

5.4.2 Solution of the equation

By analogy with the behaviour of single degree of freedom system, it will be assumed that for a multi-degree of freedom system, each free vibration motion i is harmonic, expressed as:

$$[u_i(t)] = [\beta_i] \cos(\omega_i t - \vartheta_i)$$

where

$[\beta_i]$ is the vector of the n unknown constants $\beta_{1i}, \beta_{2i}, \ldots \ldots \beta_{ni}$

ϑ_i the phase angle of the natural vibration depending on the initial conditions.

Subsituting this expression to the equation of motion, this equation becomes:

$$[k - \omega_i^2 m][\beta_i] \cos(\omega_i t - \vartheta) = 0$$

Or, as generally $\cos(\omega_i t - \vartheta) \neq 0$:

$$[k - \omega_i^2 m][\beta_i] = 0$$

This equation leads to the computation of natural frequencies and modes of vibration of the system.

5.4.3 Computation of natural frequencies and modes of vibration

Computation of these quantities requires solution of the matrix equation:

$$\boxed{[k - \omega_i^2 m][\beta] = 0}$$

As the elements of $[\beta]$ cannot be simultaneously all zero, a nontrivial solution is obtained from the frequency equation of the system:

$$\boxed{|k - \omega_i^2 m| = 0}$$

The n roots of this equation $(\omega_1^2, \omega_2^2, \ldots \omega_N^2)$ represent the squeares of the natural frequencies of the n modes of vibration of the system.

Eigen - Vectors:
For each value of ω_i a vector $[\beta_i]$ is determined and is called eigen vector.

Modes of Vibration:
Dividing all the eigen-vector elements by the largest absolute value of its elements, the modes $[\phi_i]$ are obtained.

$$[\phi_i] = \begin{bmatrix} \phi_{1i} \\ \cdot \\ \cdot \\ \cdot \\ \phi_{ni} \end{bmatrix} = \frac{1}{\max \beta_{ni}} \begin{bmatrix} 1 \\ \beta_{2i} \\ \cdot \\ \cdot \\ \beta_{ni} \end{bmatrix}$$

5.4.4 Orthogonality conditions

1) The modes of vibration are orthogonal to the stiffness of the system.

$$[\phi_i]^T [k][\phi_j] = 0 \quad (i \neq j)$$

2) The modes of vibration are orthogonal to the mass of the system.

$$[\phi_i]^T [m][\phi_j] = 0 \quad (i \neq j)$$

5.5 Damped systems

5.5.1 Equation of motion

Initial conditions: $u_i(0)$, $\dot{u}_i(0)$ (I=1,2,....N)

Equation of motion:

$$\boxed{[m][\ddot{u}] + [c][\dot{u}] + [k][u] = [0]}$$

The damped free vibration response of a multi-degree of freedom system can be studied in a similar to the undamped case process, on condition that the expression of damping matrix [c], is defined.

Following Clough's proposal, this can be done if we assume a linear relation of matrix [c] to the matrices [m] and [k].

$$[c] = a_0[m] + a_1[k]$$

where a_0, a_1 constants to be determined, or in a more general form:

$$[c] = \alpha_0[m] + \alpha_1[k] + \alpha_2[km^{-1}k] + \alpha_3[km^{-1}km^{-1}k] + \ldots\ldots\ldots = [m]\sum_{r=0}\alpha_r[m^{-1}k]^r$$

For practical applications, linear approximation (first two factors) is usually taken into account. In this case the equation of motion can be written in the form:

$$[m][\ddot{u}] + [\alpha_0 m + \alpha_1 k][\dot{u}] + [k][u] = [0]$$

5.5.2 Determination of damping matrix

In order to follow a similar process to the one of the undamped systems, the damping matrix $[c]$ has to fulfil the orthogonality conditions. It can be proved that a general expression of the damping matrix $[c]$, of the form:

$$[c] = \sum_r [c_r] = [m]\sum_r \alpha_r \left\{[m]^{-1}[k]\right\}^r \tag{1}$$

fulfils the requirements imposed, term by term:

$$[\Phi_i]^T [c_r][\Phi_j] = 0 \text{ for } (i \neq j),$$

and consequently such an expression of matrix $[c]$ can be used for the solution.

We denote:

$$C_i = [\Phi_i]^T [c][\Phi_i]$$

Substituting for matrix $[c]$ the general expression (1), and taking into account that $[k][\Phi_i] = \omega_i^2 [m][\Phi_i]$, we get the final expression for C_i :

$$C_i = M_i \sum_r \alpha_r \omega_i^{2r}$$

Alternatively, by definition, it is:

$$C_i = 2\xi_i \omega_i M_i$$

Consequently:

$$\xi_i = \frac{1}{2}\sum_r \alpha_r \omega_i^{2r-1}$$

In the forementioned series, r-terms, as many as the available number of modal damping ratios ξ_i, must be included . Then the values of constants α_r can be determined from the resulting set of simultaneous equations. After that the damping coefficients are defined from the equation (1).

The modal damping ratios ξ_i can be estimated on the basis of existing experience (experimental determination).

6. DYNAMIC LOADING

Loadings may be static and/or dynamic.The static loading do not vary with time (dead loads, etc).The dynamic loading vary with time (earthquake, blast, etc). The inertial loading due to an earthquake is one, type of dynamic loading.

6.1 Static versus Dynamic actions

The term "dynamic" generally means "time-varying". Oppositely to a static action, dynamic action is a loading whose magnitude, direction and point of application vary with time. The structural system response due to a dynamic load (i.e, resulting deflections and stresses) is also time-varying (fig18).

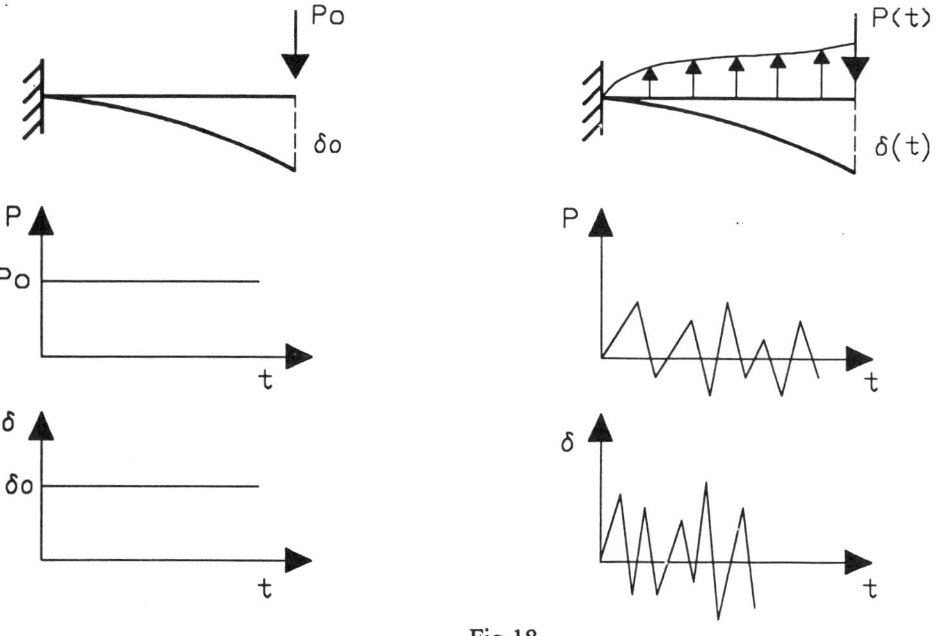

Fig. 18

6.2 Types of Dynamic Loading

The various types of dynamic loading shown below (fig.19)

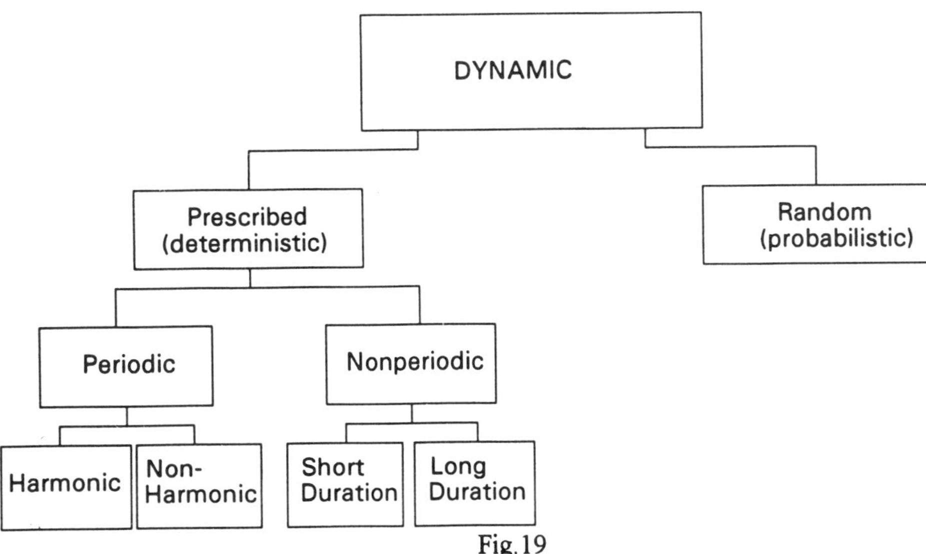

Fig.19

6.2.1 Harmonic Loading

Definition: Periodic repetitive load, exhibiting successively for a large number of cycles the same time variation of harmonic type (sinusoidal, cosinusoidal variation) (fig.20).

Example: unbalanced-mass effect in rotating machinery.

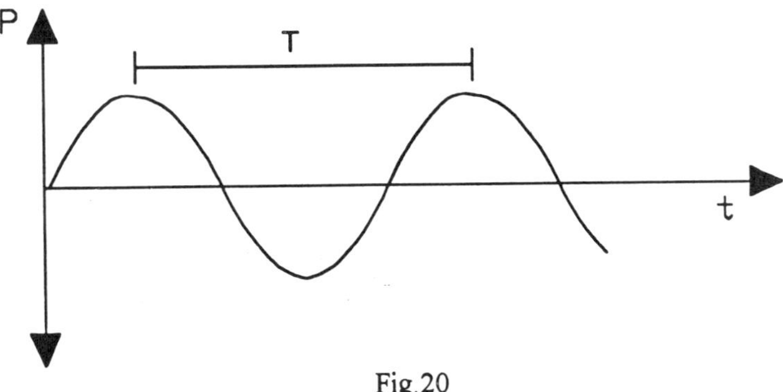

Fig.20

6.2.2 Non-Harmonic Loading

Definition: Periodic repetitive load, exhibiting successively for a large number of cycles the same time variation of random shape (fig.21).

Example: the hydrodynamic pressures generated by the propeller of a ship.

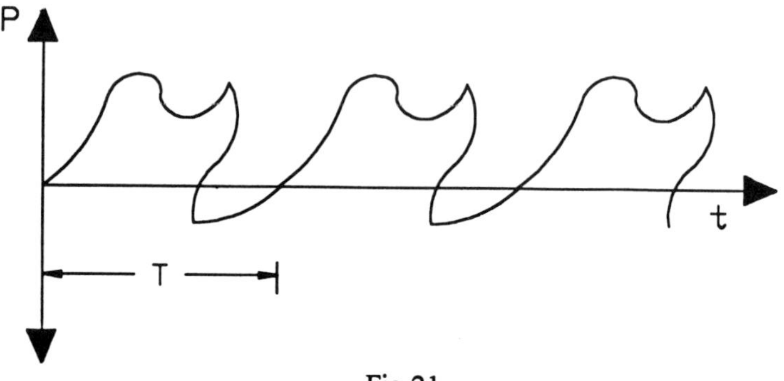

Fig.21

6.2.3 Shord Duration

Definition: Short duration loading with arbitrary law of distribution in time (fig.22).
Example: blast or explosion, impact.

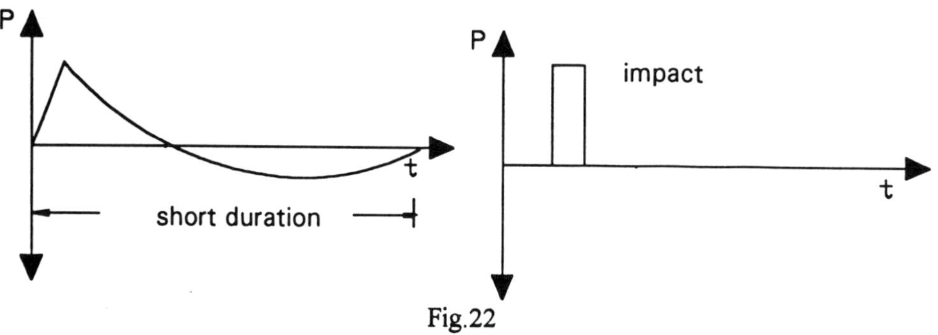

Fig.22

6.2.4 Long Duration

Definition:Long duration loading with arbitrary law of distribution in time (fig.23).
Example: traffic loading on a bridge.

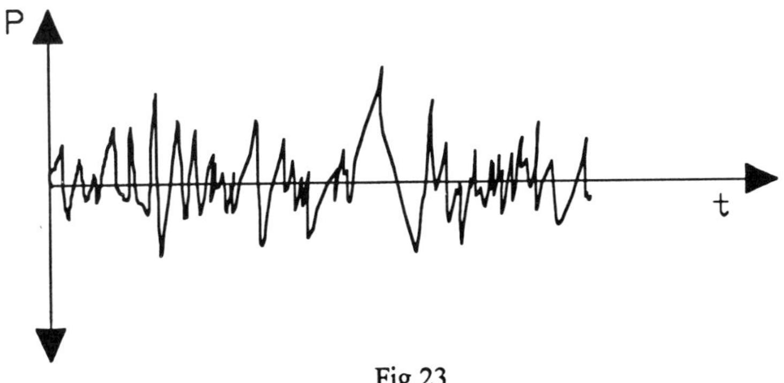

Fig.23

6.3 Earthquake Effect

6.3.1 The equation of motion of single and multi-degree of freedom systems under earthquake effect (fig.24).

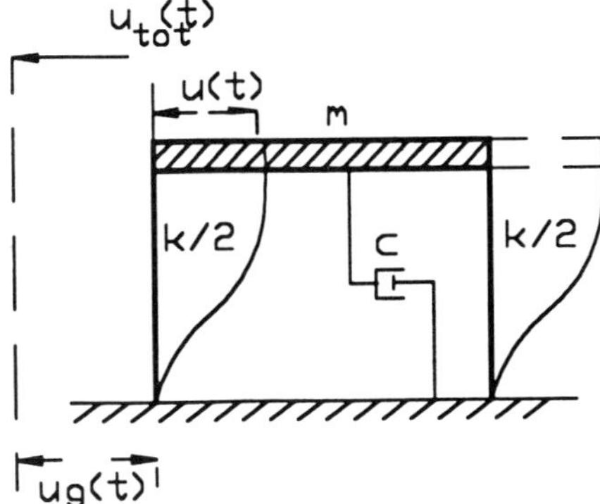

Fig.24

Conditions:

$$u_{tot}(t) = u_g(t) + u(t)$$
$$\ddot{u}_{tot}(t) = \ddot{u}_g(t) + \ddot{u}(t)$$

Equation of equilibrium :

$$m\ddot{u}_{tot}(t) + c\dot{u}(t) + ku(t) = 0$$

Equation of motion :

$$m\ddot{u}(t) + c\dot{u}(t) + ku(t) = -m\ddot{u}_g(t)$$

6.3.2 Dynamic effect of earthquake

The response of the moving due to the earthquake system (a), is identical to that of the system (b) with fixed base and subjected to external dynamic load equal to $-m\ddot{u}_g$ (fig25).

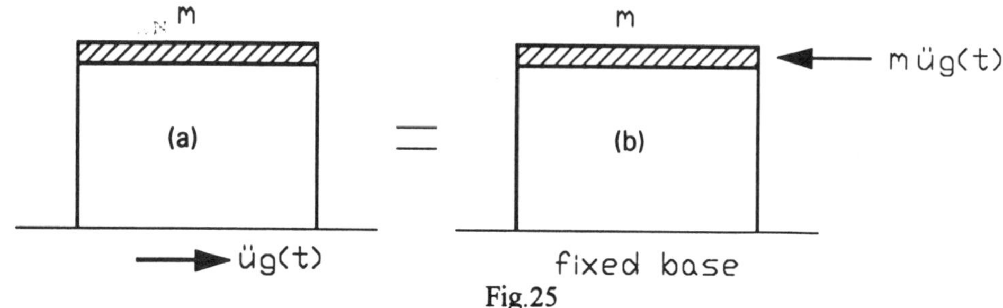

Fig.25

6.3.3 Equation of motion of a multi-degree of freedom systems (fig26)

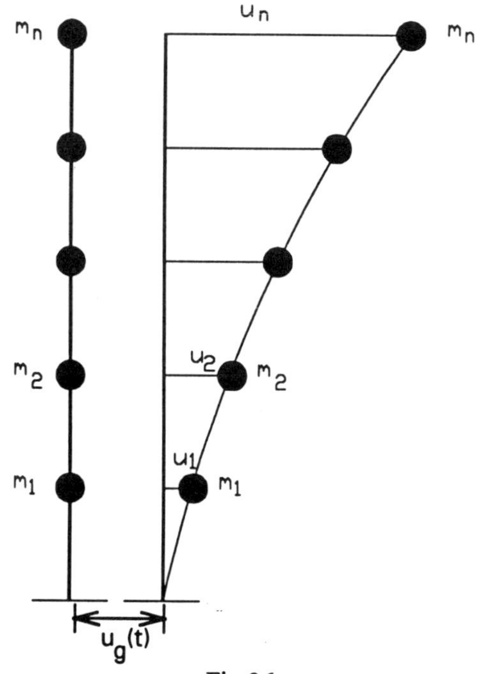

Fig.26

Conditions:

$$u_{i,tot}(t) = u_g(t) + u_i(t)$$

$$\ddot{u}_{i,tot}(t) = \ddot{u}_g(t) + \ddot{u}_i(t)$$

Equation of Equilibrium:

$$[m]\left[\ddot{u}_{tot}(t)\right] + [c]\left[\dot{u}(t)\right] + [k]\left[u(t)\right] = [0]$$

Equation of motion

$$[m]\left[\ddot{u}(t)\right] + [c]\left[\dot{u}(t)\right] + [k]\left[u(t)\right] = -[m]\left[\ddot{u}_g(t)\right]$$

7. FORCED VIBRATIONS

The forced vibrations, either of single degree, or of multi degree of freedom systems, are examined.

7.1 Basic definitions

7.1.1 Response ratio

The response ratio $R(t)$ is the ratio of the dynamic displacement to the displacement produced by the static application of the maximum value of load P_0:

$$R(t) = \frac{u(t)}{P_o / k}$$

When $u(t) \to \infty$, $R(t) \to \infty$.

7.1.2 Dynamic magnification factor

The dynamic magnification factor is the maximum value of the response ratio $R(t)$:

$$D = \max R(t)$$

7.2 Single degree of freedom systems

7.2.1 Response of undamped systems

The dynamic response of undamped single degree of freedom systems under external excitation, is presented.

Equation of motion:

$$m\ddot{u} + ku = p(t)$$

General solution of the equation:

$$u(t) = \frac{\dot{u}(0)}{\omega}\sin\omega t + u(0)\cos\omega t + \frac{1}{m\omega}\int_0^t p(\tau)\sin\omega(t-\tau)d\tau$$

The last term of the expression is called Duhamel integral for an undamped single degree of freedom system. It may be used to evaluate the response of the system under any form of dynamic loading P(t). When the Duhamel integral can be calculated analytically the solution of the equation is obtained in a closed form.

7.2.2 Applications :

Applications for undamped single-degree of freedom systems under forced vibrations.

1) Harmonic loading

The system is subjected to a harmonic loading P(t) of amplitude P_0 and angular frequency $\bar{\omega}$.

Equation of motion:

$$m\ddot{u} + ku = P_0 \sin\bar{\omega}t$$

Solution of the equation:

$$u(t) = \frac{\dot{u}(0)}{\omega}\sin\omega t + u(0)\cos\omega t + \frac{P_0}{k}\frac{1}{1-\beta^2}\sin\bar{\omega}t[\sin\bar{\omega}t - \beta\sin\omega t]$$

$$\beta = \frac{\bar{\omega}}{\omega}$$

Example:
For the case of initial conditions u(0) = u̇(0) = 0, the displacement expression becomes:

$$u(t) = \frac{P_o}{k} \frac{1}{1-\beta^2} (\sin \overline{\omega} t - \beta \sin \omega t)$$

Resonant response:

The situation where the frequency ratio β equals to unity is called resonance and the corresponding frequency, resonant frequency. In that case, for an undamped system, u(t)→ ∞.

Real example:

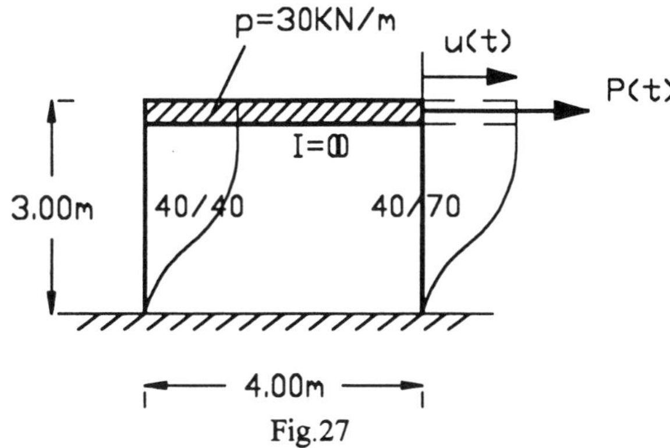

Fig.27

<u>Data:</u>
   Reinforced concrete plane frame with rigid supports and applied force P(t) (fig.27).
   Modulus of elasticity E=2.1x10^7 KN/m
   Span distributed vertical loading p=30 KN/m
   Specific weight ã=24 KN/m
   Column weight: 2x0.4x0.4x0.4x3x24=23.04 KN
   Girder weight: 4x30=120 KN
   Floor mass: m=[(120+23.04)/2]/9.81=7.29 Kgr

$$\text{Moments of inertia of columns } I = \frac{bh^3}{12} = \begin{cases} I_1 = \dfrac{0.4 \times 0.4^3}{12} = 2.133 \times 10^{-3} \\[2mm] I_2 = \dfrac{0.7 \times 0.4^3}{12} = 3.773 \times 10^{-3} \end{cases}$$

Stiffness: $k = \dfrac{12EI_1}{h^3} + \dfrac{12EI_2}{h^3} = 19908 + 34841.33 = 54749.33 \text{ KN/m}$

Angular frequency: $\omega = \sqrt{\dfrac{k}{m}} = 86.66$ rad/s

Period: $T = \dfrac{2\pi}{\omega} = 0.0725$ s

Damping ratio: $\xi = 0.05$

Damping coefficient: $c = 2\hat{i}m\dot{u} = 63.175$ KN/m.s

Equation of motion:

$$7.29\ddot{u} + 63.175\dot{u} + 54749.33u = P(t)$$

2) Rectangular impulse (fig.28)

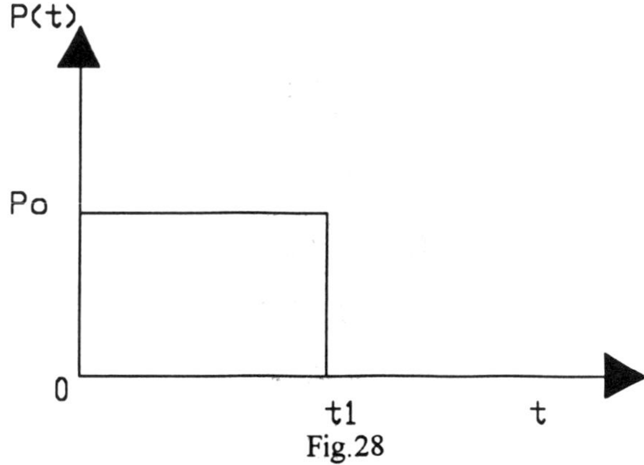

Fig.28

The response is examined in two phases: the loading phase (I) and the subsequent free-vibration phase (II)

Phase (I) : $0 \le t \le t_1$

During this phase the system is subjected to the rectangular impulse. This is a forced vibration with equation:

$$m\ddot{u} + ku = P(t)$$

The solution is:

$$u(t) = \frac{\dot{u}(0)}{\omega}\sin\omega t + u(0)\cos\omega t + \frac{1}{m\omega}\int_0^t P(\tau)\sin\omega(t-\tau)d\tau$$

For initial conditions $u(0) = \dot{u}(0) = 0$, and loading $P(\tau) = P_0$:

$$u(t) = \frac{P_0}{k}(1 - \cos\omega t)$$

Phase (II): $\bar{t} = t - t_i \geq 0$

During this phase the system is under free vibration conditions. The free vibration is described by the equation:

$$u(t) = \frac{\dot{u}(t_1)}{\omega}\sin\omega\bar{t} + u(t_1)\cos\omega\bar{t}$$

3) Constant loading suddenly applied (fig.29)

Fig.29

Equation of motion:

$$m\ddot{u} + ku = p(t)$$

Solution of the equation:

$$u(t) = \frac{\dot{u}(o)}{\omega}\sin\omega t + u(o)\cos\omega t + \frac{1}{m\omega}\int_o^t p(t)\sin\omega(t - \tau)d\tau$$

For initial conditions: $u(0) = \dot{u}(0) = 0$

$$u(t) = \frac{P_o}{k}(1 - \cos\omega t)$$

4) Pseudo-Static loading (fig30)

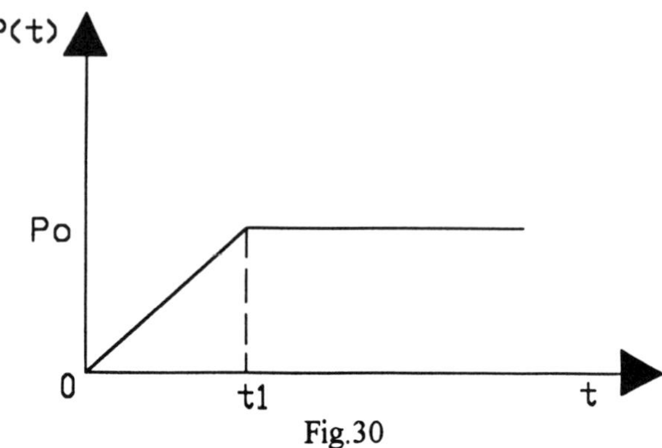

Phase (I) : $0 \le t \le t_1$

Forced Vibration with $P(t) = \dfrac{P_o}{t_1}t$ and initial conditions $u(0) = \dot{u}(0)$

$$u_I(t) = \frac{1}{m\omega}\int_0^t \frac{P_0}{t_1}\tau \sin\omega(t-\tau)d\tau = \frac{P_0}{Kt_1}\left(t - \frac{\sin\omega t}{\omega}\right)$$

Phase (II) : $\bar{t} = t - t_1 \ge 0$

Forced Vibration with $P(\bar{t}) = P_0$ and initial conditions $u_{II}(0) = u_I(t_1)$,

$$\dot{u}_{II}(0) = \dot{u}_I(t_1)$$

$$u_{II}(t) = \frac{\dot{u}_I(t_1)}{\omega}\sin\omega\bar{t} + u_I(t_1)\cos\omega\bar{t} + \frac{P_0}{k}(1 - \cos\omega\bar{t}) \qquad \bar{t} = t - t_1 \ge 0$$

7.2.3 Response of damped systems

The dynamic response of damped single degree of freedom systems under external excitation, is presented.

The system is subjected to a harmonic loading P(t) of amplitude P_0 and angular frequency $\bar{\omega}$.

Equation of motion:

$$m\ddot{u} + c\dot{u} + ku = P(t)$$

General solution of the equation:

$$u(t) = \frac{\dot{u}(o)}{\omega}\sin\omega t + u(o)\cos\omega t + \frac{1}{m\omega_D}\int_0^t P(\tau)e^{-\xi\omega(t-\tau)}\sin\omega_D(t-\tau)d\tau$$

Application for harmonic loading

Equation of motion :

$$m\ddot{u} + c\dot{u} + ku = p(t)$$

Solution of the equation :

$$u(t) = e^{-\xi\omega t}\left(A\sin\omega_D t + B\cos\omega_D t\right) + \frac{p_0}{k}\frac{1}{(1-\beta^2)^2 + (2\xi\beta)^2}\left[(1-\beta^2)\sin\overline{\omega}t - 2\xi\beta\cos\overline{\omega}t\right]$$

The first term represents the transient response to the applied loading and the second the steady state response. The constants A,B can be calculated for any given initial conditions.

Resonant response

As it was mentioned before the situation where the frequency ratio β equals to unity is called resonanse. In the case for a damped system:

$$u(t) = e^{-\xi\omega t}\left(A\sin\omega_D t + B\cos\omega_D t\right) - \frac{p_0}{k}\frac{\cos\overline{\omega}t}{2\xi}$$

7.3 Multi-degree of freedom systems

7.3.1 Undamped systems

The dynamic response of undamped multi-degree of freedom system is presented.

Equation of motion

$$[m][\ddot{u}] + [k][u] = [P(t)]$$

Solution of the equation of motion following the mode superposition method:(i=1,2,...n)

Solution of the equation of motion following the mode superposition method:(i=1,2,...n)

$$[u(t)] = [\Phi_1]Y_1(t) + [\Phi_2]Y_2(t) + \ldots\ldots + [\Phi_n]Y_n(t)$$

$[\Phi_i] =$ vector of i-mode

$Y_i(t) =$ the so called generalised coordinates of the system, are unknown function of t.

Generalised coordinates (functions $Y_i(t)$)

For each one of the n modes of vibration, the following steps lead to the determination of the generalized coordinates $Y_i(t)$ (i=1,2,...n):

1) Determination of the n circular frequencies ω_i (i=1,2,...n), and the equivalent n mode vectors $[\Phi_i]$ i=1,2,...i).

2) Determination of the generalised masses M_i and generalised loads $P_i(t)$

$$M_i = [\Phi_i]^T[m][\Phi_i]$$

$$P_i(t) = [\Phi_i]^T[P(t)]$$

3) Derivation of uncoupled equations:

$$\ddot{Y}_i + \omega_i^2 Y_i = \frac{P_i(t)}{M_i}$$

This equation for each mode i (i=1,2,...n) is derived from the general equation of motion after:

a) substitution of the general solution expression, following the mode superposition method, to the general equation of motion

b) premultaplication of the equation of motion by the transpose of the mode vector $[\Phi_i]$ (i.e $[\Phi_i]^T$)

c) use of the ortogonality conditions

4) Determination of the generalised initial conditions $Y_i(0)$, $\dot{Y}_i(0)$

$$Y_i(0) = \frac{[\Phi_i]^T[m][u(0)]}{M_i}$$

$$\dot{Y}_i(0) = \frac{[\Phi_i]^T[m][\dot{u}(0)]}{M_i}$$

5) Determination of the generalised coordinates $Y_i(t)$:

$$Y_i(t) = \frac{\dot{Y}_i(0)}{\omega_i}\sin(\omega_i t) + Y_i(0)\cos(\omega_i t) + \frac{1}{M_i\omega_i}\int_0^t P_i(\tau)\sin\omega_i(t-\tau)d\tau$$

7.3.2 Damped systems

The dynamic response of damped multi-degree of freedom systems is presented.

Equation of motion

$$[m][\ddot{u}] + [c][\dot{u}] + [k][u] = [P(t)]$$

Solution of the equation following the mode superposition method:

$$[u(t)] = [\Phi_1]Y_1(t) + [\Phi_2]Y_2(t) + \ldots\ldots + [\Phi_n]Y_n(t)$$

where (i=1,2,...n):

$[\Phi_i]$ = vector of i-mode

$Y_i(t)$ = the so called generalised coordinates of the system, are unknown function of t.

Generalized coordinates

For each one of the n modes of vibration, the following steps lead to the determination of the generalized coordinates $Y_i(t)$ (i=1,2,...i):

1) Determination of the n circular frequencies ω_i (i=1,2,...n), and the equivalent n mode vectors $[\Phi_i]$ (i=1,2,...n), as in the case of undamped system.

2) Determination of the generalised masses M_i and generalised loads $P_i(t)$ (i=1,2,...n):

$$M_i = [\Phi_i]^T[m][\Phi_i]$$

$$P_i(t) = [\Phi_i]^T[P(t)]$$

3) An assumption has to be made concerning the values of coefficients ξ_i (i=1,2,...i)

4) Derivation of uncoupled equation of motion:

$$\ddot{Y}_i + 2\xi_i\omega_i^2\dot{Y}_i + \omega_i^2 Y_i = \frac{P_i(t)}{M_i}$$

This equation is derived from the general equation of motion after:
a) substitution of the general solution expression to the general equation of motion.
b) premultaplication of the equation of motion by the transpose of the mode vector

$[\Phi_i]$ (i.e $[\Phi_i]^T$).

c) use of the orthogonality conditions.

5) Determination of the generalised initial conditions $Y_i(0)$, $\dot{Y}_i(0)$

$$Y_i(0) = \frac{[\Phi_i]^T [m][u(0)]}{M_i}$$

$$\dot{Y}_i(0) = \frac{[\Phi_i]^T [m][\dot{u}(0)]}{M_i}$$

6) Determination of the generalised coordinates $Y_i(t)$:

$$Y_i(t) = e^{-\xi_i \omega_i t}\left[\frac{\dot{Y}_i(0) + Y_i(0)\xi_i\omega_i}{\omega_i}\sin(\omega_{Di}t) + Y_i(0)\cos(\omega_{Di}t)\right] + \frac{1}{M_i\omega_i}\int_0^t P_i(\tau)e^{-\xi_i\omega_i(t-\tau)}\sin\omega_{Di}(t-\tau)d\tau$$

REFERENCES

1. Clough, R. and Penzien, J.: Dynamics of Structures, McGraw Hill, Singapore 1975.
2. Craig, R.: Dynamics of Structures, John Willey & Sons, Singapore 1981.
3. Paz, M.: Structural Dynamics, Van Nostrand Reinhold, New York 1991.
4. Chopra, A.: Dynamics of Structures, EERI California 1980.

SEISMIC BEHAVIOUR OF TRADITIONALLY-BUILT CONSTRUCTIONS

REPAIR AND STRENGTHENING

P.G. Touliatos

National Technical University of Athens, Athens, Greece

ABSTRACT

A traditional, old construction is a very uncertain subject to be precisely modelized and analysed. On the other hand, the originality and personality of each case, often developed during some centuries, does not permit an easy generalisation of its behaviour, failure modes and proper strengthening methods.

The most secure way, still seems to be, the careful recognition of each local constructional system, the understanding of its behaviour under dynamic loading and the study of the history and the present condition of each building before any intervention decision is taken.

Nevertheless, there are some basic principles that can be common for the typical cases of the traditional construction. This work tries to determine these basic principles of behaviour, pathology and intervention methods of a typical construction with masonry walls and wooden floors and roof, in seismic region.

1. INTRODUCTION

Seismic action acts on buildings with catastrophic results depending on the intensity of the earthquake, the quality of the ground, the region and the type of the construction.

But, it is generally noticed that the largest percentage of damages is observed on old "traditionally-built" constructions. This fact was once more confirmed, during the latest earthquakes in the regions of Perachora, Loutraki, Kiato (1981), and in Kalamata (1986). During this last earthquake the 77% of the traditional constructions were totally or gravely ruined.

These traditional buildings are usually constructed with vertical load bearing walls made of stones, bricks, cement or mud bricks and light wood framed structures.

The horizontal load-bearing structures consist of : wooden floors, roofs and balconies, stone or -brick constructions (e.g. marble or stone beams, arches, vaults, domes, etc.) or reinforced concrete slabs and beams.

They are buildings mainly constructed before the ample use of reinforced concrete with elements and technology based on the experience of the builders, without any structural or aseismic calculations.

All the above facts, plus ageing of constructions and wearing from different causes (as humidity, earthquake, ground-settlement) as well as the lack of maintenance, make these buildings a lot more vulnerable to earthquake action, than modern constructions.

The consequences usually are grave damages or total collapses, unfortunately followed by human victims.

Most of Greek villages and small rural towns are formed by this kind of buildings as well as a substantial part of our Architectural heritage. The memory of the total catastrophy of Argostoli in 1954 - a beautiful town in Cephalonia - is still tragically alive.

All countries, who are menaced by earthquake action, have organised research concerning the seismic behaviour of these traditional constructions.

In Greece, also, such a research is carried on with purpose to classify and explain the typical damages observed, to propose repair methods and ways of strengthening before the occurance of seismic action.

The prevention of damages by strengthening is a better and cheaper method than repairing after the damages have occured.

2. LOAD BEARING WALLS. SEISMIC BEHAVIOUR

Load-bearing walls of traditional constructions can be classified in two large categories according to their seismic reaction, in relation to the main direction of

                                        P.G. Touliatos

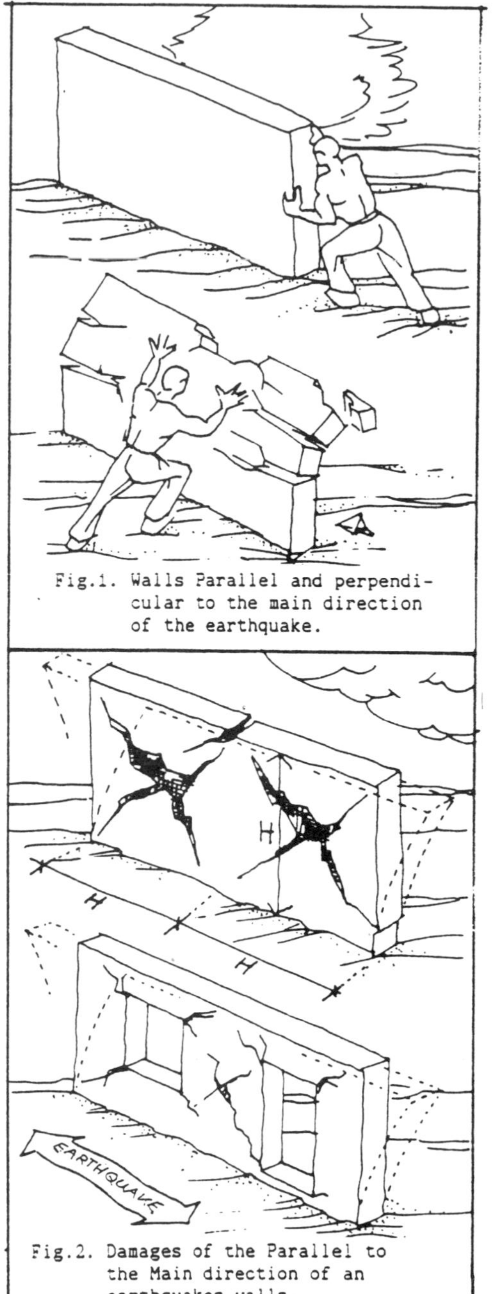

Fig.1. Walls Parallel and perpendicular to the main direction
of the earthquake.

Fig.2. Damages of the Parallel to
the Main direction of an
earthquakes walls.

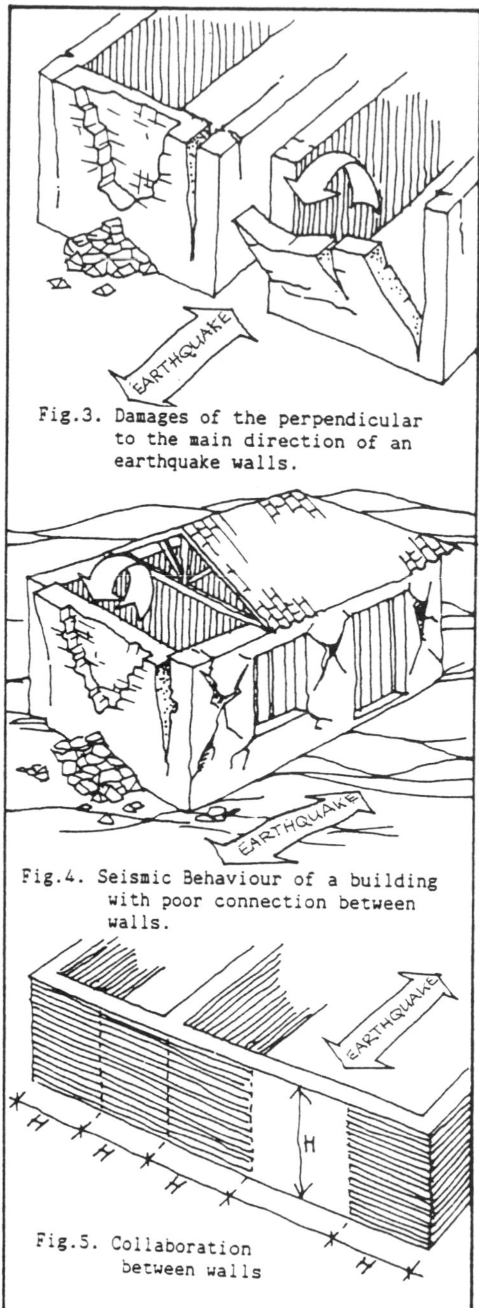

Fig.3. Damages of the perpendicular
to the main direction of an
earthquake walls.

Fig.4. Seismic Behaviour of a building
with poor connection between
walls.

Fig.5. Collaboration
between walls

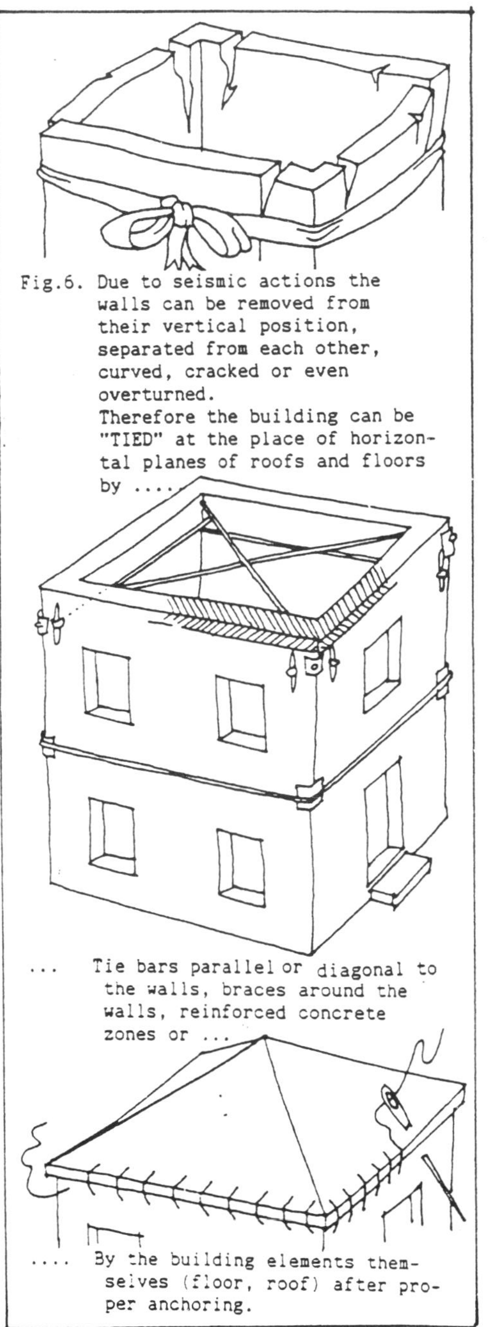

Fig.6. Due to seismic actions the walls can be removed from their vertical position, separated from each other, curved, cracked or even overturned.
Therefore the building can be "TIED" at the place of horizontal planes of roofs and floors by ....

... Tie bars parallel or diagonal to the walls, braces around the walls, reinforced concrete zones or ...

.... By the building elements themselves (floor, roof) after proper anchoring.

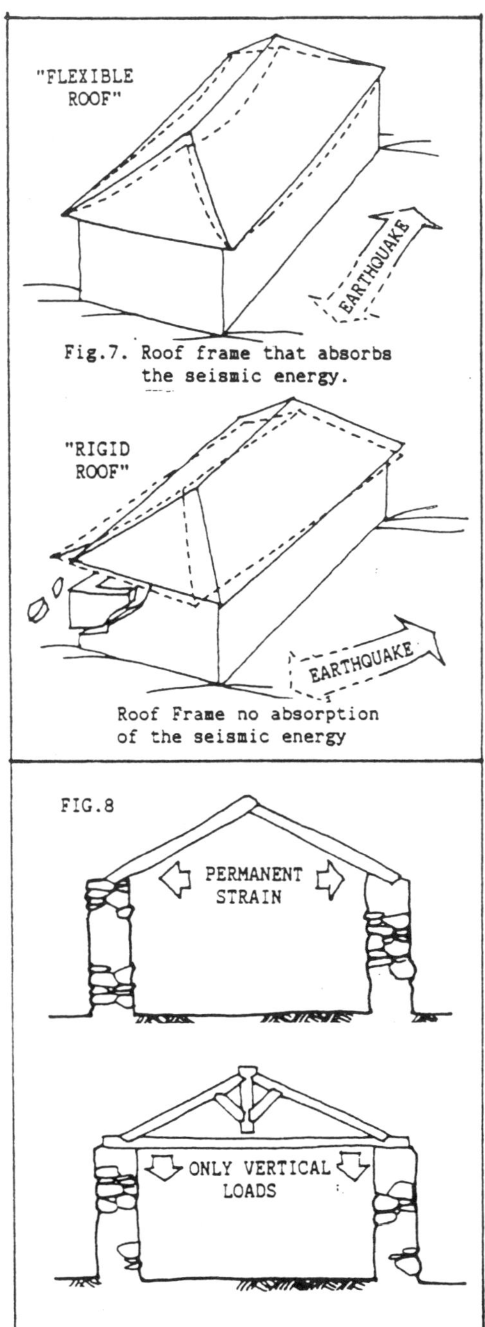

Fig.7. Roof frame that absorbs the seismic energy.

Roof Frame no absorption of the seismic energy

the earthquake.
- Walls parallel to the main direction of the earthquake (Fig. 1).
- Walls perpendicular to the main direction of the earthquake (Fig. 1).

It must be noted here for reasons of clarity, the existance of a main direction of an earthquake is accepted, though it is well known that seismic action is a very complicated phenomenon.

Walls parallel to the main direction of the earthquake have a better seismic behaviour than the perpendicular ones, where we observe cracks from bending stresses and in some cases these walls collapse after overturning.

Typical damages are :
- Walls parallel to the main seismic direction :
  Cracks of the "X" type are characteristic.
  When the walls have openings (doors, windows, chimneys etc.) cracks can also be observed around them (Fig. 2).
- Walls perpendicular to the main seismic direction :
  Cracks can be observed at the points of connection with transversal walls.
  Detachment, overturning and collapse of parts of the wall, division of the wall, division of the unity of the wall's width and even total collapse of the bending wall can ocure (Fig. 3).

Many factors play a crucial role in the determination of a wall's seismic resistance.

Basic factors are the materials which are used and the construction system(s).

Characteristic examples are wood framed walls which show very good seismic behaviour (no matter which the main seismic direction is).

Other factors are quality and shape of the construction elements of the wall (stones, bricks, etc.) existance and quality of the horizontal reinforcing zones, constructional methods, quality of plastering or other sheathing etc.

Research, observation and analysis of the damages give useful information for every different case, and point out the most adequate way to repair and strengthen the construction.

Earthquake acts on buildings in an additional way, weakening their strength step by step after every seismic action.

The weakening of the strength of a wall (or other parts of a construction) as time goes by, is due also to other factors as inadequate maintenance, humidity, ground settlements, etc.

Walls with damages from seismic action can be repaired and strengthened in several ways :
- By improving the quality of the mortar
- By repairing and strengthening the plaster
- By constructing "Jackets" of reinforced concrete or cement mortar
- By proper repair of cracks
- By confinings, embracings, horizontal tie rods or tie belts, etc.

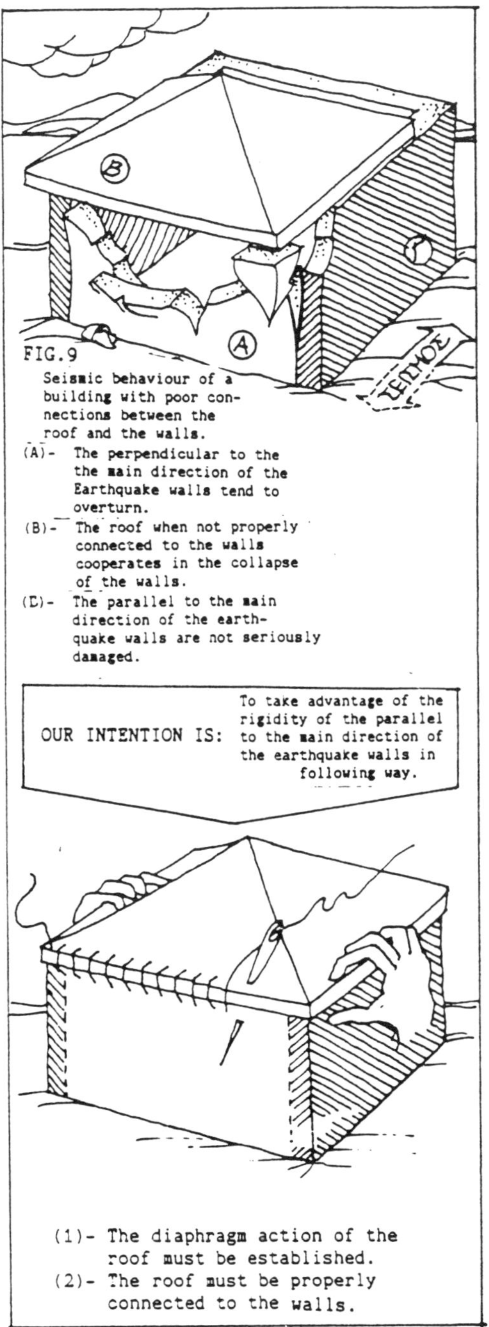

FIG.9
Seismic behaviour of a
building with poor con-
nections between the
roof and the walls.
(A)- The perpendicular to the
the main direction of the
Earthquake walls tend to
overturn.
(B)- The roof when not properly
connected to the walls
cooperates in the collapse
of the walls.
(C)- The parallel to the main
direction of the earth-
quake walls are not seriously
damaged.

OUR INTENTION IS: To take advantage of the rigidity of the parallel to the main direction of the earthquake walls in following way.

(1)- The diaphragm action of the
roof must be established.
(2)- The roof must be properly
connected to the walls.

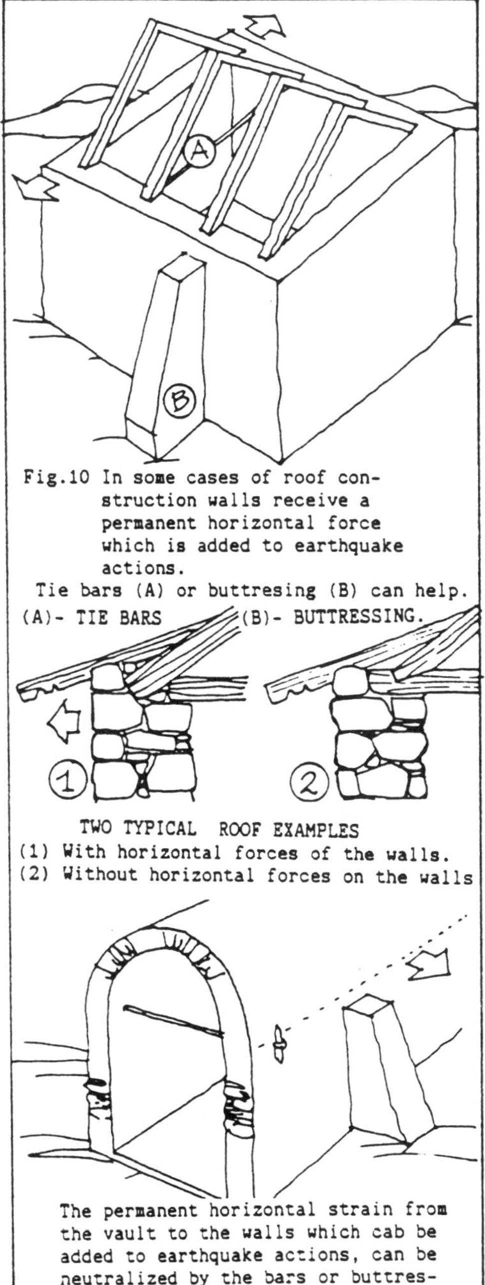

Fig.10 In some cases of roof construction walls receive a
permanent horizontal force
which is added to earthquake
actions.
Tie bars (A) or buttresing (B) can help.
(A)- TIE BARS   (B)- BUTTRESSING.

TWO TYPICAL ROOF EXAMPLES
(1) With horizontal forces of the walls.
(2) Without horizontal forces on the walls

The permanent horizontal strain from
the vault to the walls which cab be
added to earthquake actions, can be
neutralized by the bars or buttres-
sings.

3. COLLABORATION OF WALLS PARALLEL AND REPRENDICU- LAR TO THE MAIN SEISMIC DIRECTION

The building or parts of it behave like an orthogonal parallelepiped (an opened box) (Fig. 4).

Good construction and strong connection of the walls between them have the following results :

Walls parallel to the main seismic direction - rigid walls- cooperate with the perpendicular ones, retraining and strengthening them.

This collaboration is valid for a length H   (of the perpendicular wall) corresponding to a height H of this wall (Fig. 5).

Correct, strong connection of the walls between them is insured by:

- Correct entanglement and joining of stones or bricks at all connections.
- Existance and obligatory crossing of horizontal tie belts.
- Confining, embracing
- Butressing
- All existant floor and roof members after having established their diaphragm action and after using proper anchorages between them and the walls (Fig. 6).

4. HORIZONTAL LOAD BEARING STRUCTURES (FLOORS-ROOFS)

Traditional roofs and floors are usually made of wood. Sometimes we find horizontal load bearing structures of stone or bricks (vaults, domes, arches) metal beams (neo-classical buildings) even reinforced concrete.

Especially during seismic action, horizontal load bearing structures, (partially or in the whole) transfer horizontal forces to the vertical load bearing system (Fig. 7,8).

A general goal is not only to extinguish the transferring of horizontal forces but also to transform these horizontal structures into diaphragms.

In this way, with adequate anchoring and joints, the horizontal structures become rigid and connect and strengthen the walls during seismic action (Fig. 9).

The main actions in order to strengthen the roofs and floors and to establish through them the co-operation between the parallel and perpendicular to the main seismic direction walls are :

- To extinguish all existing permanent horizontal thrusts with the rods, butresses etc.   (Fig. 10).
- To establish diaphragm action of roofs/floors through sheathing and bracings or skew boards etc.
- At the same time all damaged members of roofs/floors should be restored.
- To establish connection and collaboration of horizontal and vertical load bearing structures using horizontal belts, wood-plates, tie bars etc.
- To repair and strengthen the sheathing of the roof so as to establish water-

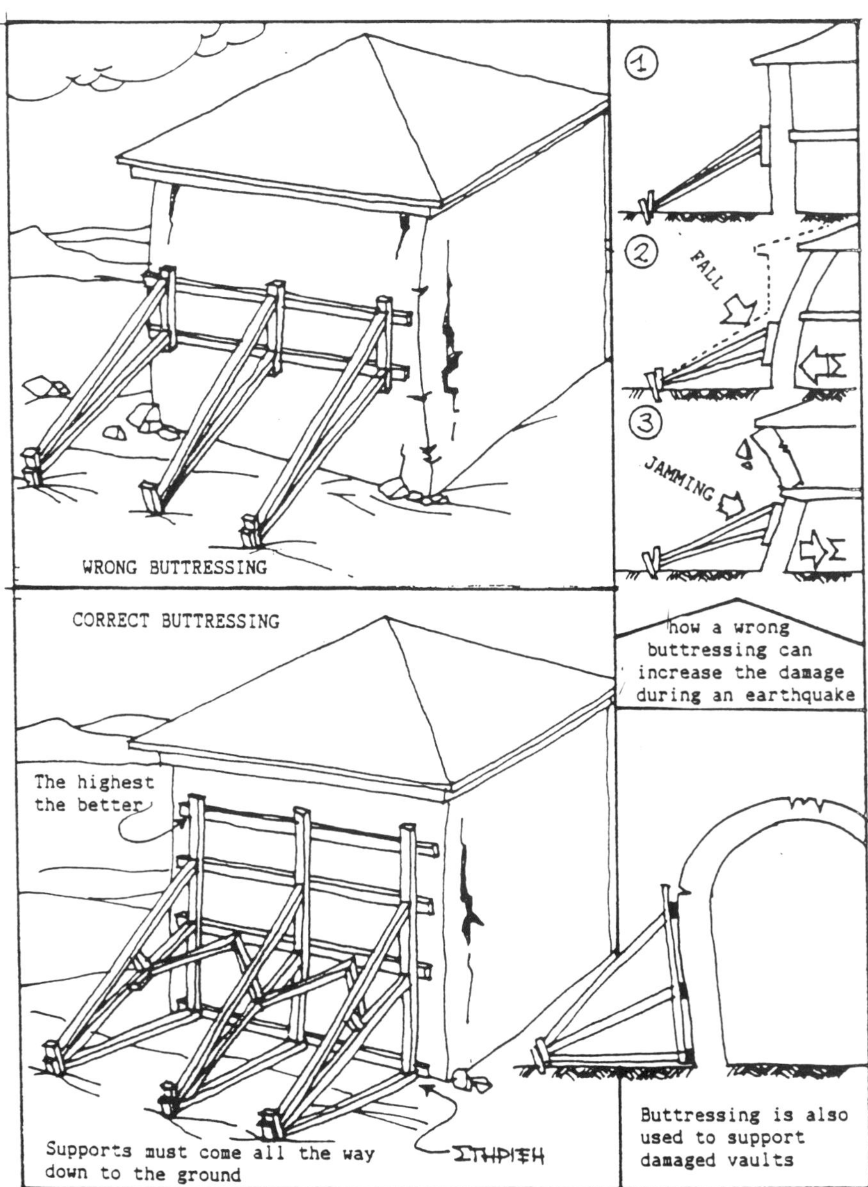

FIG.11 IMMEDIATE MEASURES AFTER AN EARTHQUAKE

proofing.

5. IMMEDIATE STRENGTHENING ACTIONS IN ORDER TO SAVE BUILDINGS DURING SEISMIC ACTION

Right after the earthquake, damaged buildings and monuments are in great danger of getting more seriously damaged or collapsing, if seismic action goes on.

Immediate actions, such as butressing protection and strengthening of the constructions could avoid further damages.

But serious attention should be paid when constructing supports and buttreses in a damaged building, because faulty constructions could cause greater damages (Fig. 11).

6. REPAIRS-STRENGTHENING AND PREVENTION OF DAMAGES FROM SEISMIC ACTION

All the above mentioned about collaboration of the more rigid (stiff) walls of a building during an earthquake, with the weaker ones using the help of floor and roof structures, can be achieved and should be achieved before seismic action.

Considering the importance of every traditional construction, the seismic strengthening can be planned and realized at different stages, even during the stages of regular works of maintenance.

Information about the appropriate actions should be given to the authorities and also to the owners and users of these buildings.

The organisation of specially informed technicians and workshops in regions of high seismic risk could prevent many disasters.

These workshops should be provided with special knowledge, equipment and materials in order to be capable of acting immediately during seismic action. Also they must be able to strengthen traditional constructions and monuments in order to prevent damages.

We must note that immediate actions demand risky decisions with criterior the saving of traditional buildings.

It is "easy" and "safe" to characterize buildings as "dangerous" or "to be demolished" but this leads us to the extinction of traditional Architecture.

7. ADDING-ALTERING A TRADITIONAL BUILDING

Traditional buildings, in order to survive through the years in their social and natural surroundings, usually need some kind of altering of their function and

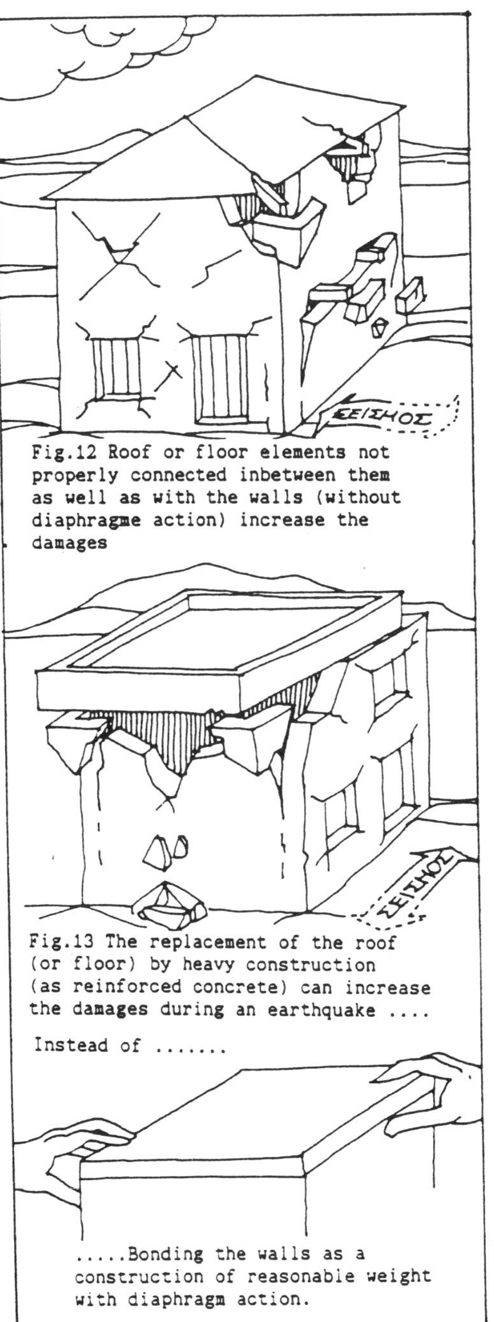

Fig.12 Roof or floor elements not properly connected inbetween them as well as with the walls (without diaphragme action) increase the damages

Fig.13 The replacement of the roof (or floor) by heavy construction (as reinforced concrete) can increase the damages during an earthquake ....

Instead of ....

.....Bonding the walls as a construction of reasonable weight with diaphragm action.

EARTHQUAKE OF KALAMATA
SEPTEMBER 1986

Damaged monument in the Cemetery of the town.

structure.

This happens with adding a story or some extra rooms or simply altering the function of the building.

But it must be noted that in order to alter an existing traditional construction, we must examine the project from the aseismic point of view.

Some of the most usual faults are :
- Adding Heavy-weight floor-roof constructions usually made of reinforced concrete.
- Adding extra rooms in a way that form., mass, and seismic behaviour of the building is completely altered and weakened.
- Adding constructions made of materials that cannot collaborate with the existing ones.

In Figures 12, 13 we can observe damages that occur from the above mentioned actions.

CONCLUSION

Our civilisation and architectural heritage was born and lived through-out the centuries, facing seismic action.

With the knowledge and skills we now have, if we pay some attention this can go on for many years yet.

Already in Greece Institutions (like the N.T.U.A.) or organisations (like the Organization of Aseismic Design and Protection) have started to work in this direction. The International collaboration, the proper research and the observation and analysis of the seismic effect concerning traditional buildings will contribute in our effort to solve the above mentioned problems.

REFERENCES

1. CHR. ARNOLD, "BUILDING CONFIGURATION AND SEISMIC DESIGN", New York 1982.
2. A. COBURN - R. SPENCE, "EARTHQUAKES PROTECTION" , New York 1993.
3. P.G. TOULIATOS, "REPORT ON THE GREEK EXPERIENCE CONCERNING THE STRUCTURAL BEHAVIOUR OF TIMBER CONSTRUCTION IN SEISMIC ZONES".
    FROM : "STRUCTURAL BEHAVIOUR OF TIMBER CONSTRUCTIONS IN SEISMIC ZONES", A.CECCOTTI, FLORENCE UNIVERSITY 1990.
4. T. TASSIOS - E. KATSARAGAKIS - P. TOULIATOS, "TIMBER IN EARTHQUAKE - SAFE HOUSES. THE GREEK EXPERIENCE", Athens

1989:
5. J. EBERHARD, "ARCHITECTS AND EARTHQUAKES", New York 1975.
6. P. CARYDIS, O. VAGELATOU, I. VLACHOS, M. XINOSTATHI, P. TOULIATOS, K. HOLEVAS , "RESEARCH FOR THE CONSTRUCTIONAL ASEISMIC REGULATION", Athens 1985.

A MECHANICAL MODEL FOR STATICS AND DYNAMICS OF HISTORICAL MASONRY BUILDINGS

A. Giuffré

University of Rome 3, Rome, Italy

From:

- A.Giuffré, *Mechanics of historical masonry and strengthening criteria,*
  XV Regional Seminar on Earthquake Engeneering, Ravello, 1989.
- A. Giuffré, C. Carocci, *Statics and dynamics of historical masonry buildings,*
  Workshop on masonry monuments, Creta, 1992.

Rearranged by:
Antonella Cecchi and Lorena Sguerri

AIM OF THE LECTURE

The aim of this lecture is to present an organic point of view on the mechanical qualities of the historical masonry, and on the mechanical behaviour of the historical buildings.

The presentation will be subdivided in seven paragraphs:

1.   *actual masonry in historical buildings*
2.   *mechanical idealization of the masonry*
3.   *the structure of house*
4.   *the seismic problem in historical centres*
5.   *collapse mechanisms of the buildings*
6.   *dynamic features of historical masonry*
7.   *technical criteria for restoration.*

But one point should be made clear before beginning.

Elapsing the centuries the masonry got many different features, often related to very different static ideas, and in different regions the constructive practise was affected by local demands and local availability of materials; so, it is evident, this presentation can't achieve a complete effectiveness. In addition, if you consider for how much time the active practise of the historical masonry has been neglected, and in the meanwhile how sophisticated we have become with the analytical approach to the structural analysis related to the modern technologies, you can understand that mechanical modeling of ancient structures must be rediscovered.

Nevertheless we need to take up a rational way in order to preserve our architectural heritage; the cultural world demands philological reasons to accept the interventions of strengthening we propose; and rationality and philology are the natural requirements of the scientific approach.

So, as you can see, I should go to present a rational, philological, and consequently scientific way for the restoration of the historical towns, and it seems a not little task. Unfortunately I'm not able to accomplish such a duty, and then you will have only an attempt to apply the scientific methodology to the analysis of the static features of our ancient architecture.

1. ACTUAL MASONRY IN HISTORICAL BUILDINGS

Looking at the masonry work it is possible to distinguish between two different realisation.

The high architecture, religious or militar, has been almost always realised with worked stones. Squared stones are in portion of Micene's walls, squared stones constitute the most important greek and roman buildings.

The name given by Vitruvio to this kind of masonry is "opus quadratum", and with this name I will refer to it.

Mechanical qualities of "opus quadratum" will be investigated later, but it is possible to observe immediately the regularity of its horizontal layers and the regular interlocking of the stones across the wall, to make it monolithic *(see Pl.1)*

In parallel with the high architecture tradition, masonry has also a popular tradition, which can be represented by walls made with raw clay (see *Pl. 2*).

In french they are called *mures pisés*, accounting for the beating of the clay concerning that construction; in italian they were called *muri formacei*, accounting for the form on which they were shaped.

In all Europe the raw clay masonry was adopted until the end of the XIX century, and it is adopted at present in many regions of the world[1]

Leon Battista Alberti mentions it as an excellent way to get comfortable houses[2], and first of him Plinio[3] and Vitruvio[4] have mentioned the raw clay masonry has the most durable work.

In this kind of masonry not horizontal lies, not transversal connections can be recognized.

As an evolution of the formed clay masonry, about at the end of the first century before Christ, the Romans invented the *opus concretum* made with alternate layers of little stones (*caementa*) and mortar (*materia*) put into forms and beaten.

Every three feet a layer of large bricks (*bipedali*) was placed across the wall, to achieve the horizontal lying, while the monolithism was assured by the cohesion of the mortar.

Everybody knows how extraordinary works the Romans built with this kind of masonry. Nevertheless it should be observed that the most of the Roman monuments have not arrived at our days: their monolithism was assured by the mortar, but if it is not a good mortar the wall is easily broken up.

The squared stone masonry (*opus quadratum*) doesn't require mortar, because its monolithism is achieved by means of the suitable arrangement of the stones, but mortar is needed by the little stones which make the *opus concretum*.

The most common masonry met in the Italian ancient buildings is made with irregular stones connected by mortar, but in spite of their unevenness, the techniques of the "raw stones" masonry is not without rules.

These rules are not so geometrically simple as for the *opus quadratum*; on the contrary they are complicated and varied as the different languages of the people.

We find in all the ancient authors of treatises, till the beginning of our century, the description of the raw stone masonry *(see Pl.3)*. Requirements are always the same: every three feet of height an horizontal layer shall be settled; big stones oblong shaped shall be disposed across the wall as

---

[1] The raw clay is used too in form of bricks, arranged like the *opus quadratum*.

[2] L.B. Alberti, *L'Architettura [De Re Aedificatoria]*, Il Polifilo, Milano 1966, book III, chap. XI, p. 220.

[3] Plinio, *Storia naturale*, Einaudi, Torino 1988, vol. V, book XXXV, chap. 48, p. 491.

[4] Vitruvio, *De Architectura*, book II, chap. III (*de lateribus*).

frequently as possible. Other rules regard as to fill the voids among the stones with little pieces of brick and mortar, how to position the stones accounting for their particular shape, ... and so on.

The trend to obtain the mechanical feature of the squared stone masonry is evident, but only the skill of the mason can succeed in getting such a goal. Building a good masonry is really an art: the artist knows and applies its rules.

2. MECHANICAL IDEALIZATION OF THE MASONRY

We have reached the conviction that the raw stone masonry, as applied in Europe since the Middle Age and until the end of last century, directly derives from the mechanical ideas implicit in the *opus quadratum*. So it is possible to formulate the following statement:

The squared stone masonry can be assumed as the mechanical idealization of the most part of the historical masonries. If those have been "perfectly done", an effective analysis of their strength can be carried out adopting the squared stones model, related to the actual percentage of big and little stones.

Referring to such a statement the first analysis to be performed in order to evaluate the mechanical quality of a masonry wall should be devoted to understand its texture and its quality.

The knowledge of the techniques usually applied in a given region, their evolution in the time and their local features, is the essential base for getting the observer able to distinguish a wall "perfectly done" from a bad masonry made by an ignorant bricklayer.

Often the wall has been built with two exterior textures, filled in the interior with little stones without any interlocking. In this case the masonry looks a good work but it is awful.

Anyway any raw stone wall in which it is possible to identify the size and position of the stones can be analyzed in statistical terms, pointing out the average size of the blocks (subdivided if necessary into categories) and the rule for positioning them implicitly applied by the mason.

The corresponding model of *opus quadratum* is thus defined, and the results of mechanical analysis carried out on the model can provide useful indications for the prototype.

Plates 4 and *5* illustrate the derivation of models from actual walls.

The rule for positioning the blocks which defines *opus quadratum* masonry emphasizes the presence of an internal structure in the material making up the wall. It cannot be reduced to an isotropic material: not only do the layers make it eminently anisotropic, but the considerable size of the components with respect to the thickness of the wall introduces unavoidable structural effects. If it is desired to pass to the continuum, masonry must be modeled as a material having a structure, according to the principle of the "continuum of Cosserat".

The approach we are following represent any unit as a rigid body with monolateral contact, and performs a limit analysis with the homogeneization method.

The research is going on with good results; but since there isn't a mathematical model able to study such a material, it is possible to approach the problem in experimental terms, constructing prototypes of *opus quadratum* masonry according to the actual rules of assembling cited above[5]. In a simple way it is possible to analyze the mechanical features of the squared stone masonry

---

[5] Such an experimentation has been carried out by Vittorio Ceradini in his doct. thesis at the Department of Structural Engineering of the University of Rome, Faculty of Architecture, 1992.

examining it stone by stone, imposing equilibrium conditions to each of them with the hypothesis that every stone transmits to the one below a linear diagram of normal stresses. Shear stresses can be transmitted too, following the Coulomb's friction theory.

In-plane action

The diagrams of the tensions show that the lack of strength in traction makes the masonry unable to diffuse the load according to the requirements of the De Saint Venant postulate.

Only the portion of masonry directly interested by the load is stressed, along the action line of the force: that is evident for both the vertical and the diagonal direction (see *Pl. 6*).

The diagonal load is transmitted from a stone to the other by means of normal and tangential stresses, according to the Coulomb's theory.

The result of this kind of analysis is in agreement with the more sophisticated theories related to the no-tension continuum body[6].

It can be observed that the interlocking between blocks produces, through friction, a pseudo-tensile resistant strength; this is a function of the compression stress on the sliding surface.

The dependence of tensile strength on the compression stress makes impossible a direct comparison between the *opus quadratum* model and a wall in raw stone with cohesive mortar.

The mortar presents a uniform tensile strength, and the interlocking between stones which are not perfectly squared does not create friction as effectively as in *opus quadratum*.

It is not difficult to specify the laws which govern the correlation between prototype and model, given that both are actually masonries of *opus quadratum* which obey the same law of geometric formation.

Plate 7 shows the basic relationships which govern this similarity.

The most significant result, as concerns the possible appearance of cracks, is the relationship between the tensile stress produced by loads, and the pseudo-tensile strength made possible by friction; relationship conditioning the formation of the cracks.

This "stability relationship" (*Plate 7* reports its essential formulas) differs between prototype and model only in the differences which may occur between them in the friction coefficients: if these are equal, both in the prototype and in the model, the two stability relationships are also equal, and the appearance of cracks in the model corresponds to the appearance of cracks in the prototype.

The experimental results obtained with the "opus quadratum" model are extremely interesting. The testing method is set down in *Plate 7*: the model is placed on an inclinable frame whose inclination introduces a component of weight parallel to the layers of the masonry.

The component normal to the layers takes the place of the vertical force (namely the own weight), while the component parallel to the layers, which is equal to the former multiplied by the tangent of the angle of inclination, corresponds to the horizontal force.

The tangent of the angle of inclination thus plays the role of a "seismic coefficient".

The angle of inclination which produces collapse (that is, the seismic collapse coefficient), as

---

[6] S. Di Pasquale's theory, related to the continuum and isotropic body, clearly shows that the stress only follows a linear path inside the body, stressing it only in compression.
It will be shown later that such characteristic concerns the building structure too: an historical masonry building doesn't behave as a whole, but every loaded portion carries its load quite independently by the next portions. That is the consequence of the lack of connections.

well as the crack pattern (that is, the collapse mechanism), are strongly influenced by the pseudo-tensile strength of the model.

It is possible to account for different values of such pseudo-tension by changing the position of the blocks so as to modify the amount of overlapping.

Plate 8 illustrates in schematic form the typologies of collapse: the sliding of the upper layers for walls of small height; for higher walls the overturning of a portion toward what might be called the "downhill" side (the size of this portion decreases with smaller values of tensile strength); and total overturning, which happens as for a monolithic block, when tensile strength reaches high values.

It is interesting to observe the inclined cracks, which are normally called "shear cracks"; experience shows that the mechanism which produces them is the overturning of a portion which detaches from the wall.

The difference in the resistance to tensile strength appears in the inclination of the crack, which tends to become vertical as the resistance tends to zero.

It is also useful to compare the form of the crack obtained experimentally with that which occurred in a wall in Reggio Calabria during the 1908 earthquake.

Plate 9/a illustrates this comparison.

Although, as has been mentioned, the *opus quadratum* model can represent the reality of stone and mortar only qualitatively, the likelyhood between these crack patterns encourages to go on with experiments of this kind.

Moreover, it may be useful to observe that the cracks tending towards the vertical are produced by the distributed horizontal force; the failure mechanism would not be the same if the wall were stressed by a concentrated force acting on one of its vertical edges. Corresponding to the clearly different distribution of internal stress, there is a different crack pattern and a different value for the collapse load.

Such differences can be observed too in a wall with openings: in *Plate 9/b* can be seen the result of the experiment carried out in both conditions.

The out-of-plane collapse of a wall braced by buttresses also assumes the classic forms observed after earthquakes: see in *Plate 10* the result obtained on the test frame, where the model reached collapse at a seismic coefficient value of 0.15, and the photograph of a facade of the Palazzata of Messina, which partially collapsed in 1908.

The agreement of these images proves the interest of the *opus quadratum* model, easy to set up and test.

But now let's go straightaway to the most challenging action on the masonry wall: the out-of-plane action. The main task of a masonry wall has always been that of closing spaces and its main weakness is the resistance to lateral push.

Out of plane action

It is not difficult to study the behaviour of a masonry wall pushed out of plane.

The first step concerns the analysis of the monolithic bloc.

A monolithic wall subjected to a lateral foce acting at its upper edge has two possibilities of collapse: rocking and sliding, depending on the height-thickness ratio. If the wall is higher than $H = b/2f$ (were f is the friction coefficient) the maximum horizontal force it is able to support is $F = \gamma b^2/2$ (see *Pl. 11*).

This value doesn't depend from the height: if the height increases, and with it the overturning moment produced by the horizontal force, the weight and its stabilizing moment increase too.

But if the wall is lower than the over-mentioned value it doesn't rock: it slides under the force F $= f \cdot \gamma bH$.

The second step concerns the analysis of a wall made of superimposed stones.

A lateral force acting on the upper stone provokes the sliding of that stone, but if there is also vertical load, and the derived friction strength exceed the value of the lateral force, the sliding is avoided. The vertical load makes the superimposed stones to behave as the monolithic wall[7].

We are used to account only for mechanically effective bonds, but it is important to remember that monolateral bonds, like the contact between two stones, is quite effective if the vertical load is enough.

Even if it is obvious, it is important to remember that a lateral force applied on the top of a wall, without an adequate vertical one, provokes the sliding of the first stone: many ineffective tie-beams would be saved.

If an *evenly spread horizontal force* acts on the wall its maximum value related to the total weight only depends on geometrical parameters: $F/P = b/H$[8].

Effectiveness of the "diatoni"

When the thickness of the wall is made by more then one stone the effectiveness of the *diatoni* is accounted for.

The lack of stones placed through the thickness of the wall leaves it to behave as two independent walls, but the presence of the *diatoni* makes the wall monolithic. The difference of the horizontal strength is evident: the monolithic wall resists twice more than one made by two independent walls.

The *diatoni* have the task of supporting the smaller stones, making their weight participant to the stabilizing moment.

Even if much part of the horizontal contact between the stones is lost, the global behaviour is that of the monolithic wall.

Plate 12/a lists some walls tested by Vittorio Ceradini; the graph illustrates the decrease in resistance to overturning as the number of diatones crossing the wall is reduced.

Plate 12/b shows a pattern of partial diatones and smaller blocks; in this case also the wall's effectiveness to overturning decreases with the reduction of the diatones, that is, with the percentage increase of small stones with respect to larger ones[9].

In Ortigia (the ancient Syracuse) we found three kind of masonry work. The "opus quadratum" model allows to evaluate their strength against lateral push. It is evident how the presence of little stones makes less resistent the wall *(see Pl.13)*. So the dependence of strength by the arrangement of the stones is demonstrated. More sophisticated analysis can be made in every cases, but the direct observation of the masonry work allows to understand its mechanical quality.

---

[7] Similar observations were first presented in Italy by Mario Como and Antonio Grimaldi, from the second University of Rome.

[8] We are often ingenuously amazed recognizing in the actual behaviour of the structures the strict parallelism between mechanics and geometry, because we are not used to deal with isostatic structures; but the ancient builders were conscious of that, and their geometric considerations were quite close to our usual mechanical analysis. Now we must recover the reliance to the geometrical analysis, if referring to that ancient masonry structures.

[9] V. Ceradini, doct. thesis 1992.

Force-displacement relationship

The force-displacement relationship, according to the rocking motion of the rigid body, can be easily derived.

If the ratio between the height and the base is not too small, such relation can be assimilate to a linear function starting from the value of the overturning force and getting zero at a displacement equal to half the thickness (see *Pl. 14*).

The same relationship attains the squared-stone masonry, if it is built with all the necessary *diatoni*; but a different arrangement could be cause of a different behaviour: the significant parameters of the force-displacement relationship, the ultimate force and the ultimate displacement, can change from the values concerning the monolithic body, and both decrease as much as the standard arrangement of the stones has been betrayed.

As conclusion of this paragraph it can be pointed out that the arrangement of the squared stone, the dimension of the partial *diatoni* with reference to the thickness of the wall, the distance between them, the dimension of the little stone, and so on, can be used as multi-dimensional parameter useful to define the actual consistence of the masonry and to explore, parametrically, its behaviour. In this way the force-displacement relationship can be defined as a function of such parameter and the indefiniteness of the geometrical description is removed.

The strength of a wall against out-of-plane forces depends on the arrangement of the stones. The regular arrangement (rule of art) allows a behaviour like the monolithic body; a less effective texture makes the mechanical parameters of the rigid behaviour to decrease.

If the parametric analysis of the masonry walls, referred to the stone texture, is available, the force-displacement relationship of the actual masonry can be evaluated by pointing out the actual value of the multi-dimensional parameter of the actual arrangement.

Cyclic behaviour of the rocking wall

It is interesting to introduce a last mechanical consideration before completing the analysis of the squared stone masonry. Of course the actual behaviour will present an initial elasticity, dependent on the elasticity of the material and on the elasticity of the foundation. So the force-displacement relationship will start from zero with a linear branch, and it will evolve as an elastic bilinear material, with a decreasing second branch.

The cyclic behaviour following this "bilinear model" doesn't present any hysteresis (see *Pl. 14*).

Nevertheless, accounting for the actual stresses that will stress the masonry, it is easy to foresee that some degradation is inevitable, and that produce hysteresis.

The well known TAKEDA-model can represent an hysteretic behaviour, but, I need to underline, it doesn't describe a clear characteristic of the masonry, it can only be applied like a mathematical way to account for not recognizable mechanical behaviours.

We will examine later the dynamic response of these kinds of models and we will discuss the difference between the two formulations.

What has been said until now is enough as an attempt to give a mechanical description of the squared stones masonry, the ancient *opus quadratum* which can be adopted as a model of usual raw stones masonry.

This presentation can be synthesized with the following words:

The well done masonry is characterized by horizontal lies and regular transversal connections.
That allows it to behave as a monolithic body rocking on horizontal straight lines.
The lack of transversal connection makes the masonry unable to a regular rocking motion.

INTERMEZZO: SOME CONSIDERATIONS ABOUT DIFFERENT KIND OF MASONRIES

It could be interesting to have a mention of different kinds of masonry[10] that can't be assimilated to the *opus quadratum*, and required a new analysis completely different from the one presented (see *Pl. 15*).

The people of Cartage, which have been taking up the little island of Mozia (on South-West of Sicily) about on the V century b.C, used to build a kind of completely irregular masonry, were no trace of horizontality can be found.

It is not infrequent to find more recent masonry made like that; but whereas the ancient african people had no idea about horizontal support as a good way to sustain vertical loads, and such kind of masonry is coherent with their culture, modern realizations in Europe indicate the ignorance of consolidate rules.

Typical of the ancient African people is the so called *opus africanum*, or *frame masonry*, adopted in Mozia too, in Selinunte and in Morgantina ,in the South-West coast of Sicily as well as in Africa.

It is evident that the static idea is completely different from the one implicit in the *opus quadratum*. Here the vertical stones play the role of the piles in a modern framed structure, but without the connections usual in the modern materials. In some way the vertical stones can remember the wood architecture, and the irregular texture of the wall the non-structural partitions.

An other ancient constructive tipology, used since the Roman time, is the *opus craticium*: wooden frame mixed to the masonry as can be find in Ercolano, preserved under the ash after the famous eruption of the Vesuvium on 62 b.C.[11]

It has been adopted all over Europe, improving the stiffness with crossed trusses in the frame. From this model all the wooden architecture in North Europe derives, and to this model, after the earthquake of 1783, Vivenzio made reference offering to the Government of Naples the typical design of the "antiseismic house"[12].

Of course the mechanical considerations derived from the *opus quadratum* can't be applied to this kind of masonry. It must be clear that now and in the following of lecture 1 only concern with italian historical centres.

---

[10] See, on this subject, J.P. Adams, *L'arte di costruire presso i Romani*, Longanesi & C., Milano 1988.

[11] Vitruvio didn't love this kind of technology. He wrote: *Craticii vero, velim quidem ne inventi essent* (I would have preferred that *opus craticium* had been never invented); Vitruvio, *De Architectura*, book II, chap. VIII, p. 11.

[12] G. Vivenzio, *Istoria de' tremuoti avvenuti nella Provincia di Calabria ulteriore, e nella città di Messina nell'anno 1783, e di quanto nella Calabria fu fatto per lo suo risorgimento fino al 1787, preceduto da una Teoria ed Istoria generale di Tremuoti*, Stamperia Reale, Napoli, 1788.

3. THE STRUCTURE OF THE HOUSE

The study of masonry is but one step, albeit an important one, towards the study of the historical house.

The house is a simple organism made up of juxtaposed elements, but their assembly generates an organism which offers itself as safe and resistant to external actions.

The construction of the wall respects general rules that come from that ideal common matrix which is *opus quadratum*, but nevertheless there exist local interpretations conditioned by the materials of the place and by the building culture of a certain period. In the same way the house-organism is articulated according to general rules, but offers particular aspects closely connected to a geographical area and an historical period. This means that there exists, and there was known to the builders of old, an unwritten project, an implicit model and they made constant reference to it.

There existed, therefore, a concept of house which included its intrinsic characteristics: functional-distributive, formal and static.

Observing examples of houses in different Italian regions, it is easy to understand that there exists a common matrix translated into reality by using particular interpretations. The comparison of these organisms is possible because the elements making them up are the same, and perform the same rules with respect to the whole.

All the houses are composed of elementary cells aggregated differently in plan and superimposed in order to form units of more than one floor. The differences depend on the urban texture in which the house is built: pre-existing elements, available spaces; every new unit is adapted to its neighbours maintaining constant the tendency to realize, wherever possible, the configuration of the elementary cell with its dwelling surface.

The walls which form partitions support the horizontal elements: floors and roofs, and between these is placed a connecting structure between the various levels of the dwelling: the staircase.

In some geographical areas the first floor is realized with a vaulted masonry structure, reminder of a Roman practice.

This is a less simple structure than the wooden floor. It requires a greater degree of construction skill, and its thrust must be countered by increasing the thickness of the surrounding walls.

Some times the house is organized so that the lower part does not communicate with the upper, the latter being reached via an external staircase: this is the "proferulum house", found in many European areas.

Plate 16 shows some photographs of "proferulum houses" in different parts of southern Italy.

In *Plate 17/a* are reproduced the schemes of the vaulted house and the wooden floor house, models of reality which are useful for theoretical analyses.

However, in the generality of cases, both the parts and their assembly are defined elementarily, and certain tangled, urban fabrics, crisscrossed by small, tortuous dead-end streets, reveal themselves upon closer observation to be nothing more than a system of elementary cells, side by side and one atop the other.

What becomes inextricable, in such aggregates of dwellings, is the ownership situation. The addition of rooms in adjoining cells which expanding families have made, especially in the last century, are such that, taking each dwelling by itself, it is impossible to find its structural thread.

This thread must be sought in the logic of development of the inhabited area, instrumentally disregarding its present use; only from this point of view does it regain its underlying essential simplicity.

The process of growth of the building texture in the historical city has an important structural aspect. The elementary cells lean against one another, often using the walls of preceding cells.

Plate 18 shows the urban texture of two quarters in historical Italian cities, very different from one another. One is La Mattonata in Città di Castello[13]. This is a quarter laid out in prevision of expansion outside the city walls in the 13th century, while the other, La Graziella in the historical centre of Syracuse on the isle of Ortigia, derives from spontaneous building in an area which had been the site of the ancient Greek settlement[14].

Such evolving processes can be modeled in general terms: *Plate 17/b* reports the model derived by the linear texture of Città di Castello. The scheme illustrates the possibilities of inserting a new cell between preceding ones, pointing out the resulting non-connected angles, potential weak points in case of an earthquake. *Plate 17/c* models the spontaneous growing of the district "La Graziella" in Ortigia. It lists the positions in the urban texture which involve different exposure: a single exterior wall, two walls for corner houses, three walls for end houses.

Generally the unit of the masonry grid varies little from the dimensions of 6x6 m; wall thicknesses are at times conditioned by the dimensions of the stones, but are often close to three palms, or 60 cm. Nevertheless local conditions and material can modify such a trend.

In Città di Castello the masonry cell is usually elongated (*Pl. 19*) because it is built within a preconstituted subdivision.

This structuring of the cell characterizes the floor, made up of large main beams, parallel to the street facade and placed at rather close intervals, and smaller secondary beams orthogonal to them, placed at a distance which allows for a covering of thin bricks, the traditional "pianelle", which form the structure of the floor above.

The staircase, at right angles to the street front and located along one of the lateral walls, is here an independent structure: there is always present a second wall which provides the second support for the steps and for the beams of the floor, and which terminates where the staircase ends; the roof also has beams parallel to the street front, supported on gable walls, and therefore naturally has no thrust.

The lateral walls are of greater thickness with respect to those facing the street, and are often prolonged towards the exterior leaving to the front wall the function of a filler between them[15]. This way of building the house, realizing the lateral walls disconnected from the front wall, can also be observed in the houses on the Lungarno in Florence; from a strictly mechanical point of view it also recalls the houses of Venice, though their appearance is very different (*Pl. 20*). But, in Venice, there are reasons concerning the foundations which explain this usage: the masonry structure "had to" be articulated in independent walls to avoid the consequences of different settling between the foundation built on pilings of the facade and the foundations in masonry of the internal walls[16].

On the other hand in Venice the facade walls are systematically tied to the floors; but also in Città di Castello steel anchorages can be seen in the facades, tied to the floor beams, as if to demonstrate a common static concern of builders in widely separated and at first glance

---

13 *Manuale del recupero di Città di Castello*, Editor F. Giovanetti, urban lavoratory, town council of Città di Castello 1992.

14 VV.AA., Techniques of preservation of the historical centre of Ortigia, A. Giuffrè editor, to be printed.

15 R. Argalia, degree thesis at the Architecture Faculty of the University of Rome, 1991.

16 For these and other observations about the venetian house see: G. Creazza, *Aspetti del degrado strutturale a Venezia*, in Atti della giornata di studio: "A vent'anni dall'evento di marea del novembre 1966", Venezia 1987, p. 83.

unconnected areas.

At Barbarano the cell is nearer to the square form (*Pl. 21*), the walls are very thick, at times increased by buttresses added in later periods; the beams of the floors, in single or double patterns, are almost always oversized[17].

Here too the roofs are generally not thrusting.

In Barbarano as in Castelvetere the staircase which reaches the second floor from the street is external and made in stone: these are proferulum houses (*Pl. 22/a*); in Castelvetere the two superimposed floors of a house on three levels have different orientation for the beams, which seems to be an excellent rule for holding the walls; moreover there is not a single roof here that does not have an attic floor, so that the inclined beams always have a horizontal match. But inclined and horizontal beams are never connected so as to form a truss; strange improvidence of the builders: incompetence or conscious choice guided by reasons unknown to us? (*Pl. 22/b*)[18].

The houses of Ortigia are characterized by extreme lightness of the horizontal structures: light floors of small beams which today we judge insufficient to support standard overloads, extremely light roofs in which the tiles often rest directly on the beams without a continuous board surface (*Pl. 23*).

Cane and plaster are used for the partitions and for the characteristic vaults which cover the rooms and create a providential air space for the insulation of the rooms under the roof.

The houses of Ortigia provide the occasion for citing other constructive details in the walls: the treatment of openings[19].

Instead of the brick lintels of Città di Castello (*Pl. 24*), realized with well-organized masonry, Ortigia uses poor, elementary wooden lintels, but on the other hand surrounds the opening with high-quality stonework (*Plates 25, 26*).

The quality is not only architectonic, but also static, for the stones of these frames are intimately connected to the wall and create a point of strength where the opening introduces an element of weakness. The door-balcony complex which characterizes the facade is a completely organized stone structure.

As can be seen, each community has worked out in the course of centuries its own particular techniques for the realization of the elements of the house. This makes explicit the nature of the construction language diversified from area to area, in close dependence on the local resources and on the conditioning derived from long years of indigenous experience.

A noteworthy feature of the historical house is its ability to sustain modifications – its being not an object defined by the moment of its first building, but a ductile, evolving organism, capable of changing to solve the new necessities of its users, the new situations in which it may find itself within the urban texture. The houses of our historical centres are all the fruit of an evolution carried out over the centuries: the aspect which presents itself to us today is the result of slow, but at times radical transformations.

Plate 27 shows a house in Città di Castello deriving from a truncated tower, transformed and enlarged, and *Plate 28* a house in Barbarano, the various phases of whose transformation can be

---

[17] For the constructive techniques of Barbarano see the degree thesis of Valentina Jappelli, Faculty of Architecture of the University of Rome, 1992.

[18] A. Giuffrè and al., "Centri storici in zona sismica, analisi tipologica della danneggiabilità e tecniche di intervento conservativo a Castelvetere sul Calore", in C. Gavarini, A. Giuffrè, G. Longhi, *Ingegneria antisismica*, editorials ESA, Milano 1991, p. 267.

[19] For this item see the degree thesis of Caterina Carocci, Faculty of Architecture of the University of Rome, 1991.

followed in the stonework and alignments.

Great transformations which deeply change an already consolidated aspect: walls are knocked down, built anew or built alongside others; windows are closed, opened or transformed and so on.

It is worth noting here, returning to the theme of mechanics, that such modifications, if executed competently, do not alter the structural consistency. It is not so much the lateral interlocking that must be provided for, as the correctness of the path of the vertical loads. The oldest houses of our historical centres are often palimpsests containing the rules for transforming a facade, closing a loggia, occupying a courtyard or inserting a new staircase.

Plate 29 contains an extremely modified facade from Città di Castello: the old windows have been closed and new ones opened, offering to the internal forces paths different from those previously followed. Yet the masonry shows no sign of suffering and seems to have acquired the homogeneity of a newly-built structure.

All this is possible thanks to that intrinsic characteristic of the masonry organism which is mentioned above: its being an articulated system of simplified forms, easily replaceable and modifiable.

This capability is inherent in the nature of the components, which can all be dismantled and substituted by parts, including the walls, and this is fundamental in explaining the habit of maintenance of the houses, undertaken through repair and substitution of the pieces as they gradually deteriorate.

In conclusion, another point should be made relative to the seismic behaviour of the house: not local characteristics determine behaviour in the event of an earthquake; rather, it is a question of more generalized situations, encountered in every region.

Particularly vulnerable are:

1. houses which are in particular positions with respect to the urban texture (corner houses, end houses).
2. houses which for some time have not had vital ordinary maintenance, and whose walls have been left to the washing-out action of the rain, which penetrates them from above, breaking up the mortar and loosening the stones.
3. houses built with intrinsic defects and with major divergence from the type, from the constructive norm which is recognized as the basis of local building practice.
4. houses whose transformations have not been correctly executed, whose load-bearing capabilities have been diminished by hollowing out or disconnecting the masonry walls.

These observations, derived from examination of historical houses, are very usefull to provide good conservative interventions: this argument will be the subject of the last section.

4. THE SEISMIC PROBLEM IN HISTORICAL CENTRES

It is true that strong earthquakes have caused tragic destruction of masonry-built cities in the past, and recently many surviving buildings of the past have suffered other severe earthquake damage, but these events must be critically re-examined in order to avoid unfairly penalizing an entire construction category.

To declare masonry construction as earthquake-inefficient (on account of the damage which some of it – much of it – has suffered during earthquakes) would be comparable to condemning reinforced-concrete constructions because of the generalized collapses which have occurred on the ex Yugoslavian coast, as well as in the towns of Friuli and Irpinia.

If historical cities are still in existence in spite of earthquakes, it is thanks to their basic resistance and repairability, a typical feature of their masonry technology. One measure of the seismic vulnerability of urban areas is contained in the definition of the degrees of the macroseismic MCS scale:

→ 8th degree: some partial collapses, and some buildings cracked;
→ 9th degree: some total collapses, and numerous buildings seriously damaged so as to be uninhabitable;
→ 10th degree: total destruction.

Obviously these definitions have a statistical value which cannot be applied to the individual building; moreover, they seem to be intrinsically inadequate. The effect of an earthquake on a building depends not only on the intensity of its action, but also on the quality of the building's construction.

Nevertheless, as we have seen, referring to the structure of houses, their intrinsic nature varies little from place to place, or from one building to another.

The basic rules for their construction are constant, the lack of connections is generalized and the local typological characteristics are limited to accessories which are unessential from the point of view of earthquake resistance.

We can conclude that in the load-structure binomial, implicit in macroseismic definitions, the structure is – with statistical approximation – constant, and the effects of the earthquake can correctly be considered a measure of the intensity of load: the seismic action.

What can differ from one building to another is the quality of their parts, or particular features of urban aggregation. Houses with constructive details distinct from the usual rules must be considered with special care; it is that they could be singled out by 8th degree earthquakes, which produce some partial collapse.

Localized damage will occur in those cases – and only in those cases – in which the general rules have not been respected, or in which deterioration has made them useless.

On the other hand, it is not by chance that the Italian norms issued after the 1908 earthquake, and periodically updated, still require new masonry constructions to have a maximum wall interaxis of seven metres and a maximum height of three floors.

Making some concessions to modern construction technology, including the use of reinforced concrete, the norms allow the realization of cells a bit larger than the traditional ones, but the legislation has basically codified the pre-existing urban texture.

An examination of the seismic behaviour of the buildings in our historical centres can be carried out by identifying the average norm of construction and comparing the individual building to it. Any difference which reduce the anti-seismic qualities will be evident, as will be evident the criteria for adapting the building to the correct rule. So the comprehension of the weakness suggest the design of the intervention. To bring the building back to its correct original rule. Of course such intervention is made with the original technique and its effect is clearly measured: it make the building able to resist the 8° degree.

The damage of the 8th degree earthquake is thus avoided, but greater intensities involve the overall nature of the structure rather than local chance factors.

Intensity above the 8th degree involves the fundamental concept of masonry construction: its nature as a system easy to assemble and disassemble.

Preventive action is essential, but it calls for understanding of the mechanics of the structural behaviour.

This will be discussed in the next section.

5. COLLAPSE MECHANISMS OF THE BUILDING

Since houses built according to the traditional norms resist the 8th degree of the macroseismic scale, but are in large part damaged by the 9th degree, the 9th degree can be taken as their resistance threshold.

It should be noted that, although the use of tie-rods was already known in the 18th century, and all writers, particularly Leon Battista Alberti[20], recommend special care in the connections between walls, the basic urban fabric of historical cities is characterized by poorly connected walls.

As has been mentioned, the house is an assembly of juxtaposed parts; the best interlocking between transverse walls will never make effective a tension strength that the regular continuity of the masonry provides only in a very limited degree.

The collapse of such houses due to the forces of inertia of an earthquake normally occurs with the fall of the perimeter walls, directed outwards.

This mechanism was identified as the basic movement of the masonry wall by Rondelet in the late 18th century, and is encountered in all seismic events of considerable intensity.

In *Plate 30* can be seen Rondelet's drawings[21], and a street in Messina after the 1908 earthquake. The fall of the facade walls is evident.

This phenomenon occurs in all buildings when the 10th degree is reached; the 9th degree produces detachment, but collapse occurs only in the more precarious situations. (In fact the definition of the 9th degree says: *numerous buildings seriously damaged so as to be uninhabitable*).

The studies carried out to date on historical centres have produced a series of concordant features of collapse: considering the overall structure of the building and the particularities of the phase of construction and the position within the urban fabric, a range of possible forms of collapse has been identified.

Plate 31 shows a scheme developed for the small town of Castelvetere. The aggregation pattern of the cells which have subsequently occupied the free space is schematically pointed out. With reference to the cell as originally built, the aggregation can present perimeter walls with one edge connected and the other juxtaposed, or with both sides simply placed near pre-existing walls.

The forecast of the collapse mechanism is evident; the various cases are shown in the figure. In a certain sense it can be predicted that non-connected corners or walls constructed between two pre-existing buildings will suffer damage at the 8th degree. Of course the prediction is not rigorously deterministic, but it is extremely realistic; without calling for impossible measures of probability it could be considered highly probable, or, at least, sufficiently probable to decide to intervene. In

---

[20] L.B. Alberti is explicit in declaring: "Other kind of bond (made of larger stones), this fundamental, is represented by that ones that go around the walls, along all their length in order to keep tied the corners and to chain the structure of the work". L.B. Alberti, *L'Architettura (De Re Aedificatoria)*, Il Polifilo, Milano 1966, book III, chap. XI, p. 208.
[21] G. Rondelet, *Trattato teorico e pratico dell'arte di edificare*, (first italian translation by Soresina) Mantova 1834, vol. IV.

more complicated urban texture, like in Ortigia (see *Plate 18*), the foundamental element of earthquake vulnerability is the position of the house. *Plate 32* shows a frame with the most characteristic failure mechanisms as function of the positions in the urban texture.

The formulation of the frame is an operative instrument: based on it the foreseeable damage in a portion of urban fabric has been pointed out, observing the buildings one by one and recognizing the failure mechanism concerning each of them. *Plate 33* gives a sketch of the group of houses and the damage predicted, by the application of the frame. A scenario like the 9th degree derives, as it is evident, since we have put into effect all the detachments we know to be primed by the 9th degree.

The examination of each individual building can be carried out on the basis of the frame specifications, and a prediction can be made as to which mechanism the building will be subject to.

Inevitably, due to the nature of intrinsic disconnection of the structural organism, collapse occurs as a result of local kinematisms like the type illustrated, and if numerical analyses are to be developed, reference must be made to these local mechanisms.

But what type of analysis should be carried out?

Seismic legislation suggests in many cases a static test, almost as an assertion that if a structure resists certain horizontal forces, it will be able to withstand the dynamic action transmitted by an earthquake.

This formulation obviously involves a drastic simplification, but it is not mistaken from the qualitative point of view. Seismic motion produces horizontal accelerations, and therefore forces of inertia, and there is no doubt that structures which are more resistant to such forces will suffer less damage.

Moreover, whereas elastic structures can amplify the dynamic effect in function of their vibrational characteristics, masonry construction does not have an elastic behaviour capable of producing a significant influence on the response, as it will be seen in the next section.

It should also be pointed out that the intensity of the static forces prescribed by the code for seismic check is based on the concept of acceptable damage: it is predicted that the dynamic action will push the structure beyond its limit of resistance, but the passing of this limit is accepted in light of the ductility of the structure. This allows for a certain range of damage without reaching ruinous collapse.

Masonry structures are in no way inferior to those in reinforced concrete as regards available ductility, that is, capability for deformation beyond the threshold of resistance. This is clearly demonstrated by the cracks of many centimetres' width which can be observed in earthquake-damaged buildings that have not collapsed, comparing them with the few millimetres of deformation which accompany the stress at the ultimate resistance.

This consideration authorizes the use for masonry structures, within the convention accepted by seismic regulations for modern buildings, of the same values of design forces prescribed for modern buildings.

Once a system of horizontal forces distributed in constant ratio on the mass of the building has been established, tests can verify the equilibrium, one by one, on all the failure mechanisms which the previous qualitative survey predicted.

It should be noted that these analyses rarely provide truly significant information. For example, the seismic coefficient $\alpha = 0.07$ prescribed by Italian code for seismic areas where earthquakes only slightly greater than the 8th degree are expected, requires, for a wall six metres high and completely free from transverse walls, the thickness of 0.42 m.

It is truly unusual for the external wall of a historical building six metres high to be less than 50 cm thick, nevertheless, if we were to wait for an 8th degree earthquake we could not accept leaving it disconnected, in spite the good result of the check.

In effect, a six-metre high, 42-cm-thick wall would easily withstand an earthquake superior to the 8th degree if it were a monolith freely rocking on its base, as has occurred for the obelisks of Rome, which have survived major earthquakes unharmed, and as has been confirmed beyond the shadow of a doubt by numerical analysis. But the presence of floors and roofs, which inhibit oscillation towards the interior and make the response to dynamic action unsymmetrical, worsens the situation.

Under no circumstances would we want to leave such a wall disconnected if the expected earthquake were to exceed the 8th degree, even if its thickness were greater than that suggested by conventional numerical analysis.

The collapse mode considered until now, the outward overturning of the walls, can be called "first mode of damage": it is this type that the structure is most subject to. The intensity of the earthquake which produces it is similar to Euler's first value of critical load in a beam with axial load. If "first mode" collapse is avoided, the structure moves on to a different form of collapse, and to a much higher critical load value.

The task of those who provide for the safety of houses in historical centres is to seek devices to impede "first mode" damage. Such devices are easily found in the usual technical literature relative to such constructions. But the literature is not necessarily to be found in libraries; it is present in houses which have been damaged in previous earthquakes, and which intelligent restorers have repaired.

The use of tie-rods to bind the external walls to those orthogonal to them was common throughout the 19th century, and is without doubt the most efficient anti seismic device, consecrated by tradition and still completely valid today. This system avoids the overturning which above the 8th degree of the Mercalli scale is to be feared in any wall; the building can face the 9th degree without fear of becoming one of the "numerous buildings seriously damaged so as to be uninhabitable".

Another problem now arises for the bound walls: are they able to withstand distributed seismic forces if they are held at points by scattered anchorages? And at what interval should these be placed?

Once again the earthquake concerns the out of plane resistance of the wall, and the internal masonry structure conditions it.

Of course, if the external walls are "tied" to the buttressing walls, the forces which would have overturned them now pass to the buttressing walls. Indeed, the right-angle walls are called on to withstand more than the overturning resistance of the external walls. The inertia forces carried by the whole seismic acceleration on all the present masses now act on the transverse walls, a portion distributed and a portion brought to them by the tie-rods, to the limit of their resistance.

Limit, as it has been seen, dependent on the internal tensile strength but not easy to predict.

The experimentation available in the literature is normally limited to panels loaded with concentrated vertical and horizontal forces, but the study of the entire wall does not follow from the simplicity of such results.

The parameter "resistance to tensile stress" which seems to condition its behaviour is extremely ambiguous: difficult to evaluate, and even to define. The parameter "resistance to shear stress" is even more ambiguous because it assumes a horizontal sliding mechanism and a distribution of stresses systematically disproved by the experiments.

Anyway a limit-case consideration can certainly be expressed as follows: the crack pattern of masonry walls acted upon in their plane, such as can be observed after earthquakes, demonstrates the high ductility of the structure. After cracking the wall does not leave the plane, does not move towards an unstable configuration, but remains in any case upright, still available to support the weights it is loaded with.

Only if the texture of its stones is severely out of norm, the local disconnection which follows cracking can cause a detrimental disgregation, but this possibility can be verified with a preliminary survey.

If the masonry is of good quality, we can be confident that, the external walls being held, collapse is avoided even at the expense of wide cracks.

There is only one case in which this confidence would be poorly placed, and it is when the transverse walls are so limited in number or in dimension that they themselves suffer partial or global overturning under the action of the inertia force produced by the seismic acceleration.

Nevertheless only in extremely rare cases a situation of overturning can be reached, and making reference to the usual distribution of masonry walls in the traditional house it can be affirmed that no test is necessary if in each of the two fundamental directions of the masonry grid there is present a percentage of masonry section equal to 0.5% of the built surface. For a ground acceleration of 0.25 g this limitation leads to an average shear stress of about 1 kg/cm^2 (though this parameter is hardly significant).

The seismic safety of the buildings of historical centres must thus be controlled by identifying the first mode mechanisms to be avoided, introducing devices capable of inhibiting their appearance, and making sure that the quantity of resistant masonry in the direction of the earthquake does not fall below a certain level. But together with these overall controls, a whole series of local controls is indispensable, which it is the case to define as "analytic" insofar as they must examine analytically the characteristics of detail of the structure in order to be able to intervene upon it with mechanical correctness and efficiency.

What is meant by "mechanical correctness" and how its criteria are respected is the subject of the last section.

6. DYNAMIC FEATURES OF HISTORICAL MASONRY[22]

The dynamic behaviour of a masonry building is conditioned by the monolateral nature of the bonds. The stabilizing action of the weight can be exceeded by the overturning action of the lateral acceleration: in all the cases in which the friction is enough to avoid sliding a rocking motion is triggered off.

The kinematic mechanisms arising under strong earthquakes are often one degree of freedom systems, which can be suitably generalized and effectively described by the force-displacement relationship derived from the behaviour of the rocking wall.

In many cases, of course, the mechanisms are dissymmetrical, and sometimes more than one degree of freedom can be recognized in the collapse configuration. Nevertheless, the elementary model deserves to be studied: in some cases it directly gives a reliable result, and in general it provides a qualitative indication of the dynamic behaviour.

In this section, through the response to a set of synthetic accelerograms, we discuss the

---

22 Written by A. Giuffrè, C. Baggio, R. Masiani.

parameters of the symmetric one degree of freedom model (which briefly will call *elastic rocking model*) and the way to account for dissipation[23]. Then, the generalization procedure will enable to apply the results to more complex systems, and to point out some interesting trends of the dynamic effects connected with the seismic action[24].

Twenty accelerograms "a (t)" have been generated as samples of a non-stationary random process, with the duration of 20 seconds, normalized to the peak of acceleration[25]. The usual average response spectrum is reproduced in the left of *Plate 34/a*.

The value of the peak ground acceleration has been indicated by A.

The one-body model

The *elastic-rocking* model reproduced in *Plate 34/b*[26] is characterized by:
a) K: the initial elastic stiffness;
b) F_0: limit value of the restoring force depending from the weight P;
c) S_u: ultimate displacement.
The yielding displacement is $S_0 = F_0/K$.

The horizontal force, effect of the dragging motion produces, at first, the bending of the beam and the elastic displacement of the mass; but when it reaches the limit value F_0 the body starts overturning and the displacement increases further on S_0. Beyond this limit the lateral force able to make equilibrium to the restoring action produced by the weight decreases as the displacement increases.

The dynamic equilibrium in terms of rotation θ around the edge of the base may be written, for $\theta \ll 1$, as:

$$J \cdot \ddot{\theta} + F(\theta) + M \cdot h \cdot a(t) = 0$$

where:
J   is the mass moment of inertia around the rotational axis,
M   is the mass of the body,
A   is the peak ground acceleration,
$F(\theta)$ is the restoring moment around the centre of rotation produced by the weight.

If the structure is slender enough the function $F(\theta)$ is quite linear, as it is reproduced in *Plate*

---

[23] For this section see: C. Baggio, A. Giuffrè, R. Masiani, *I parametri del comportamento sismico di strutture murarie elementari*, Proc. 4th INAEE conference "L'Ingegneria Sismica in Italia", Milano 1989, vol. 2, pp. 666-673.

[24] For this section see: C. Baggio, A. Giuffrè, R. Masiani, *Seismic Response of mechanisms of masonry assemblages*, 9th ECEE, Moscow 1990, (to be published).

[25] The characteristics of the process have been inferred from statistical analyses on italian strong motion records. See: R. Masiani, *Accelerogrammi sismici per l'analisi non lineare nel dominio del tempo*, Proc. 3th INAEE conference: "L'Ingegneria Sismica in Italia", Roma 1987, pp. 753-764.
F. Sabetta, R. Masiani, A. Giuffrè, *Frequency non-stationarity in italian strong motion accelerograms*, Proc. 8th ECEE, Lisbon 1986, pp. 3.2/25/32.

[26] The mechanical model adopted is made by a concentrated mass M connected to a rigid base measuring b by an elastic beam high h/2. Such a model is more general than the elastic wall rocking on its base, because the mass is independent from the dimensions. In the figure the force-displacement relationship is reproduced: the horizontal force F(S), acting on the mass, produces the bending of the beam and then the rocking around B. Its value depends on the weight P. Of course, the second branch of the function can be assumed linear for h very greater than b.

34/b.

The bilinear force-displacement relationship can be expressed as follows:

$$F\,(\theta) = K \cdot \theta \qquad\qquad\qquad\qquad\qquad (\,|\theta| < |\theta_o|\,)$$

$$F\,(\theta) = F_o\,(1 - \frac{\theta - \theta_o}{\theta_u - \theta_o}) \cdot \text{sign}\,(\theta) \qquad (\,|\theta| < |\theta_o|\,)$$

F_o is the limit restoring moment: $F_o = P \cdot b/2$; and $\theta_o = b/h$ is the ultimate rotation.

Since some degradation is certainly present a new parameter C, measuring an hypothetical viscous damping, has been introduced into the equation:

$$J\,\ddot{\theta} + C\,\dot{\theta} + F\,(\theta) + M \cdot h \cdot A \cdot a\,(t) = 0$$

This differential equation can be easily solved by a step by step procedure. It represents the behaviour of the rocking wall (*Plate 34/b)*, but gives some interesting results useful for more complex cases as those described further.

In the following the results of a parametric analysis are critically examined to study the importance of the terms of the equation. The results in terms of the maximum displacement all over the duration of excitation are showed.

The initial elasticity

The value K of the initial elasticity can be expressed by the natural period concerning the first branch of the elastic rocking model: $T = 2\pi\,/\,\sqrt{K\,/\,M}$.

The *Plate 34/c* reports the results of the calculations made changing the value K in both the bilinear and the Takeda model, taking constant the value of the damping factor C.

The first model shows that the maximum displacement is independent from the initial elasticity, and the second one presents a slight variation. Nevertheless it must be observed that the Takeda model, decreasing the K value[27],, offers a decreasing hysteretic cycle: for this reason the response increases.

Both models show that the initial elasticity doesn't affect the maximum displacement produced by the earthquake. This is a very useful result because it enables to avoid the impossible evaluation of the elastic parameter.

The ultimate displacement

The ultimate displacement S_o geometrically depends on the thickness of the wall, but it can attain a lower value if the quality of the masonry is not good.

The decreasing of such parameter changes the slope of the second branch in the bilinear model.

The computation has been carried out for different values of S_u, and the results are reproduced in *Plate 34/c*. It is very interesting to observe that no perceptible difference appears in the response if the ultimate displacement decreases, but when its value is less than a certain threshold (about one third of the geometrical one) the collapse is quite certain.

---

[27] And then increasing the natural period.

This result enable to set that defects in the quality of the masonry, reducing its possibility to rock until the geometric limit, doesn't affect the response unless such reduction is extremely serious; and in this case the collapse can be given for unavoidable.

The viscous damping

In *Plate 34/d* the results concerning different values of the parameter C are reported. The bilinear model shows a significant dependence from such parameter: the response decreases if the viscous damping increases. The Takeda model, which contains hysteretic dissipation, is less affected by the viscous damping factor.

First annotations

The results presented with reference to the elastic rocking model allow to derive the following considerations:
a) in order to foresee the seismic resistance of a masonry building we have not to study its elastic characteristic, but its possible geometric kinematisms, and perhaps from those we will be able to evaluate the maximum response in terms of displacement.
b) In order to check the seismic strength of a masonry we have not to evaluate its compression strength, which has never been taken into account in the previous analyses, but its quality. If masonry is "well done" the result of the dynamic analysis (if we'll be able to perform it effectively referring to the actual kinematism) is correct; but if the "correct rule" has been strongly disregarded the failure is quite probable.
c) Since the elastic branch is completely ineffective it can be eliminated by the model. In the following the theoretic rigid rocking model (which we'll call simply *rocking model*) will be adopted.

Besides, the viscous damping factor can't be accepted as representative of the dissipation: it has no mechanical meaning and it would be impossible to give it a significant value; the Takeda model is unsuitable as well. It seems more effective to control directly the energy dissipation: we have introduced a reduction of kinetic energy every time the rocking motion crosses the rest position.

The parameter of such reduction, ρ,[28] could be experimentally evaluated, measuring the amplitude reduction of subsequent oscillations during the free rocking motion.

After this first commentary other developments can be done concerning the *rocking model*, adopting a parameter ρ related to a very little degradation[29]. But before starting the presentation of a second level of results the generalization needed to transform a kinematic mechanism into a one degree-of-freedom model should be illustrated.

The generalized multi-bodies model

In order to make simpler the presentation, we can refer to the case reported in *Plate 35/a* where the kinematic chain contains two rigid bodies.

The generalized displacement has been expressed with reference to local axes with origin in the centres of mass of the bodies:

---

[28] It takes into account all the possible dissipation mechanisms.
[29] The low value of degradation produces an amplification of the response which is in favor of safety.

$$\{S\} = \{X_1, Y_1, \vartheta_1, X_2, Y2_1, \vartheta_2\}^T$$

where X_1 and Y_1 are the displacement of the centre of mass of the body i in the direction x and y, and ϑ_1 is the rotation of the body i.

In order to evaluate a set of coherent values for $\{S\}$ the kinematic matrix of the constraints must be written. With reference to the figure:

$$[V'] \cdot \{S\} = \{q\} \quad \rightarrow \quad
\begin{vmatrix}
1 & 0 & -y_{1,1} & 0 & 0 & 0 \\
0 & 1 & x_{1,1} & 0 & 0 & 0 \\
1 & 0 & -y_{2,1} & -1 & 0 & y_{2,2} \\
0 & 1 & x_{2,1} & 0 & -1 & -x_{2,2} \\
0 & 0 & 0 & 0 & 1 & x_{3,2} \\
0 & 0 & 0 & 0 & 0 & 1
\end{vmatrix}
\cdot
\begin{Bmatrix}
X_1 \\ Y_1 \\ \vartheta_1 \\ X_2 \\ Y_2 \\ \vartheta_2
\end{Bmatrix}
= q \cdot
\begin{Bmatrix}
0 \\ 0 \\ 0 \\ 0 \\ 0 \\ 1
\end{Bmatrix}$$

The values $x_{i,j}$ and $y_{i,j}$ are the coordinates of the constraint i relative to the local axes of the body j.

Into the kinematic matrix [V] has been added a last row [0,0,0,0,0,1] in order to evaluate $\{S\}$ when $\vartheta_2 = q$:

$$\{S\} = [V']^{-1} \cdot \{q = 1\} \rightarrow \{S'(q)\} = \{S\} \cdot q$$

The configuration of the mechanism is expressed by the vector $\{S\}$ and measured by the parameter q.

The previous relation is in linearized terms: of course it could be possible to adopt a step-by-step procedure to follow the finite displacement of the system, up-to-dating the position of the constraints. Let be Δq the elementary increment of the parameter q; through the initial [V'], we can evaluate a vector $\{S\} \cdot \Delta q$. Accounting for such displacements, the coordinates of the constraints should be updated; a new matrix [V'] can be build up, and so a new vector $\{S (\Delta q)\}$ can be evaluated, which contains the displacements of the bodies starting from the previous position. The procedure can be repeated, updating again the coordinates and evaluating a further vector $\{S (2\Delta q)\}$, and so on.

The displacement of the kinematic chain is so expressed as a function of the displacement parameter q: $\{S (q)\}$.

In most cases the linear formulation $\{S (q)\} = \{S\} \cdot q$ can be accepted.

The limit value of an horizontal force, proportional to the masses and acting on the system, can be evaluated using the virtual work principle:

$$
\begin{aligned}
&\{P_y\}^T = \{0, -P_1, 0, 0, -P_2, 0\} && \text{weight} \\
&\{F\}^T = f_0 \cdot \{P_1, 0, 0, P_2, 0, 0\} = f_0 \cdot \{P_x\} && \text{horizontal force} \\
&\{S\}^T \cdot \{P_y\} + \{S\}^T \cdot \{F\} = 0 && \text{equilibrium} \\
&\{S\}^T \cdot \{P_y\} + f_0 \cdot \{S\}^T \cdot \{P_x\} = 0 &&
\end{aligned}
$$

$$f_o = - \frac{\{S\}^T \cdot \{P_y\}}{\{S\}^T \cdot \{P_x\}}$$

The parameter f_o measures the limit horizontal force. It can be interpreted as the horizontal acceleration expressed in terms of g, which provokes the starting of the motion in the mechanism.

Following the evolution of the kinematic motion, a parameter f can be evaluated as a function of q through all the vectors $\{S\}$, $\{S\,(\Delta q)\}$, $\{S\,(2\Delta q)\}$, $\{S\,(3\Delta q)\}$, ... and a monodimensional force displacement relationship f (q) is so found. The *Plate 35/b* reproduces such relationship evaluated for the second example further illustrated.

The value q_u relative to the condition (f (q_u)) = 0 is the ultimate displacement.

In most cases the monodimensional force-displacement relationship f (q) can he made linear:

$$f\,(q) = f_o \cdot (1{-}q/q_u) \cdot sign\,(q).$$

The dynamic analysis of the multi-bodies model

The procedure of the dynamic verification implies:

a) to point out a possible collapse mechanism able to rock under the seismic action;

b) to evaluate two parameters on the basis of massis and geometry of the mechanism;

c) to find in the response spectrum the maximum generalized displacement produced by an earthquake of given pick ground acceleration;

d) to evaluate the actual displacement from the generalized one and to look if the structure is able to rock untill that displacement without reaching the collapse.

The whole procedure is reported in the following.

The equilibrium condition under the inertia forces and the restoring action exerted by the weight can be formulated applying again the virtual works principle:

$$-\{S\}^T \cdot [M] \cdot \left(\{\ddot{S}\} + \{T\} \cdot \ddot{x}\right) + \{S\}^T \cdot \{P_y\} = 0$$

where $\{T\}^T = \{1,0,0,1,0,0\}$ is a vector which defines the horizontal dragging motion, and

$$\{\ddot{S}\} + \{T\} \cdot \ddot{x} = \{S\} \cdot \ddot{q} + \{T\} \cdot A \cdot a(t)$$

is the absolute acceleration, sum of the kinematic motion and of the dragging ground motion[30].

[M] is the diagonal mass matrix, which containts, for each body, the translational mass and the mass moment of inertia about the centre of mass:

$$[M] = \begin{vmatrix} M_1 & 0 \\ 0 & M_2 \end{vmatrix}; \quad \rightarrow \quad M_1 = \begin{vmatrix} m_i & \cdot & \cdot \\ \cdot & m_i & \cdot \\ \cdot & \cdot & J_i \end{vmatrix}.$$

---

[30] No problem if the non linear function $\{S(q)\}$ is adopted: during the step-by-step integration it can be updated at every step.

Remembering that:

$$\{S\}^T \cdot \{P_y\} = \{S\}^T \cdot \{P_x\} \cdot f(q)$$

and stating the following positions:

$$\{S\}^T \cdot \{P_x\} = F \qquad \text{generalized horizontal restoring coefficient}$$
$$\{S\}^T \cdot [M] \cdot \{S\} = M \qquad \text{generalized mass}$$
$$\{S\}^T \cdot [M] \cdot \{T\} = T \qquad \text{generalized dragging coefficient}[31]$$

a monodimensional equation can be formulated:

$$M \cdot \ddot{q} + F \cdot f(q) + T \cdot A \cdot a(t) = 0$$

completely similar to the one previously examined.

Some normalized parameters are very useful:

$$u = q/q_u \qquad \text{displacement}$$
$$\beta = (f_o/q_u) \cdot (F/M) \qquad \text{resistant acceleration (seismic strength)}$$
$$\gamma = (A/q_u) \cdot (T/M) \qquad \text{acting acceleration (seismic action).}$$

Dividing for M and for q_u the dynamic equilibrium condition becomes:

$$\ddot{u} + \beta \cdot (1 - u) \cdot \text{sign}(u) + \gamma \cdot a(t) = 0$$

where the parameters β and γ allow to explore a large range of possible structure, without attention to their features which will be normalized through the displacement vector $\{S\}$.

Neither hysteretic nor viscous dissipation is present in this equation, and nor initial elasticity, according to the first results: the *rocking model* is adopted.

The results of the parametric analysis have been expressed in terms of the characteristic value (referred to the 20 earthquake accounted for) of the displacement "u":
$$u_k = \bar{u} \cdot (1 + 2\sigma)$$

where $\bar{u}$ is the mean value (referred to the 20 accelerograms) of the maximum displacement produced by the earthquakes, and σ is the standard deviation.

In *Plate 36/a* some level curves in terms of u_k, referred to the *far* earthquake, have been reported in the plane of the two parameters β and γ; the *near* earthquake gives analogous results with lower values of the response.

The adopted value for the restitution factor is $\rho = \sqrt{0.8}$.[32]

The diagram show three zones. The lower one, under the thick diagonal line, represents the

---

[31] Of course, if the non linear formulation has been adopted, such coefficients shall be updated at every step according to the actual value of q and the corresponding $\{S\}$.

[32] That means that the amplitude of the free oscillations decrease, at every half-cycle, by a factor 0.8.

situations in which the formation of the mechanism does not occur. It can be defined *no-damage situation*.

If, in favor of safe, we want consider collapsed the structure when its displacement has reached the 50% of the ultimate one; we must delimit the collapse domain with the line

$$\gamma_L = 0.886 \cdot \beta + 6 \cdot 9 \text{ (Far earthquake)}$$
$$\gamma_L = 0.872 \cdot \beta + 11 \cdot 8 \text{ (Near earthquake)}$$

The upper zone can be defined as *collapse situation*.

The intermediate region concerns displacements always less than $0.5 \cdot q_u$. It can be defined as *damage situation*.

If, given β, the value of γ is contained between β and γ_L:

$$\beta \leq \gamma \leq \gamma_L$$

the mechanism will start up, but its maximum displacement will be less than the 50% of the limit. The structure will be damaged but the collapse will be avoided.

This results allow to check the seismic safety. The only condition is the possibility of defining a clear mechanism to analyse.

If the structure could rock without any degradation during the motion and if the kinematism could be perfectly described, the result would be correct. Unfortunately, the mechanism must be pointed out with a help from fantasy and the possible degradation must be qualitatively accounted anly looking at the effectiveness of the masonry.

But it must be remembered that we are used to study reinforced concrete buildings, modeling their structure without accounting for non-structural elements, that is to say we use an inactual model only to apply a standard theory.

With the same apprximation we can use the theory I have illustrated in order to foresse the dynamic response of the masonry building. But i only suggested to accept no more then the first result: if the structure statically resists the pick ground acceleration, no damage arrives.

This is an extremely important result for two reasons:

a) the statical resistence to horizontal force can be evaluated by equilibrium conditions, referring to the possible mechanism;

b) no elastic amplification must be foreseen and no amplification factor must be applied to the pick ground acceleration.

With these considerations, in agreement with the old seismic code which only required static analysis for reinforced concrete buildings, we can adfirm that static analysis for masonry buildings is significant of the dynamic behaviour too.

Some example

1. *The single wall rocking on its base*

Given a wall with base b and height h, the kinematic matrix is the following:

$$[V] = \begin{vmatrix} 1 & 0 & h/2 \\ 0 & 1 & b/2 \end{vmatrix}$$

In order to find the displacement vector {S} we can give an arbitrary value to any component; choosing the first one:

$$\begin{vmatrix} 1 & 0 & 0 \\ 1 & 0 & h/2 \\ 0 & 1 & b/2 \end{vmatrix} \cdot \begin{Bmatrix} X \\ Y \\ \theta \end{Bmatrix} = \begin{Bmatrix} 1 \\ 0 \\ 0 \end{Bmatrix} \cdot q \quad \rightarrow \quad \{S\} = \begin{Bmatrix} 1 \\ b/h \\ -2/h \end{Bmatrix} \cdot q.$$

In this case the ultimate value of q is easily evaluated: it is the value of q which makes the component $X = b/2$: $q_u = b/2$; the limit horizontal force, derived from the previous formula is $f_o = b/h$. The mass matrix is the following:

$$[M] = \begin{vmatrix} M & 0 & 0 \\ 0 & M & 0 \\ 0 & 0 & M \cdot i^2 \end{vmatrix} \quad \text{where} \quad i^2 = \frac{b^2 + h^2}{12}$$

Carrying out the previous procedure we obtain the generalized parameters:

$$F = P; \quad M = \frac{4}{3} \frac{b^2 + h^2}{h^2} \cdot M; \quad T = M$$

$$\beta = \frac{f_o}{q_u} \cdot \frac{F}{M} = \frac{3}{2} \frac{h}{b^2 + h^2} \cdot g;$$

$$\gamma = \frac{A}{q_u} \cdot \frac{T}{M} = \frac{3}{2} \frac{h}{b} \frac{h}{b^2 + h^2} \cdot A;$$

$$\eta = \frac{\beta}{\gamma} = \frac{b}{h} \frac{g}{A}.$$

We would have obtained the same results with a different choice of the displacement vector[33].

---

[33] The displacement vector could be found given the value −1 to the last component:

$$\begin{vmatrix} 1 & 0 & h/2 \\ 1 & 0 & b/2 \\ 0 & 0 & 1 \end{vmatrix} \cdot \begin{Bmatrix} X \\ Y \\ \theta \end{Bmatrix} = \begin{Bmatrix} 0 \\ 0 \\ -1 \end{Bmatrix} \cdot q \quad \rightarrow \quad \{S\} = \begin{Bmatrix} h/2 \\ b/2 \\ -1 \end{Bmatrix} \cdot q$$

For a wall with h = 3 m and b = 0.3 m, and A = 0.35 · g, we have:

$$\beta = 4.856; \qquad \gamma = 16.998; \qquad \gamma_L = 1.2$$

since $\gamma > \gamma_L$, the displacement will exceed the 50% of its ultimate value and the collapse could be expected.

If the thickness is increased to b = 0.50 m the parameters become:

$$\beta = 4.772; \qquad \gamma = 10.02; \qquad \gamma_L = 11.13$$

since $\gamma > \gamma_L$ the collapse will not happen.

If we design the wall to resist a static horizontal force equal to 20% of the weight, it can be considered completely safe against an earthquake of intensity 0.35 g:

$$\text{design criteria: } b/h = 0.2 \qquad \rightarrow \qquad b = 0.2 \cdot h = 0.6 \text{ m}$$
$$\beta = 4.716; \quad \gamma = 8.23 < \gamma_L = 11.08$$

The maximum displacement can be obtained by the expression[34]:

$$u_k = -0.045 - 0.070 \cdot \beta + 0.079 \cdot \gamma = 0.28$$

It is interesting to observe that the coefficients of β and γ in the expression of u_k are quite similar, but with opposite sign.

The maximum displacement depends from the difference between γ and β, that is to say from the difference between the seismic action (γ) and the seismic strength (β). The safety is not measured by the ratio between strength and action, but by the difference, whatever the strength is.

$$u_k \sim -0.045 + 0.0745 \cdot (\gamma - \beta)$$

From the u_k value the actual displacement can be obtained:

$$\{S_k\} = \{S\} \cdot q_k = \{S\} \cdot u_k \cdot q_u$$

In this cases examined we have:

b = 0.30;	$\gamma - \beta = 12.141$;	$u_k \sim 0.860$	$S_k = 0.860 \cdot (30/2) = 12.9$ cm
b = 0.50;	$\gamma - \beta = 5.248$;	$u_k \sim 0.346$	$S_k = 0.346 \cdot (50/2) = 8.65$ cm
b = 0.60;	$\gamma - \beta = 3.514$;	$u_k \sim 0.217$	$S_k = 0.217 \cdot (60/2) = 6.51$ cm

The S_k is, in this case, the horizontal displacement of the barycentre because it has been obtained by the value of the first component of the vector $\{S\}$.

---

in this case the ultimate displacement, able to make X = b/2 is q_u = b/h which is the ultimate rotation too.

[34] *Valid for the Far earthquake*; C. Baggio, A. Giuffrè, R. Masiani, 9th ECEE, Moscow 1990.

It is interesting to examine a very high wall: h = 9.8 m.
For different values of the thickness we can evaluate:

$$b = 0.49\ (0.05 \cdot h); \quad \beta = 1.498; \quad \gamma = 10.476; \quad \gamma_L = 8.227$$
$$b = 0.686\ (0.07 \cdot h); \quad \beta = 1.494; \quad \gamma = 7.470; \quad \gamma_L = 8.224$$
$$b = 0.98\ (0.10 \cdot h); \quad \beta = 1.487; \quad \gamma = 5.199; \quad \gamma_L = 8.217$$

It can be observed that the values of β are small; but since the γ value are small too, the displacement, in percentage of the ultimate value, are similar to those of the case h = 3 cm. It follows that the safety also is quite similar.

The maximum horizontal displacement of the centre of mass will be:

$$b = 0.49; \quad \gamma - \beta = 8.978; \quad u_k = 0.624$$
$$b = 0.686; \quad \gamma - \beta = 5.976; \quad u_k = 0.400 \qquad S = 13.7\ cm$$
$$b = 0.98; \quad \gamma - \beta = 3.712; \quad u_k = 0.232 \qquad S = 11.3\ cm$$

In spite of the greater height, the second wall seems safer than the first one even with a thickness proportionally less.

2. *The wall bonded at the top*

If the top of the wall is bonded by orthogonal chains its resistance increases. The collapse mechanism implies an intermediate hinge, as it is illustrated in *Plate 35*. The wall is so transformed into a two bodies kinematic chain.

Referring to the figure, the displacement vector will be evaluated through the following matrix:

$$
\begin{vmatrix}
0 & 0 & 0 & 0 & 0 & 1 \\
+1 & 0 & +h_1/2 & 0 & 0 & 0 \\
0 & +1 & b/2 & 0 & 0 & 0 \\
+1 & 0 & -h_1/2 & -1 & 0 & -h_2/2 \\
0 & -1 & b/2 & 0 & +1 & -b/2 \\
0 & 0 & 0 & 1 & 0 & -h_2/2
\end{vmatrix}
\cdot
\begin{Bmatrix} X_1 \\ Y_1 \\ \theta_1 \\ X_2 \\ Y_2 \\ \theta_2 \end{Bmatrix}
=
\begin{Bmatrix} 1 \\ 0 \\ 0 \\ 0 \\ 0 \\ 0 \end{Bmatrix} \cdot q; \quad
\{S\} =
\begin{Bmatrix} 1 \\ b/h_1 \\ -2/h_1 \\ 1 \\ 2b/h_1 + b/h_2 \\ 2/h_2 \end{Bmatrix} \cdot q.
$$

In order to account for the mass Q acting on the top of the wall, it is useful to add the displacements of the point D. The displacement vector becomes:

$$\{S\} = \begin{Bmatrix} 1 \\ b/h_1 \\ -2/h_1 \\ 1 \\ 2b/h_1 + b/h_2 \\ 2/h_2 \\ 0 \\ 2b/h_1 + b/h_2 \end{Bmatrix}; \quad P_x = \begin{Bmatrix} P_1 \\ 0 \\ 0 \\ P_2 \\ 0 \\ 0 \\ Q \\ 0 \end{Bmatrix} \quad P_y = \begin{Bmatrix} 0 \\ -P_1 \\ 0 \\ 0 \\ -P_2 \\ 0 \\ 0 \\ -Q \end{Bmatrix} =$$

and the mass matrix:

$$\begin{vmatrix} M_1 & \cdot & \cdot & \cdot & \cdot & \cdot & \cdot & \cdot \\ \cdot & M_1 & \cdot & \cdot & \cdot & \cdot & \cdot & \cdot \\ \cdot & \cdot & M_1 i_1^2 & \cdot & \cdot & \cdot & \cdot & \cdot \\ \cdot & \cdot & \cdot & M_2 & \cdot & \cdot & \cdot & \cdot \\ \cdot & \cdot & \cdot & \cdot & M_2 & \cdot & \cdot & \cdot \\ \cdot & \cdot & \cdot & \cdot & \cdot & M_2 i_2^2 & \cdot & \cdot \\ \cdot & \cdot & \cdot & \cdot & \cdot & \cdot & M_Q & \cdot \\ \cdot & \cdot & \cdot & \cdot & \cdot & \cdot & \cdot & M_Q \end{vmatrix}$$

The generalized parameters are the following:

$$q_u = b/2; \quad f_o = \frac{(b/h_1)P_1 + (2b/h_1 + b/h_2)(P_2 + Q)}{P_1 + P_2}$$

$$M = M_1 + \frac{b^2}{h_1^2} \cdot M_1 + \frac{4}{h_1^2} i_1^2 \cdot M_1 + M_2 + \left(\frac{2b}{h_1} + \frac{b}{h_2}\right)^2 (M_2 + M_Q) + \frac{4}{h_2^2} i_2^2 \cdot M_2$$

$$T = M_1 + M_2; \quad F = P_1 + P_2.$$

The wall has been examined for several values of the thickness, with the aim to compare the effect of the upper bond on the seismic safety[35].

---

[35] The position of the intermediate hinge depends on the weight Q acting on the top of the wall. It can be found imposing to the ultimate lateral force the condition of minimum.
The partial heights of the two portions shall be expressed as function of the total height H:

$$h_1 = (1 - \alpha) \cdot H; \qquad h_2 = \alpha \cdot H.$$

The lateral force f_o, reported in the text, is so function of the parameter α. The condition of minimum is expressed letting equal to zero the derivative of f_o with respect to α:

$$\partial f_o / \partial \alpha = 0$$

The function f (q) has been carefully explored and it is plotted in *Plate 37/b*. The slight non linearity weakly affects the values of the parameters β and γ, reported in *Plate 37*. The ultimate displacement q, for the three values of the thickness, are the following:

$$b = 0.49 \text{ m}; \qquad q_u = 0.207 \text{ m};$$
$$b = 0.686 \text{ m}; \qquad q_u = 0.295 \text{ m};$$
$$b = 0.98 \text{ m}; \qquad q_u = 0.429 \text{ m};$$

The maximum displacement q_k can be easily evaluated multiplying the adimensional value u_k by q_u.

$b = 0.49$ m; $\beta = 5.828$; $\gamma = 12.116$; $\gamma_L = 12.064$; $u = 0.423$; $S = 8.7$ cm
$b = 0.686$ m; $\beta = 5.114$; $\gamma = 8.262$; $\gamma_L = 11.431$; $u = 0.212$; $S = 6.3$ cm
$b = 0.98$ m; $\beta = 4.569$; $\gamma = 5.565$; $\gamma_L = 10.948$; $u = 0.029$; $S = 1.2$ cm

The superiorly bonded wall is safer that the free wall, and the maximum displacement is significantly reduced.

3. *The effect of suspended masses*

An assemblage like the one show in *Plate 36/b* can be considered again a single degree of freedom system, with hinges located in the point A, B, C, D or A', B', C', D' alternatively. Owing to the geometric symmetry the two mechanisms are also symmetric.

We want to know what the effect of variations in the mass of the transverse would be.

A simple kinematic analysis shows that the two piers have the same rotation θ and the architrave undergoes a translational motion. It must be observed that the initial motion of the point G_3 is equal to the initial motion of the point B or C, so in view of the evaluation of the implied parameters the assemblage in *Plate 36/b* is equivalent to the "reduced" model shown in *Plate 36/c*, where the mass of the transverse is concentrated in the point B.

Now we make the following statement:
a) On increasing the mass of the architrave the ratio β/γ remains constant: $\beta/\gamma = K$.

To prove the statement a) we must recall some of the previous exposed relations. Since $\beta = (f_o/q_u) \cdot (F/M)$ and $\gamma = (A/q_u) \cdot (T/M)$, results: $\beta/\gamma = (f_o/A) \cdot (F/T)$; the ratio F/T is always equal to the

---

The result provides the following expression:

$$\alpha = \frac{1}{x}; \qquad x = 1 + \sqrt{1 + \frac{2P + Q}{Q}};$$

where P is the weight of the whole wall. In the calculations over reported the value $Q = 1000$ kg has been adopted, and the weight of the wall has been evaluated according to $\mu = 1800$ kg/m^3. The following results have been found:

$$b = 0.49 \text{ m}; \qquad h_1 = 7.982 \text{ m}; \qquad h_2 = 1.818 \text{ m};$$
$$b = 0.686 \text{ m}; \qquad h_1 = 8.198 \text{ m}; \qquad h_2 = 1.602 \text{ m};$$
$$b = 0.98 \text{ m}; \qquad h_1 = 8.409 \text{ m}; \qquad h_2 = 1.391 \text{ m};$$

gravity acceleration g, since the terms in the vector $\{P_x\}$ are the same as those of the vector $[M] \cdot \{T\}$ multiplied by g; it follows: $\beta/\gamma = (f_0/A) \cdot g$. Now, referring to the reduced model we can guess that f_0 remains constant, whatever the value of the mass m_3 may be[36], and so prove the statement.

Now we make the statement b):

b) On increasing the mass of the architrave β decreases; we do not prove this statement: simple calculations will show that it is true.

Moreover we assume the hypothesis c):

c) The ratio β/γ is lesser than 1, that is to say the seismic action is greater than the seismic strength and the mechanism will start up; this is the case of interest, of course.

If so, in reference with *Plate 36/d* we can easily conclude that the increment of the mass m_3 results in a better behaviour of the structure, since the point Z, representative of the structural response as a function of the two parameters β and γ, shifts towards the lower part of the diagram, (along the straight line of equation $\beta - K \cdot \gamma = 0$) that is to say toward minor u_k values.

The numeric example in *Plate 36/e*, where the mass of the architrave is increased from 0 to three times the mass of one pier, shows that the earned benefit is not very high, nevertheless we can affirm that as far as mechanisms of the this type are concerned, a reduction of the suspended masses (a restoration solution sometimes applied) is not only useless, but could even resolve in a pejorative intervention.

7. ANTISEISMIC INTERVENTIONS FOR THE HOUSE AND THE URBAN FABRIC

The houses of the historical centre make up a cultural heritage which deserves to be preserved, for it bears witness to the civilization which is at our roots. It deserves to be preserved also because these durable houses serve as a brake to the explosion of the ephemeral which characterizes the present day. With reference to the historical city, preservation is not a static concept, as it is for archaeological findings, whose documentary nature must be safeguarded above all; since the city must be lived in, preservation must be inscribed in the dialectic between modern life and the past.

Such a dialectic, of course, has always guided the development of the city; but today it is posed with a completely new interpretation, characteristic, I would say, of modern culture: that is, with a request of recognition of the past and of respect for its manifestations, which was unknown in past centuries.

Living in a house of the past today arouses the pleasure of making explicit the awareness of past history, and suggests lifestyles different from those which would be realized in a newly built quarter.

Also, in assigning new uses to old buildings, there is an increasingly conscious attempt to identify possibilities proper to them, so that rooms conceived by the original builders for purposes now obsolete do not appear out of tune with uses which differ excessively from the original ones, to the detriment both of their nature and of the new use.

In this cultural climate is inscribed the work of the engineer and the architect, who intervene on

---

[36] The parameter f_0 can be interpreted as the horizontal acceleration which provokes the starting of the motion; since the ratio between the horizontal and vertical displacement of the point B is equal to the same ratio measured for the point G, center of mass of the pier, f_0 is independent from the value of the mass located in the point B.

the bearing structures of the building to save them from deterioration, to reconstitute lost resistance, to cure any intrinsic weaknesses.

The requirements of preservation demand that the intervention should be dialectically respectful of the nature of the work. The first step is therefore that of knowledge.

Not only is it necessary to know the fundamental mechanical characteristics of masonry; in order to be able to act correctly in an environment with historical value, a much more subtle awareness is required: it is necessary to recognize local particularities in constructive details, because it is with these that the dialogue of the modern restorer is carried on.

Floors with simple or complex beam systems, the use of boards or "pianelle", roofs which project or which are hidden by cornices, cantilever staircases or contained between two walls, must be known because they constitute the object of the intervention in a local situation, and because with restoration they must regain as much as possible of their original nature.

The modern expert must have a cultural level superior to that of his predecessors, not contrary to it, and being superior today's culture must contain all the culture of the past, and not deny it.

In order to intervene in the historical house, it is necessary to understand the character of wooden floors and their static nature, and thus identify the mechanical advantages that a system of wooden floor beams can offer to the masonry organism.

We must realistically evaluate the role that can be entrusted to existing structural elements – wooden beam floor or sloping roofs – and we must be able to appreciate the nature of a wall ably and enduringly assembled by a master, and distinguish it from the intrinsic inadequacy of disorganized masonry put up by an inexpert peasant.

This approach to understanding is necessary because "mechanical correctness" is obtained by operating on the masonry structure with masonry techniques.

The dialectic of intervention is carried on by replacing the insufficient detail with what modern competence judges to be more efficient, but always within the original language.

In this way the construction evolves towards a more consciously safe state, but remains itself.

It would not be difficult to list the possible defects which constructions present and to suggest remedies for them; some weaknesses, which concern the logic common to all types of construction, are so generalized that they admit the formulation of general criteria of intervention, valid anywhere; others are proper to a local construction formula, and must be solved in their own way.

The general rules, valid for any masonry construction, are the modern version of the classic treatises on th art of masonry, but they must be backed up by other, particular rules: local codes are necessary which define the special features of the constructions, and suggest remedies for damage or inadequacies.

The most common necessities, as we have said, are first of all those which regard the masonry, that is, the judgment on the mechanical qualities of the wall; the second involves the "structural" aspect, and concerns control of the overturning of the external walls.

Apropos of the first subject, the scientific debate is far from over: some insist on researching experimental methodologies which make it possible to measure the value of mechanical parameters such as resistance to compression or resistance to shear in terms of internal stress. Impossible experimentation, which in any case would produce no more than non-generalizable local information.

On the other hand it seems more realistic to judge the mechanical quality of the wall by checking its construction, and to intervene if it is found to be unsatisfactory.

The real problem is "how" to intervene when it is observed that the rules of good building, objectively defined as we have said at the beginning, have not been respected. Poor builders have always existed, and those of the past have left us poor walls.

It is not infrequently that a mistaken building practice, such as the realization of double walls with poor quality filler and without transverse connections, has become, in a certain period, common practice in a given geographical area.

The defect in this case is generalized, and in these areas an earthquake can have tragic consequences, because such walls are seriously affected even by 8th degree earthquakes; moreover these consequences are easy to predict, for the inadequacy of a wall without transverse connections is objectively known.

How to intervene when the walls are seriously insufficient is an as yet unresolved problem. Our predecessors knew that a poor wall had to be rebuilt, and performed reconstructions. Today we can, perhaps we "must" make this choice in some cases; *Plate 37* show how to realize walls in squared stone of different thicknesses, inserting, if necessary, portions of raw stones between the blocks.

Plate 38 shows some possible ways of intervening on walls of inadequate quality: when the wall has been realized working the two faces continuously, and filling the space between them more or less haphazardly, it is necessary to introduce the missing connectors, that is, the diatones, whose presence is essential to realize transverse solidity. Proposals range from the reconstruction of portions of wall which cross the entire thickness of the existing one, to the insertion of artificial diatones realized in reinforced concrete.

The second problem, of "structural" importance, involves the very nature of historical construction techniques and concerns, as has been mentioned, out-of-plane overturning of the exterior walls.

This danger is general; there is not a single building in the history of construction that does not present this problem. Even if all of the masonry walls have been built together, and all of the intersections have natural masonry continuity, we know that a 9th degree earthquake detaches them and a 10th degree inevitably reduces them to rubble.

This knowledge is more realistic than any computer analysis which it is possible to carry out today. It has been demonstrated by experience – of a destructive kind – which provides information more significant that can by gained otherwise – that the resistance to tensile stress of masonry is always inferior to that requested during a 10th degree earthquake, and on the average does not withstand 9th degree intensity.

In conclusion, intervention is inevitable if greater than 8th degree earthquakes are expected.

Is there a "proper way" to correct a weakness which seems almost to characterize the original nature of all historical building?

It is opportune to recall the insistence with which Alberti recommended the continuity of the cornices which form a "belt" half-way up and at the top of the building[37]; but it is Rondelet who, at the end of the 18th century, suggests placing metal ties along the walls so that they will not be thrust outwards by the vibrations produced by traffic[38].

---

[37] L.B. Alberti, *op. cit.*, p. 232.

[38] It is interesting to cite textually the words of Rondelet written on this subject: *It is not enough to build the walls of a construction with the wanted dimensions and with all the convenient attention, since they must be charged by the weight of the floors and the roofs which naturally tend to push them toward the void, effect which more increases for the quake derived by the moving of the carriages in the big towns, it usage to take at each floor some precautions at this regard during the building of the walls to prevent every thickness horizontal ties of flat or square steel, well placed, and firmly joined at their ends with anchors, which tie together the walls in such a way the one can't act without the other and they will give help each other. These ties will be placed in the walls during*

"Rooting" – placing beams or boards horizontally into the masonry between the stones – is a part of many building traditions, and throughout the 19th century the damage produced by earthquakes, the inevitable out of plumb situations occurring in facades after they had been near collapse, were repaired with metal tie-rods. The use of chains – metal ties placed within or along the wall to confer continuity to a structure in which continuity is intrinsically lacking – is a recent complement to the art of masonry, but one whose aim is to satisfy a long-felt need.

Consistent with this ancient practice seismic code proposes the insertion of tie-beams of reinforced concrete .

The reference to tradition is implicit in this prescription, but unconscious. In fact, if there had been historical awareness, the prescription would not have been for a tie of reinforced concrete, virtually a beam where there is no bending stress to be resisted; it is much more natural and consistent with the nature of masonry work to realize ties with metal bars and fix them to the exterior of the walls with anchorages.

At the top of the wall it is possible to reconstruct a portion of masonry, using an accurate interlocking between the stones or using bricks if they are present in the local masonry language, and insert an iron bar within the thickness of the wall. This creates a "masonry tie-beam which combines the recommendations of Alberti with the counsels of Rondelet; and at the same time it follows, even though not to the letter, the indications which today's technical norms provide us with.

The roof beams can be anchored to this metal bar by means of clamps, assigning to the wooden structure the task of holding the wall.

Plates 39 and *40* illustrate these interventions.

It is not possible to insert a tie-beam on the lower floors, but transverse anchoring must be realized. Here the job is done by the floor beams, or by further ties.

The plates which follow are dedicated to these subjects, with reference to the characteristics of the constructions of Ortigia. *Plates 41, 42, 43* and *44* concern anchoring the ties through the wooden floor. When possible the ends of the ties will be anchored directly or through a Y-system to the buttress walls; in other cases transmission can take place through the floor beams, fixing the ties to the floor boards – subsequent to solidly connecting the floorboards together, and to the beams.

Also the sockets of the beams in the walls must be accommodated, regularizing the masonry with bricks or squared stones.

It is important for the floor, though it is made of wood, to adhere to the wall so as not to leave any play which would set up vibrations. This can be achieved through the use of border beams or large boards tightly nailed to the beams, close against the wall. The transmission of forces to the buttress walls must also take place without play.

The floor will, in effect, no longer be merely a wooden structure resting on the walls to support the contents of the house, as it was in the past. Rather it is a diaphragm grafted into the four walls almost as if to guarantee their reciprocal distances apart.

If this instruction is respected, it takes little to prevent any tendency towards out-of-plane movement before it arises.

Plates 45, 46 and *47* refer to anchorages.

These are, for the masonry wall, points of support against the action of horizontal forces. It is evident that a stone structure can carry out resistance only through an arch behaviour into its thickness, and therefore the quality of its texture is once again involved, as has already been pointed

---

the building.
Rondelet, *op. cit.*, vol. III, part II, p. 77.

out.

The size of the anchorages plays an important role: the end-pieces of the original anchorages, iron bars arranged like gudgeon pins at the end of the chain, have often been found bent by the strain which pulls them towards the wall. Metal plates offer greater resistance. Prefabricated anchors in reinforced concrete have produced good results thanks to the ductility with which they follow the deformation of the wall.

Of course the technical examples illustrated here are no more than generic proposals, whose only purpose is to stimulate the research of useful details with reference to the actual nature of the construction in which we are working.

And we should conclude this panorama of problems with one last observation: in the search for seismic safety in cities it is necessary, in addition to studying individual buildings, to examine street patterns as they affect escape routes from the houses, and open them up as necessary so that each house can be reached by at least two routes). This we did in Ortigia, creating new access to the dead-end streets that characterize the Graziella quarter, but without altering the character of the old internal courtyards.

CONCLUSIONS

In conclusion, our remarks can be summed up as follows: masonry structures have excellent possibilities of resistance to earthquakes, if they are built according to traditional rules and are strengthened with connecting elements which tie the walls as has been shown.

To obtain this result it is essential to acquire a thorough knowledge of the behaviour of masonry, in order to take maximum advantage of its potential and avoid the temptation to substitute with other structures those which already possess sufficient efficiency to save us the trouble of studying and understanding them.

Much of the understanding of masonry structures comes by studying their historical aspects, involving the experimentation done over the centuries by their builders.

Interventions must respect consistency of language, using masonry techniques analogous to the original ones, but must also provide all that modern science can suggest in order to improve its mechanical characteristics.

The urban texture must be treated with the same care as the individual buildings, for it too is a system whose functions are severely put to the test by an earthquake, and it is necessary, for reasons of seismic safety, to control its functionality.

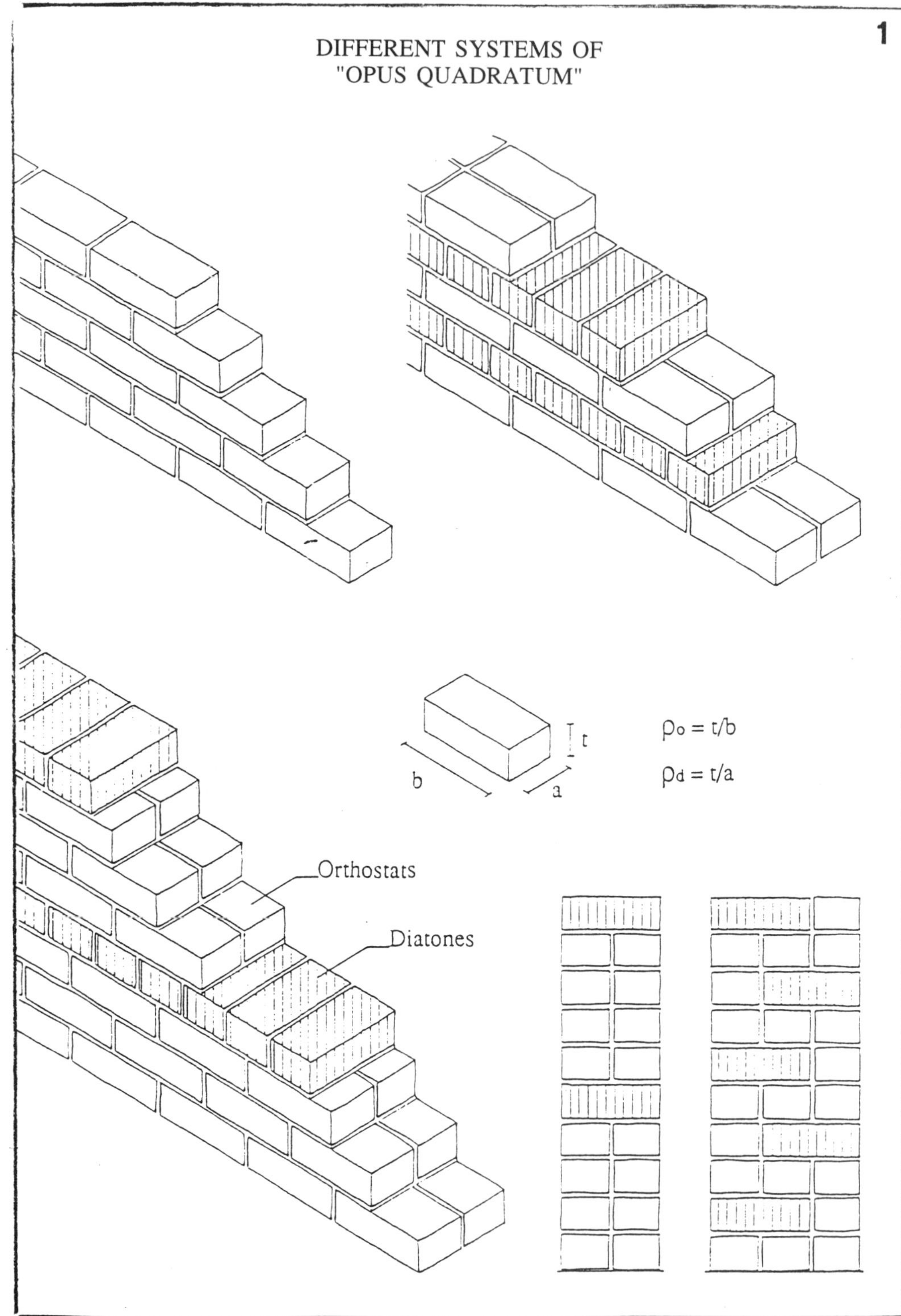

$$\rho_o = t/b$$

$$\rho_d = t/a$$

2

a

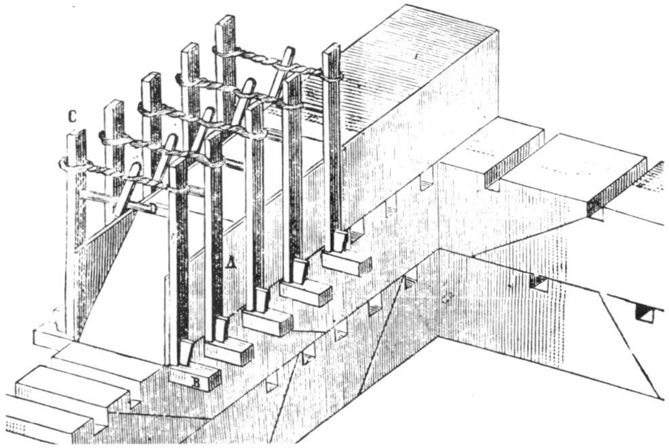

Form for
raw clay walls
(from: *Sacchi*)

b

Form for *opus concretum*
(from: *Lugli*)

c

Opus caementicium

d

Opus concretum
in the Temple of
Venere e Roma,
Rome

3

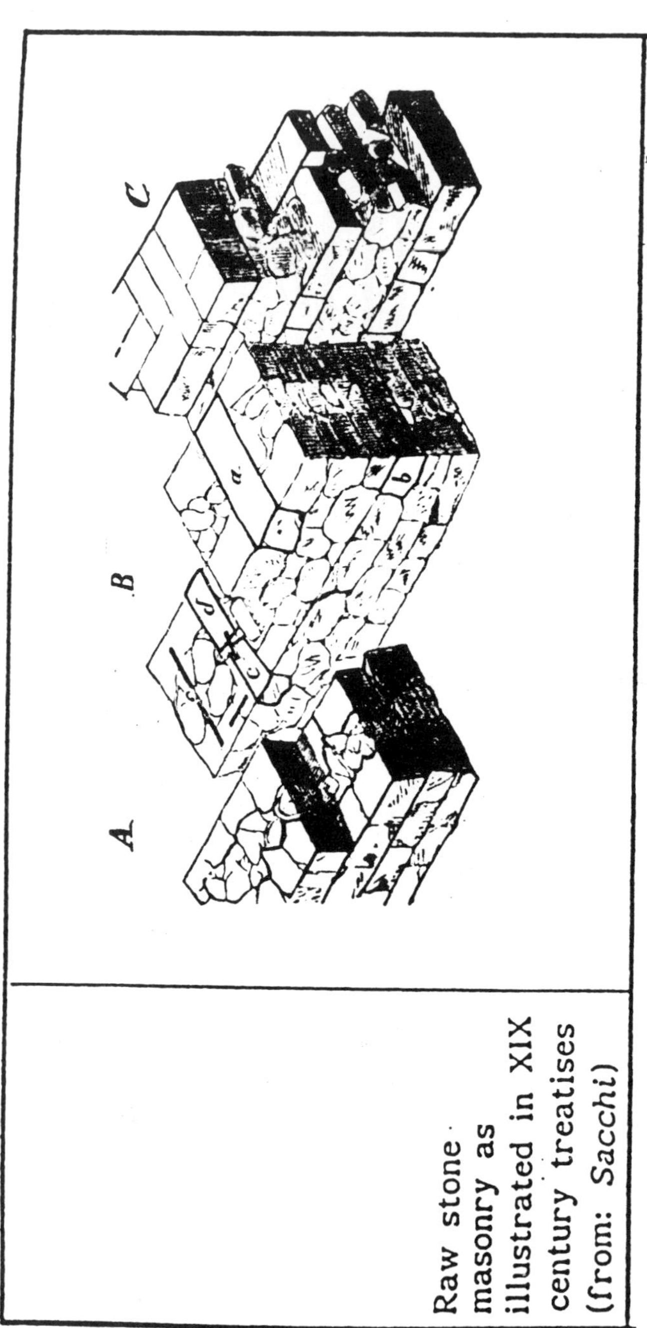

Raw stone masonry as illustrated in XIX century treatises (from: *Sacchi*)

4

MODELING AN ACTUAL MASONRY AS "OPUS QUADRATUM"

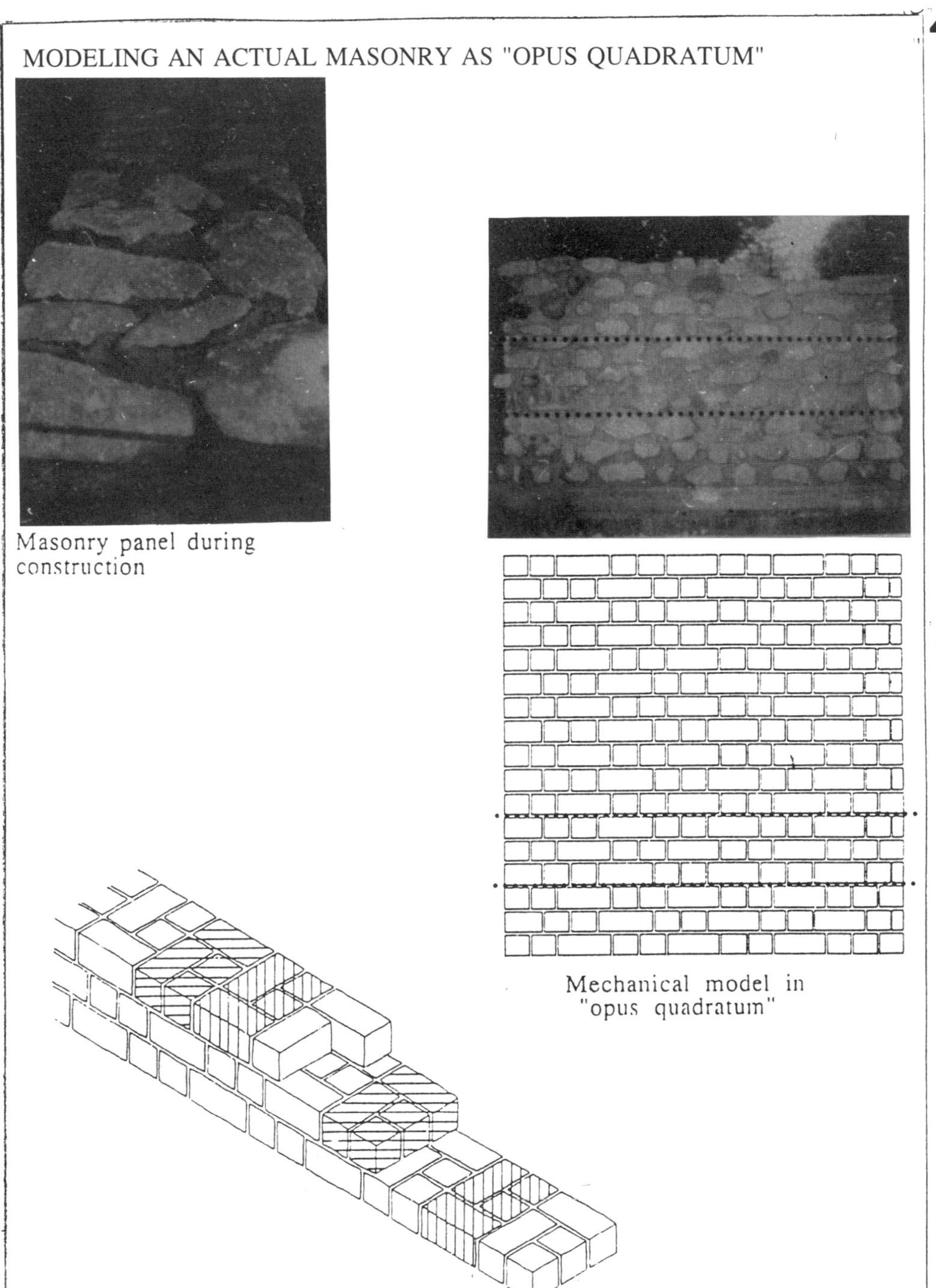

Masonry panel during
construction

Mechanical model in
"opus quadratum"

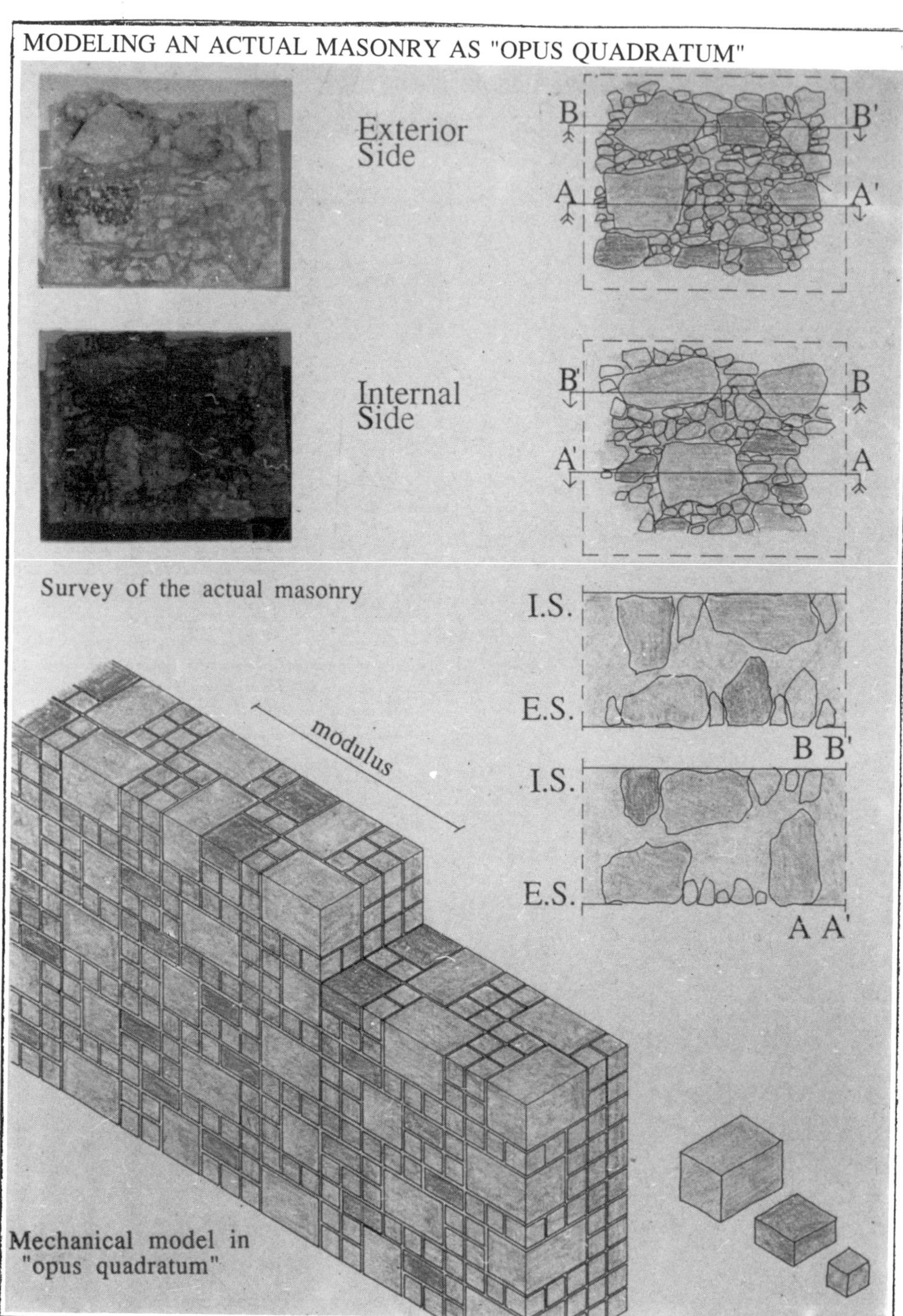
MODELING AN ACTUAL MASONRY AS "OPUS QUADRATUM"
5
Exterior Side
B
B'
A
A'
Internal Side
B'
B
A'
A
Survey of the actual masonry
I.S.
E.S.
B B'
I.S.
E.S.
A A'
modulus
Mechanical model in "opus quadratum"

6

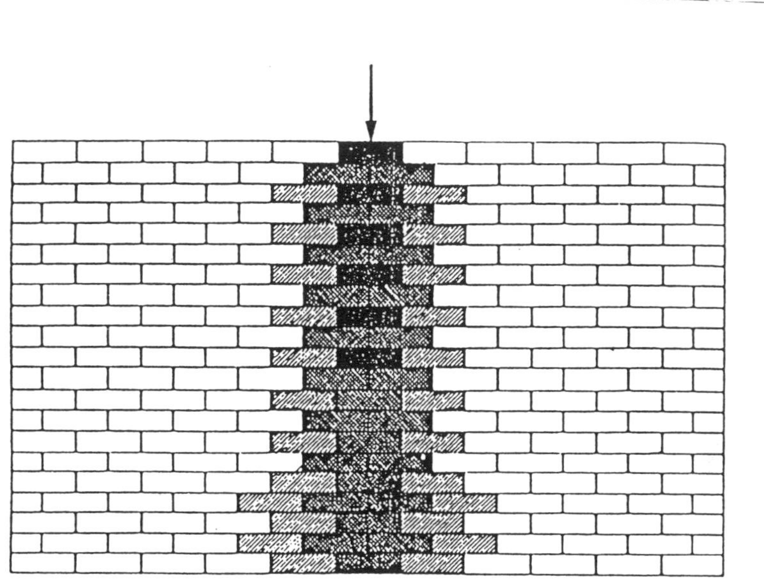

"In-plane action": vertical and diagonal load on the *opus quadratum* masonry wall

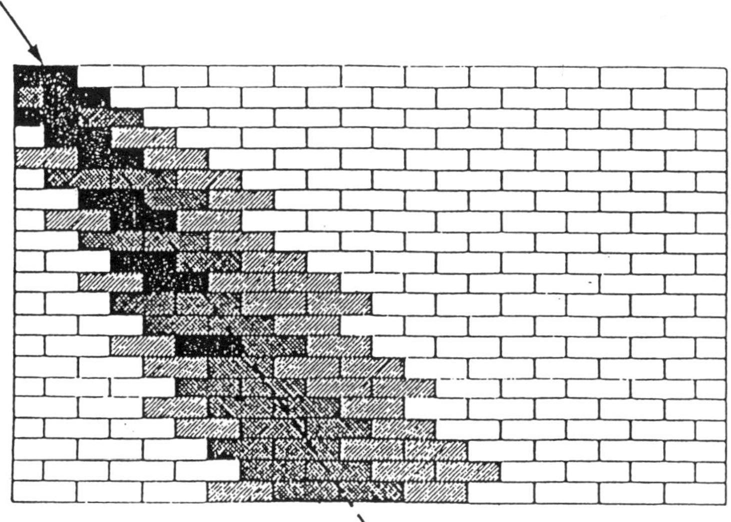

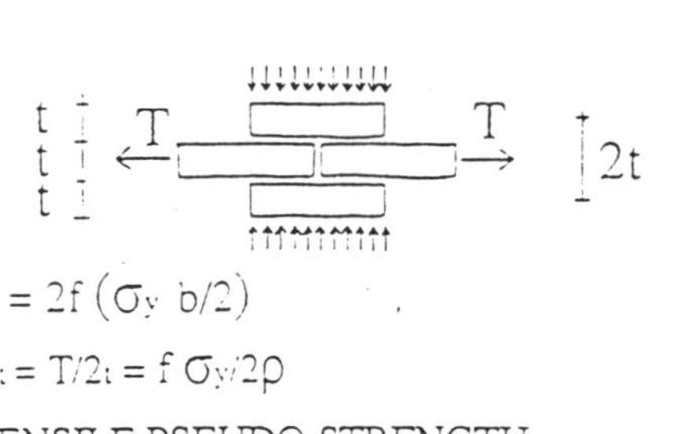

$T = 2f\left(\sigma_y\, b/2\right)$

$\sigma_t = T/2t = f\,\sigma_y/2\rho$

TENSILE PSEUDO-STRENGTH

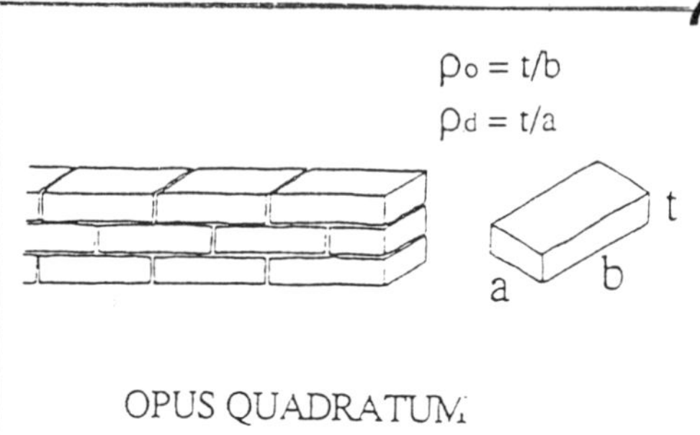

OPUS QUADRATUM

SCALE COEFFICIENT

$$\frac{\text{Prototype}}{\text{model}}$$

specific gravity $g/g' = \gamma$

geometric dimensions $L/L' = \lambda$

friction $f/f' = \varphi$

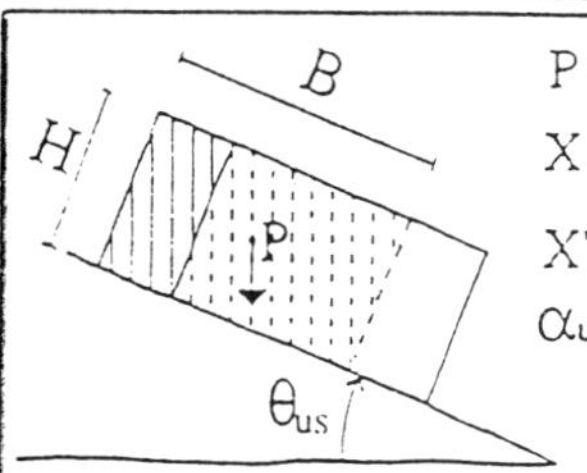

$P = gBH$

$X = fY \quad \alpha_{us} = f$

$X' = f'Y' \quad \alpha'_{us} = f'$

$\alpha_{us} / \alpha'_{us} = f/f' = \varphi$

SLIDING COLLAPSE

$X = B/H\cdot Y \quad \alpha_{ur} = B/H$

$X' = B'/H'\cdot Y' \quad \alpha'_{ur} = B'/H'$

$\alpha_{ur}/\alpha'_{ur} = 1$

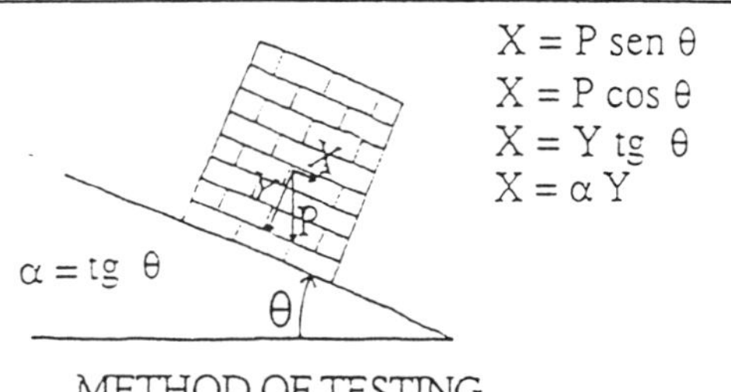

$X = P \operatorname{sen} \theta$
$X = P \cos \theta$
$X = Y \operatorname{tg} \theta$
$X = \alpha Y$

$\alpha = \operatorname{tg} \theta$

METHOD OF TESTING

ROCKING COLLAPSE

NORMAL STRESS $\rho = \rho'$

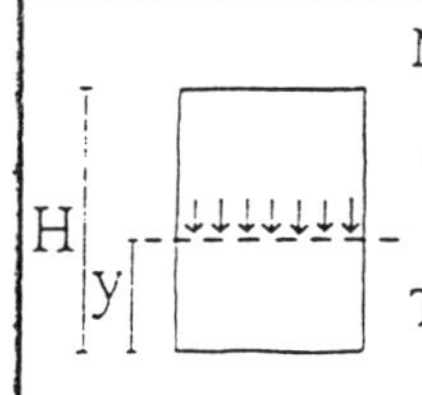

$\sigma_y (y) = g (H-y)$ $\sigma'_y (y') = g' (H'-y')$ $\sigma_y (y) / \sigma'_y (y') = \gamma \lambda$

TENSILE PSEUDO-STRENGTH

$r_t (y) = f\sigma_y (y)/2\rho$ $r'_t (y') = f'\sigma_y' (y')/2\rho$ $r (y)/r' (y') = \varphi \gamma \lambda$

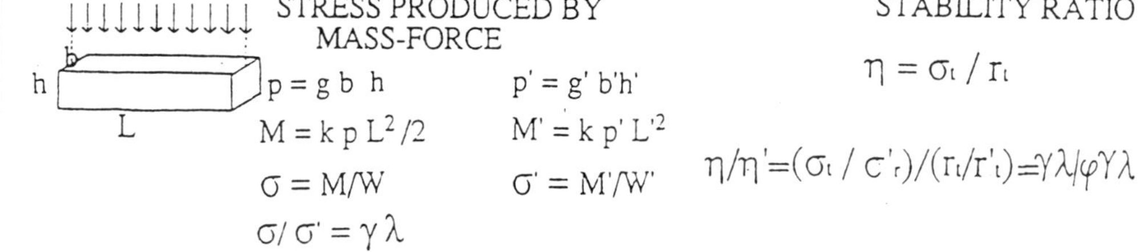

STRESS PRODUCED BY MASS-FORCE

$p = g\, b\, h$ $p' = g'\, b'h'$

$M = k\, p\, L^2/2$ $M' = k\, p'\, L'^2$

$\sigma = M/W$ $\sigma' = M'/W'$

$\sigma/\sigma' = \gamma \lambda$

STABILITY RATIO

$\eta = \sigma_t / r_t$

$\eta/\eta' = (\sigma_t / \sigma'_t)/(r_t/r'_t) = \gamma\lambda/\varphi\gamma\lambda$

8

SLIDING OF UPPER LAYER

PARTIAL OVERTURNING

NO TENSILE STRENGTH

LOW TENSILE STRENGTH

MEDIUM TENSILE STRENGTH

HIGH TENSILE STRENGTH

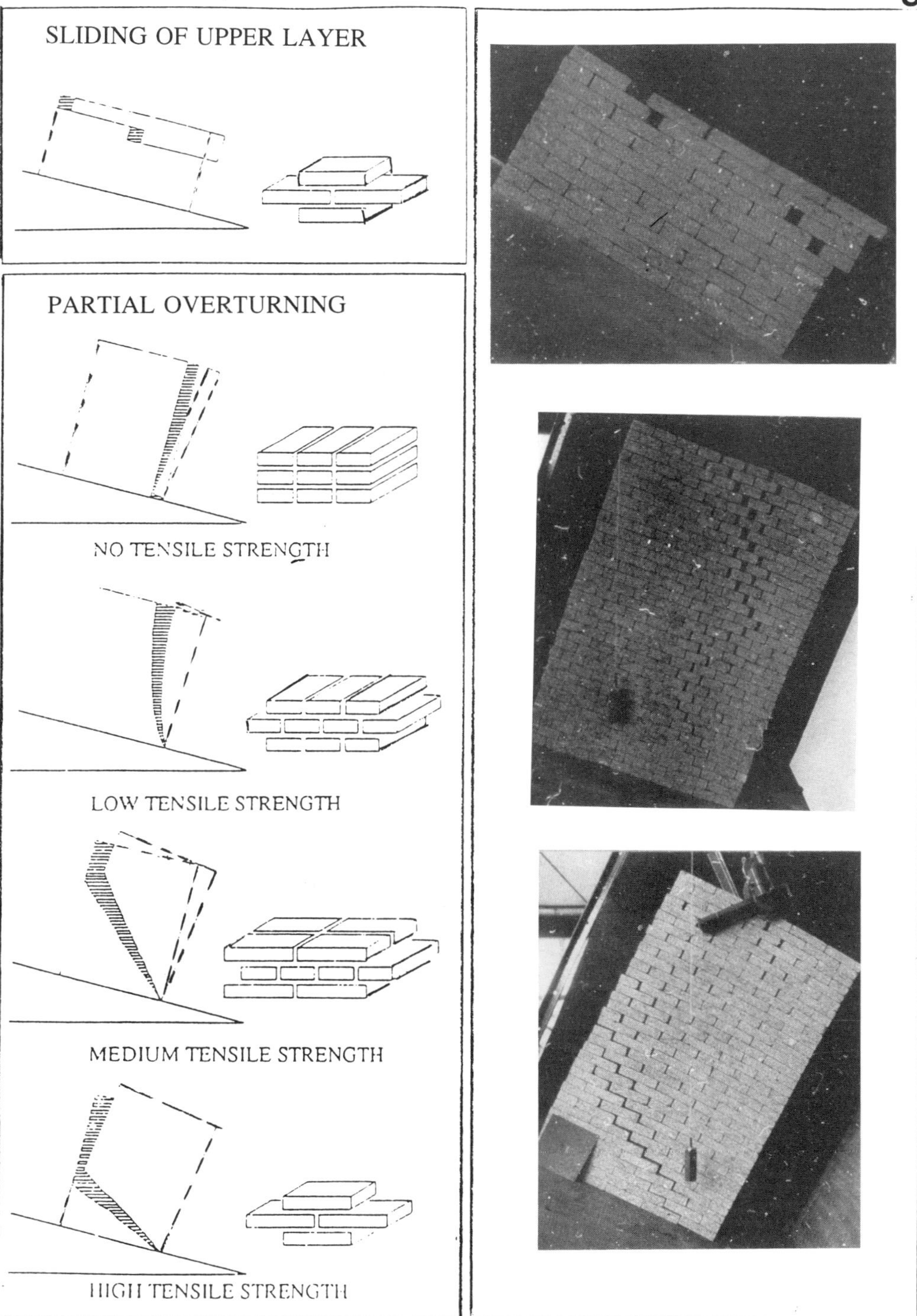

9

 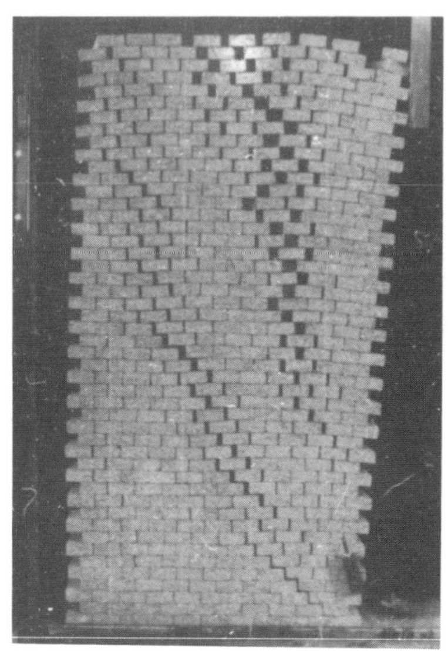

A.   Vertical crack due to low tensile strength.   Model and a house
     of Reggio Calabria hit by the 1908 earthquake

B.   Failure mechanisms for concentrated and distributed horizontal load

10

Partial collapse of the facade model and the "Palazzata" of Messina after the earthquake of 1908

11

a
"Out-of-plane action": sliding and rocking collapse modes
F_S
A
F_R
B
P
P'
P
P'
F
F_R
H_L
H
A
B
b
The wall made of superimposed stones
F
p_l
F
N
p_l
F
N
P
h
p_l
F
H
b
F
F
P
H
b
c
Effectiveness of the diatoni
C D C' D'
B A B' A'
C D
B A

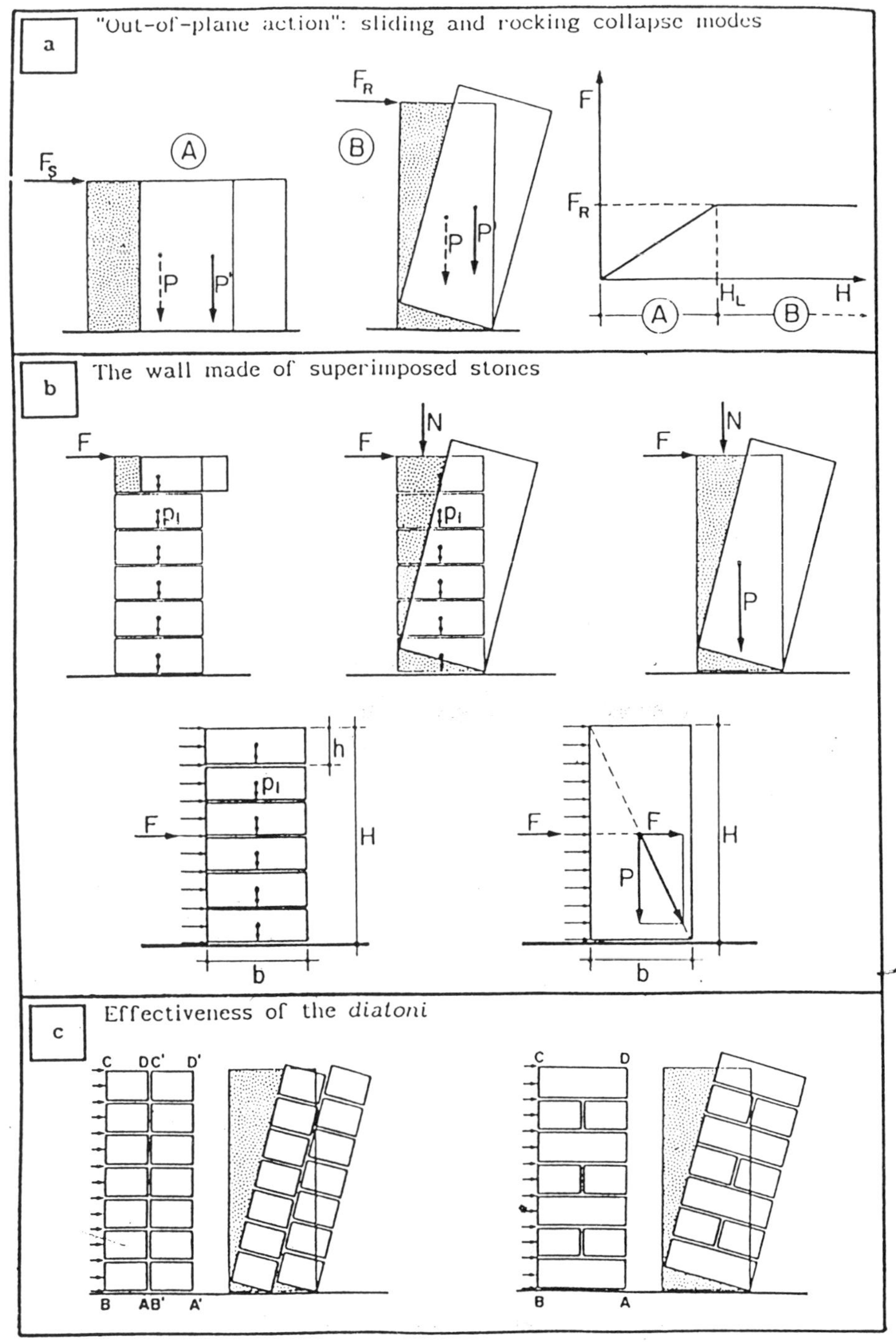

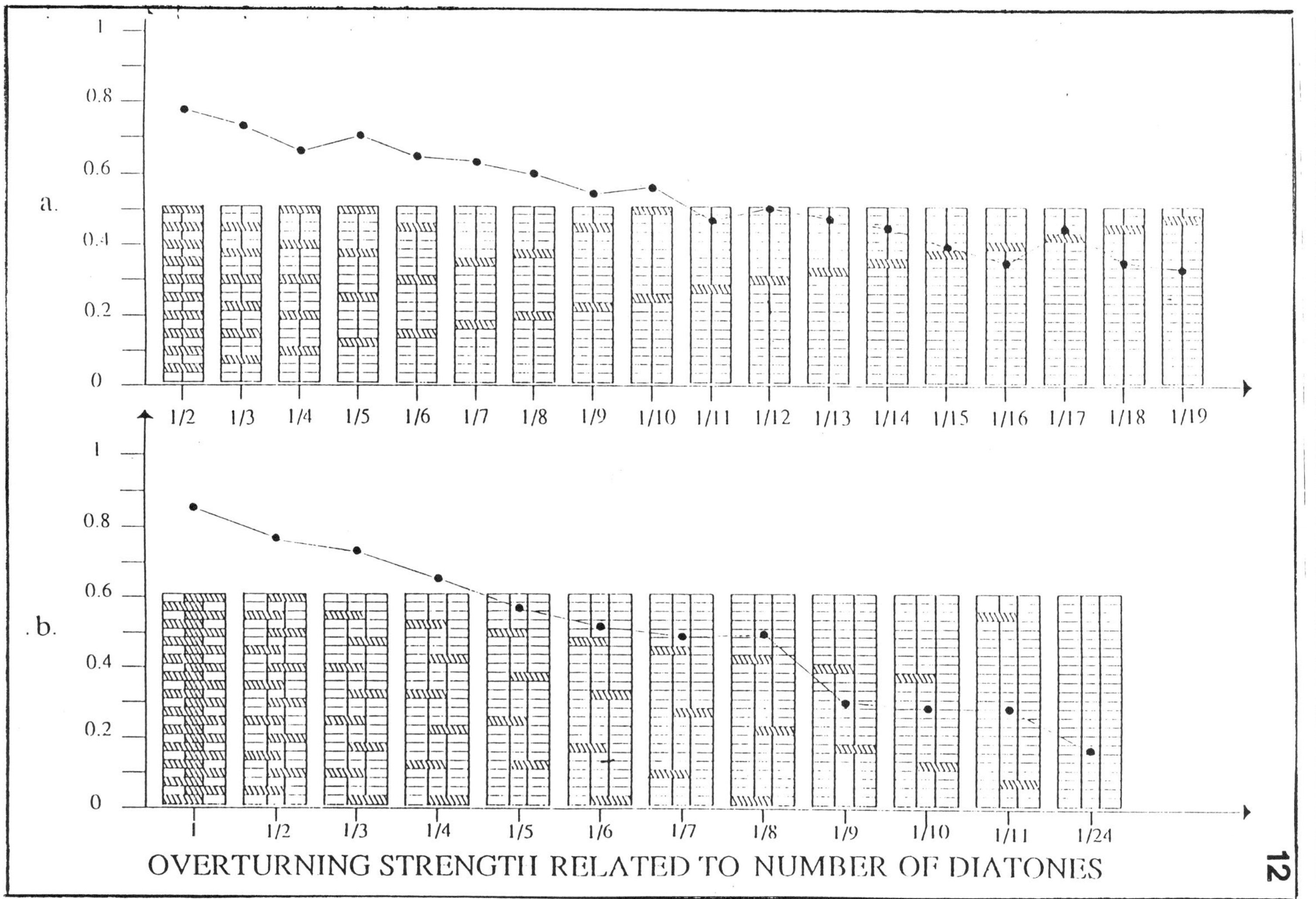
12
a.
1 0.8 0.6 0.4 0.2 0
1/2 1/3 1/4 1/5 1/6 1/7 1/8 1/9 1/10 1/11 1/12 1/13 1/14 1/15 1/16 1/17 1/18 1/19
b.
1 0.8 0.6 0.4 0.2 0
1 1/2 1/3 1/4 1/5 1/6 1/7 1/8 1/9 1/10 1/11 1/24
OVERTURNING STRENGTH RELATED TO NUMBER OF DIATONES

13

a
Disposal of the stones in historical masonry walls
of Ortigia (Syracuse, Italy)
A
B
C
b
Stress distribution in differently arranged masonry walls
A
B
C
F = 0.245 P
F = 0.205 P
F = 0.175 P

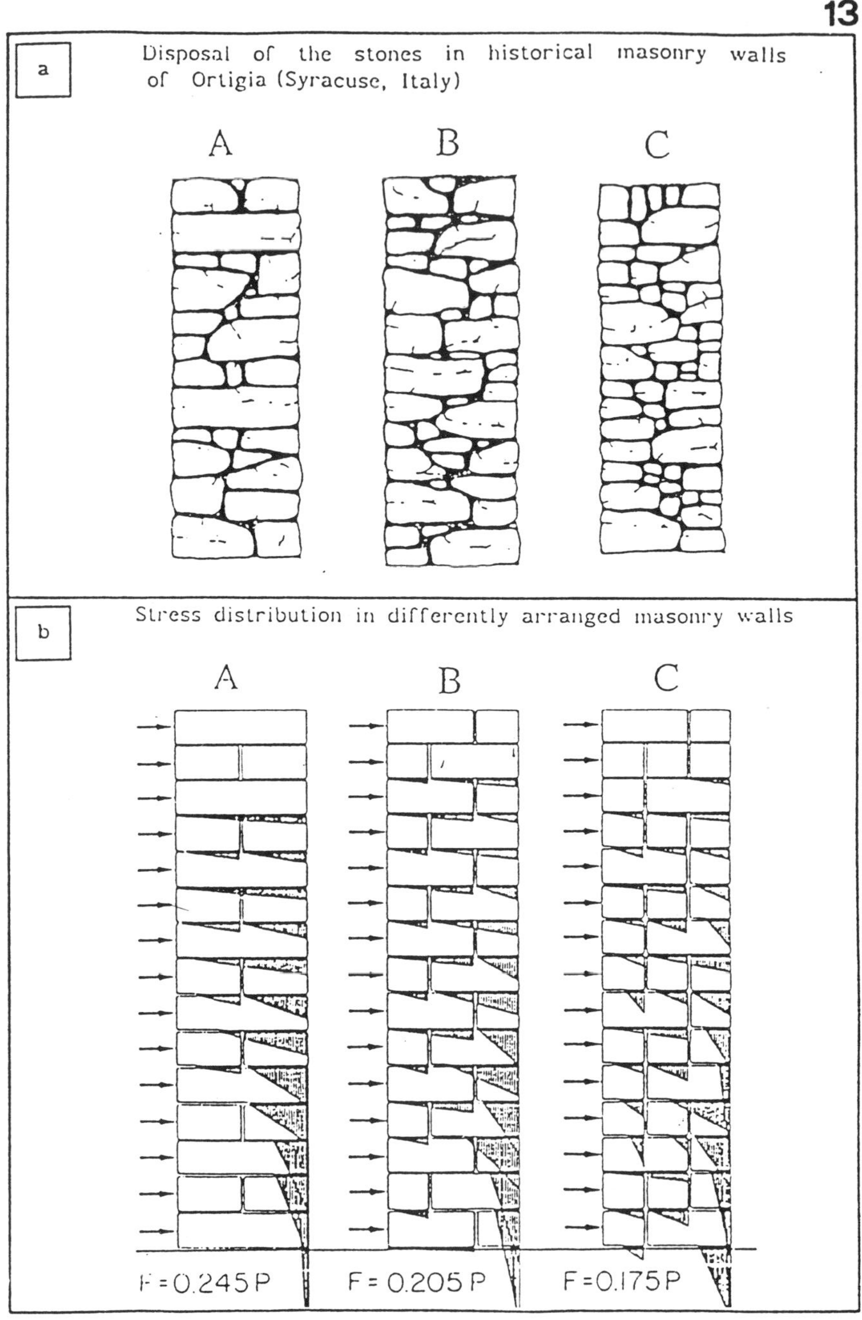

14

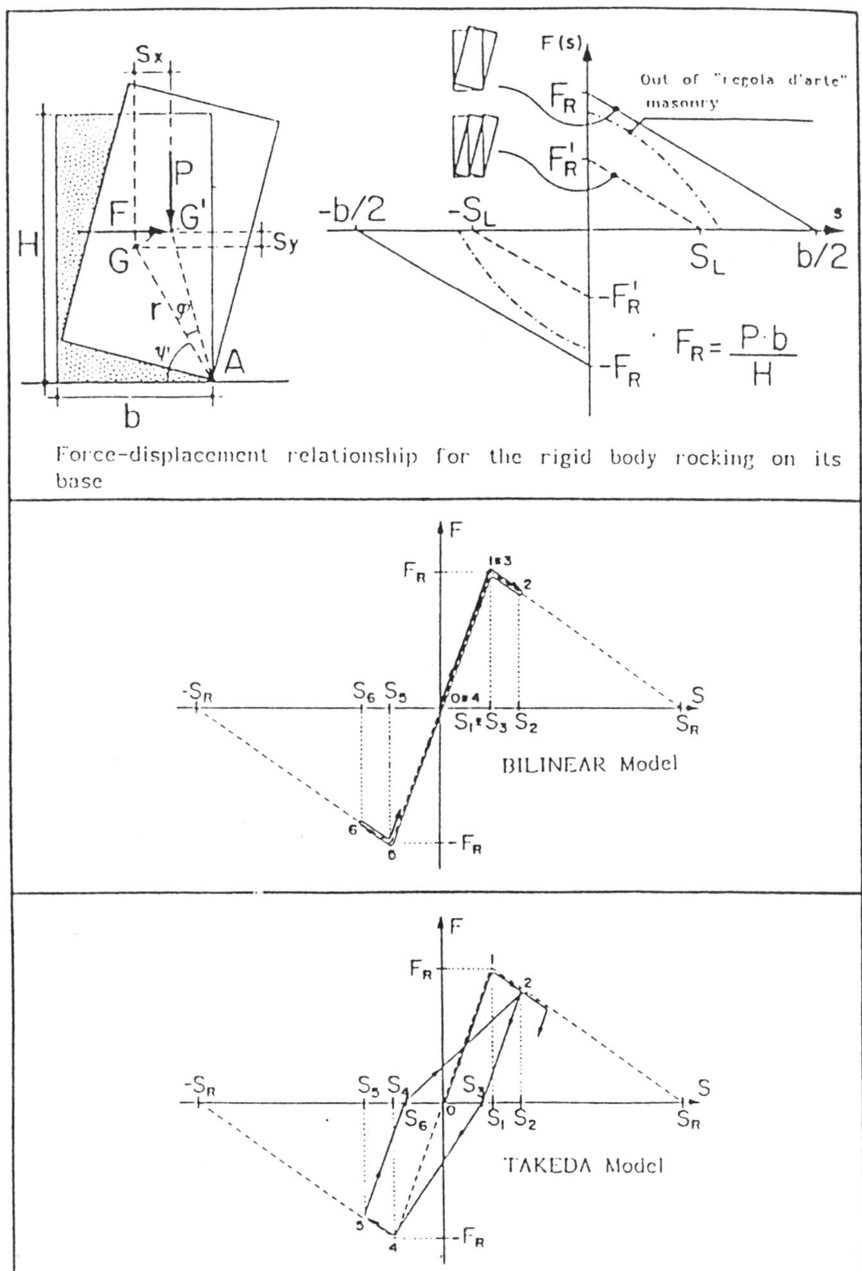

Force-displacement relationship for the rigid body rocking on its base

15

a

Irregularly
arranged
stone masonry

b *Opus craticium* (Pompei, Italy)

c Turkish popular houses

d

Opus africanum

Barbarano

Matera

17

vaulted house
wooden floor house
A   C   B   A   B'   B'   C
b. Difference in construction stage:
A. original cell
B. second stage cell
B' second stage double cell
C. third stage cell
~6 m
~6 m
C   A
B   A
c. Differences in position:
A. Inner house: only one wall exposed
B. Corner house: two walls exposed
C. Head house: three walls exposed

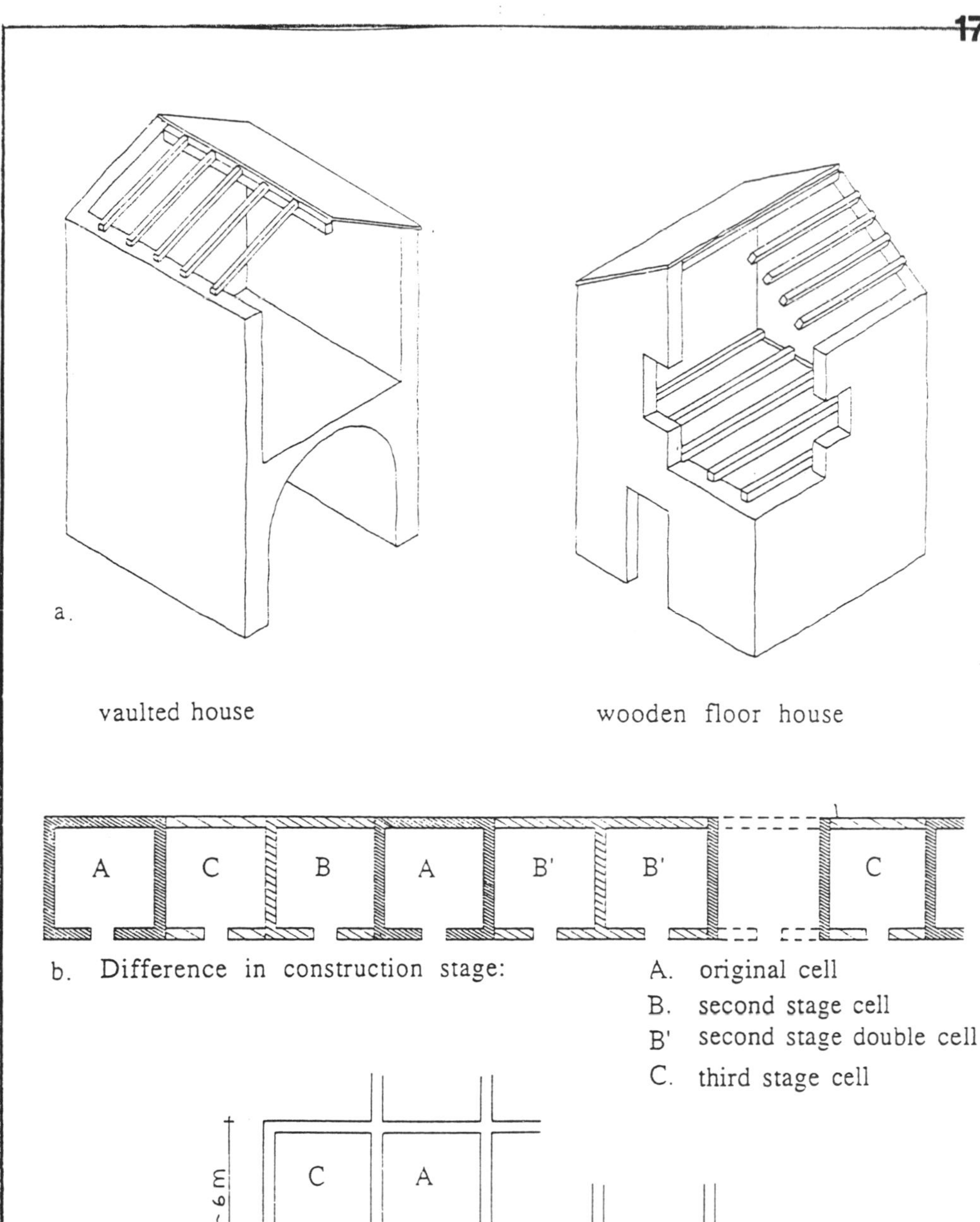

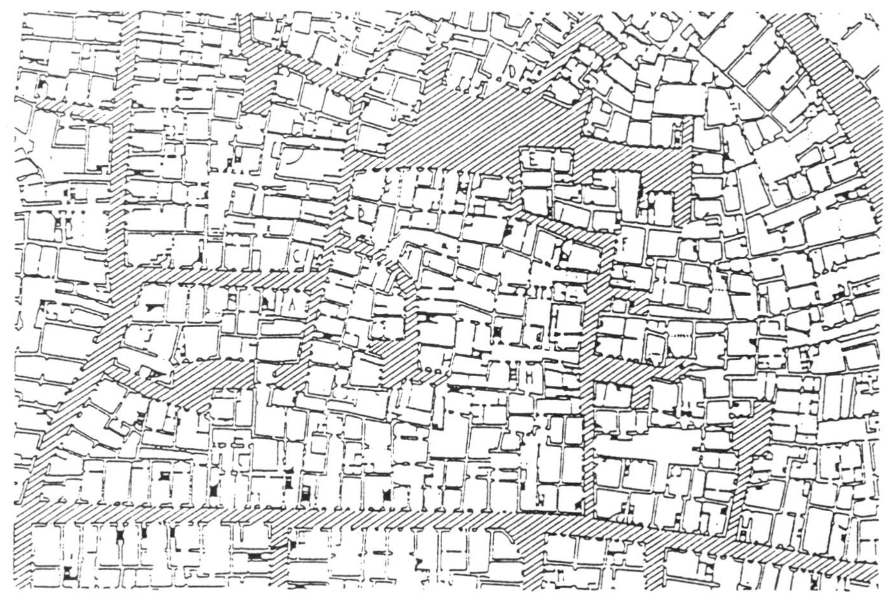

Ortigia

Città di Castello

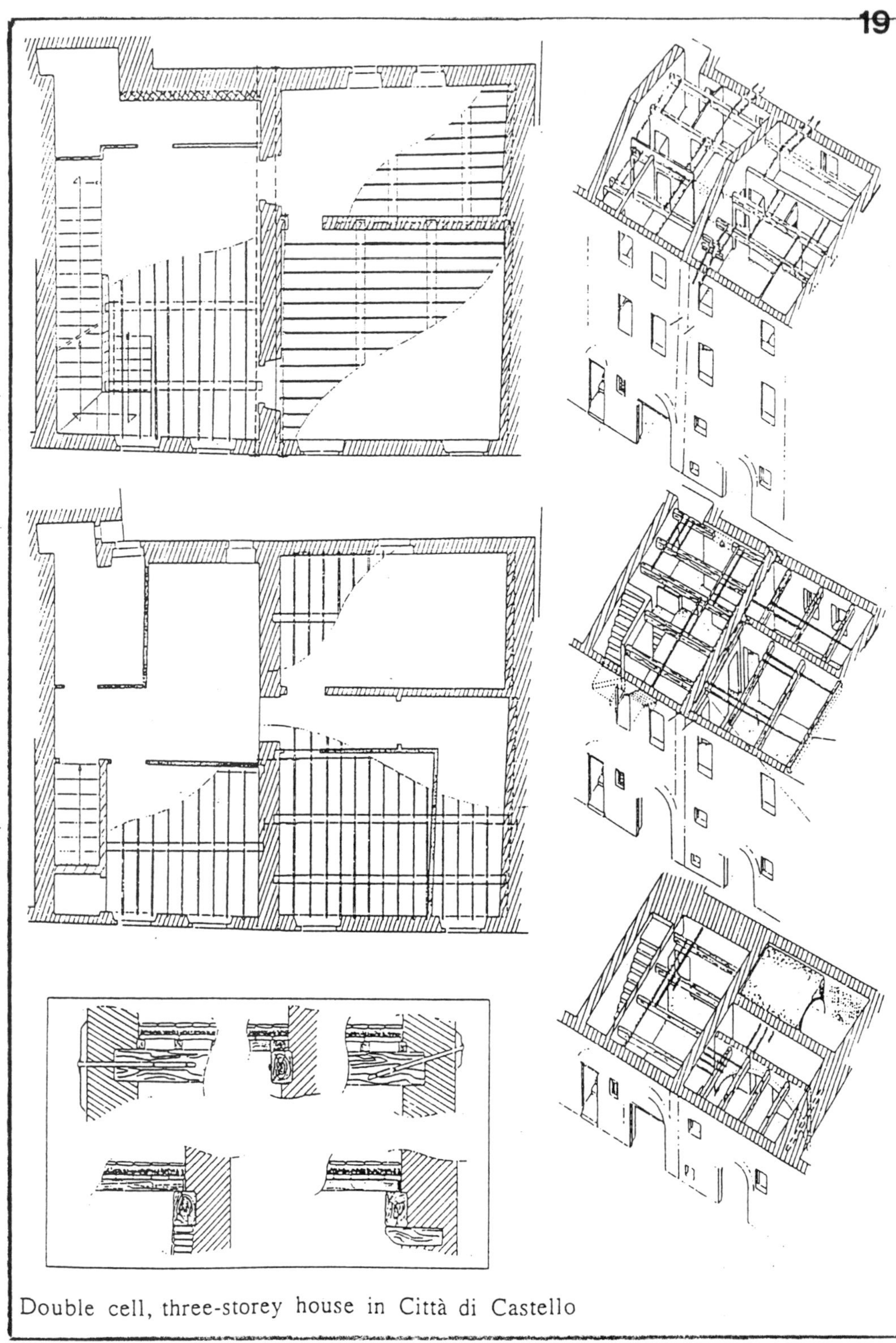

19

Double cell, three-storey house in Città di Castello

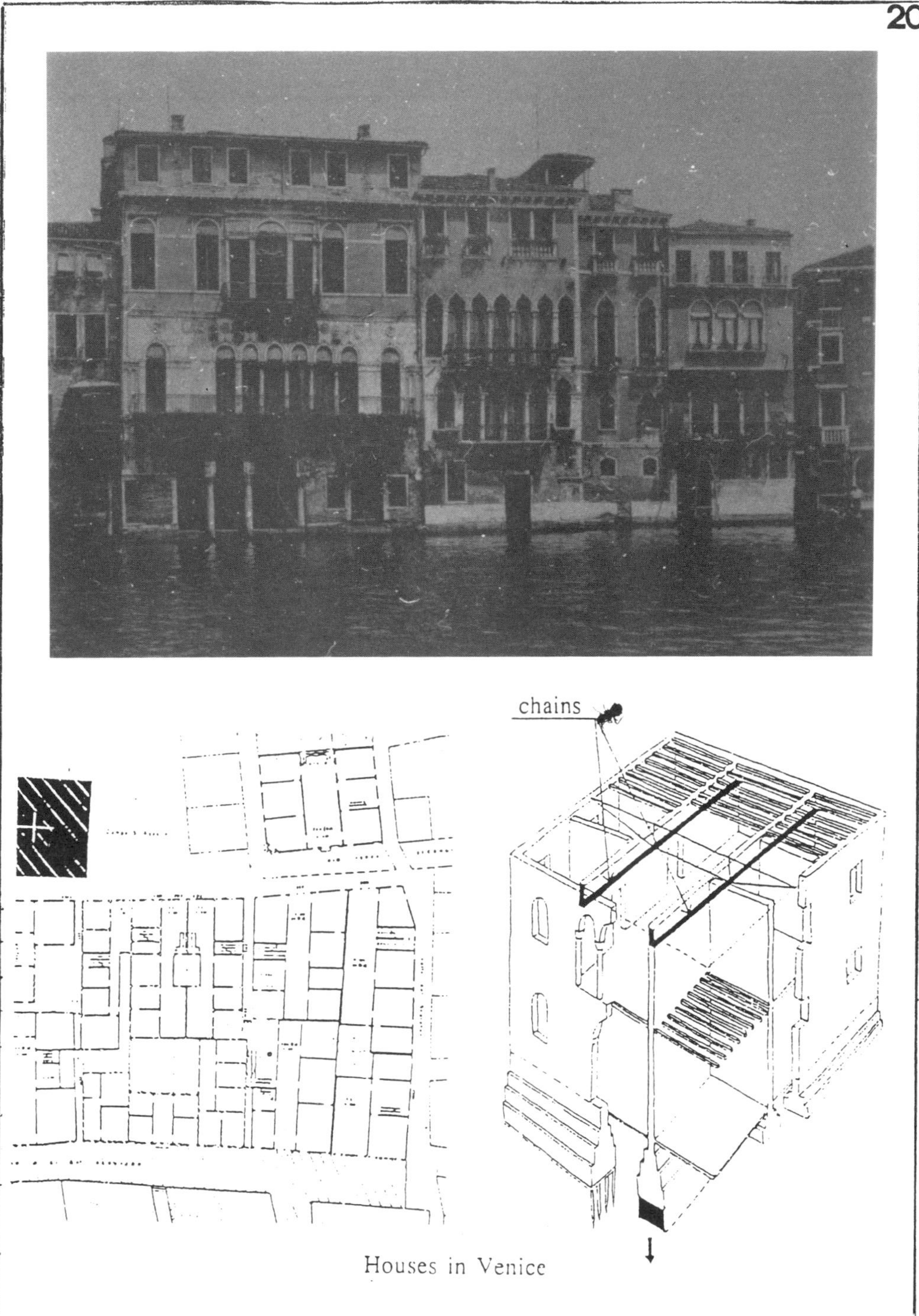

Houses in Venice

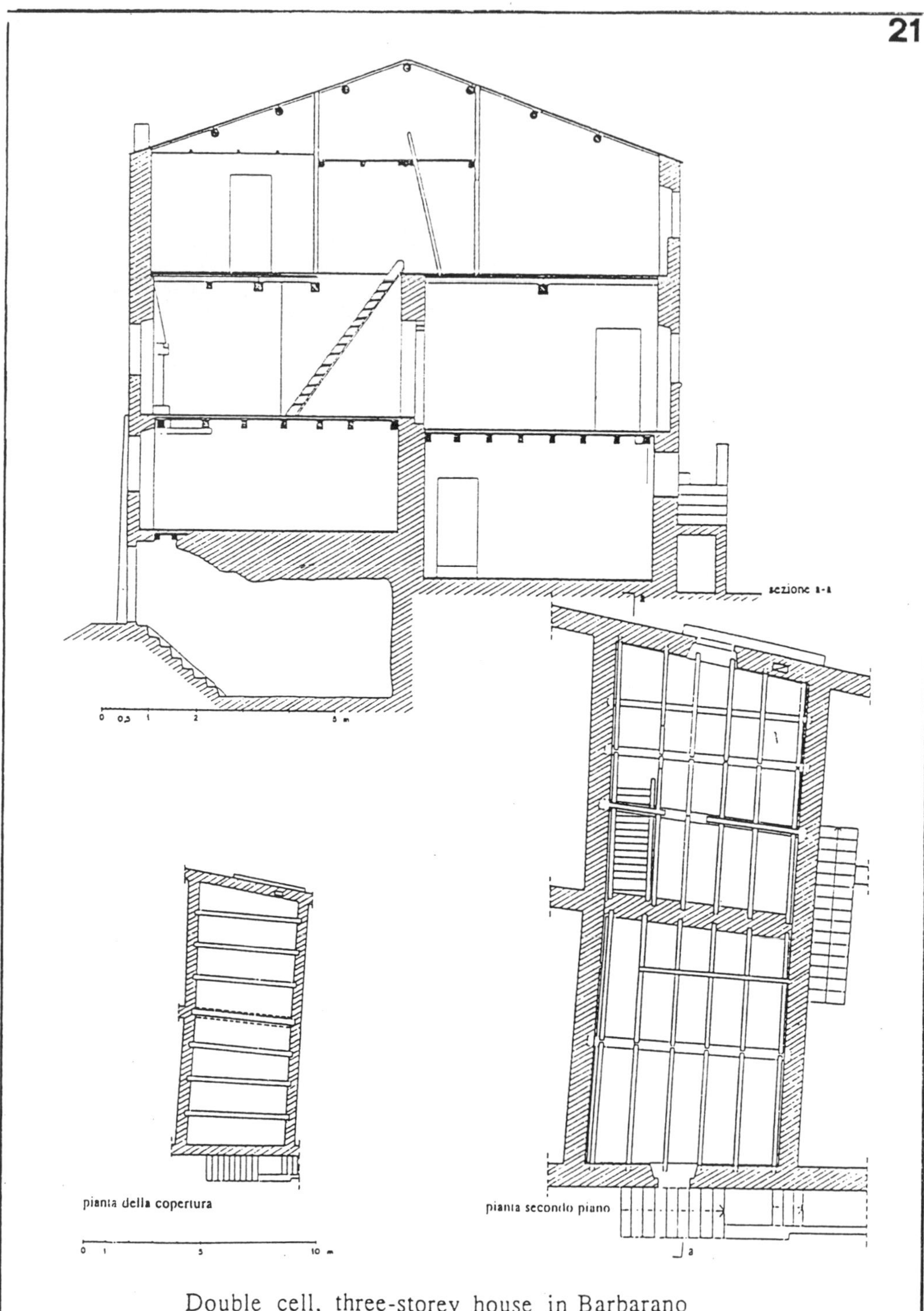

Double cell, three-storey house in Barbarano

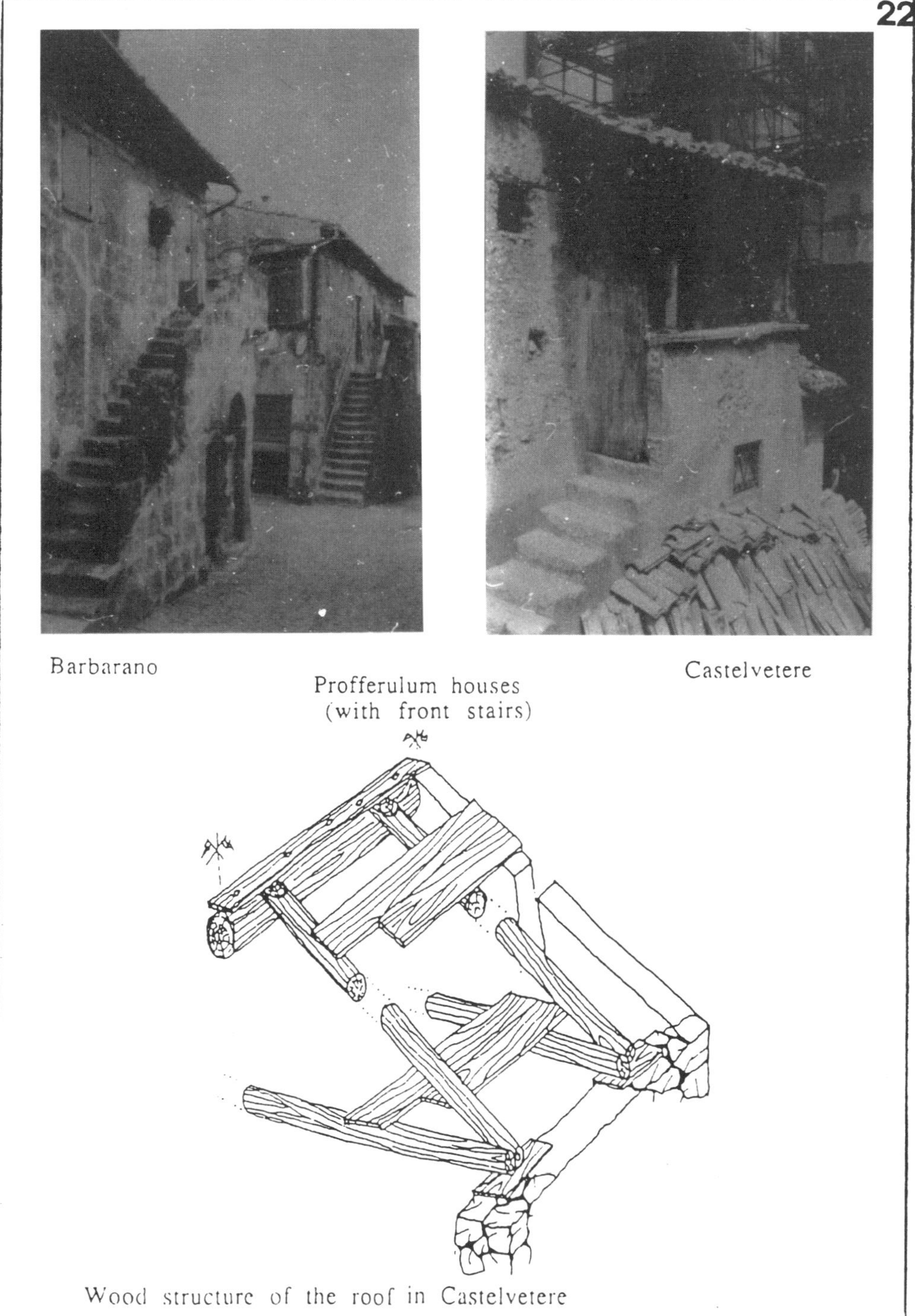

Barbarano

Castelvetere

Profferulum houses
(with front stairs)

Wood structure of the roof in Castelvetere

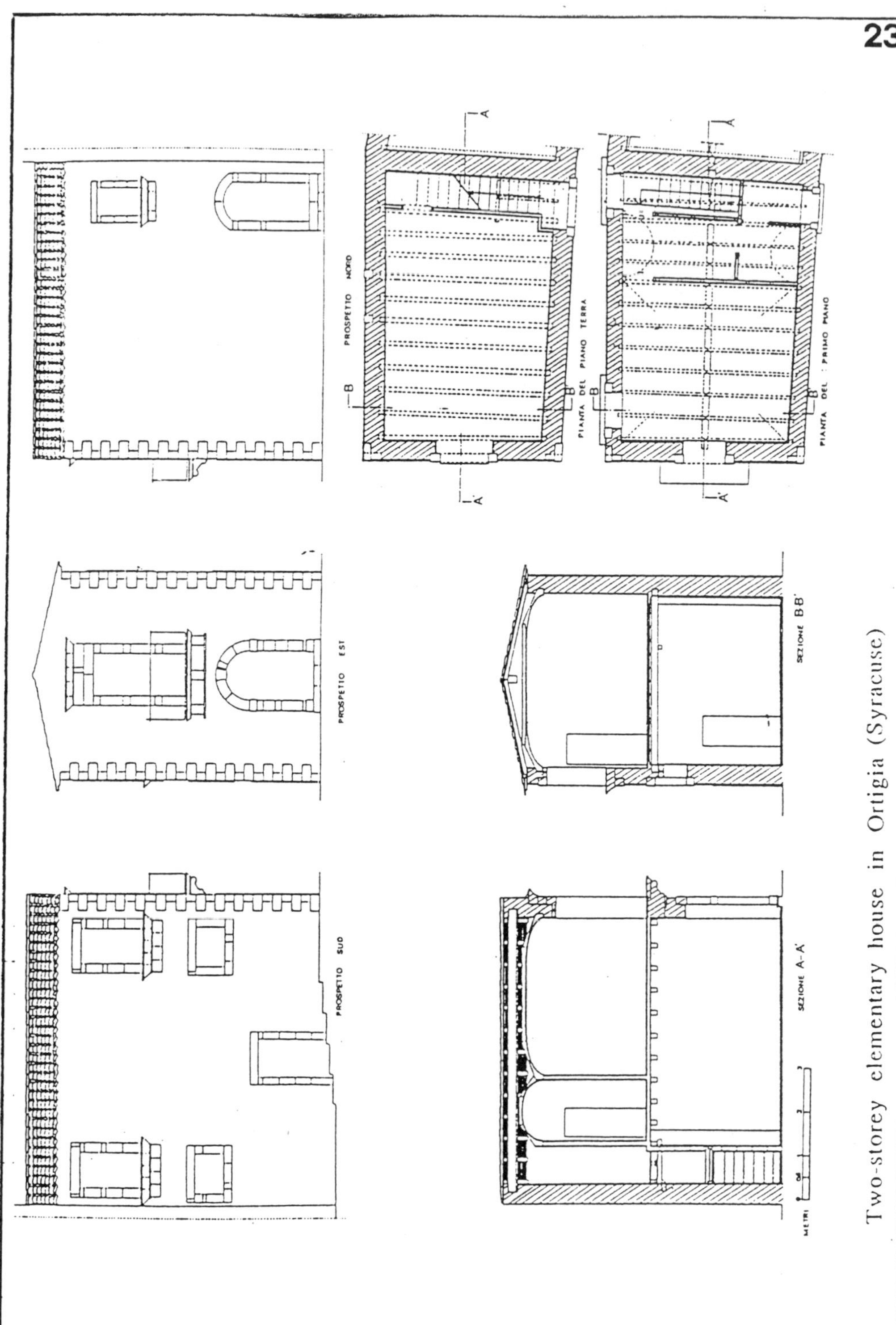

23
PROSPETTO NORD
PROSPETTO EST
PROSPETTO SUD
PIANTA DEL PIANO TERRA
PIANTA DEL PRIMO PIANO
SEZIONE B-B'
SEZIONE A-A'
METRI
Two-storey elementary house in Ortigia (Syracuse)

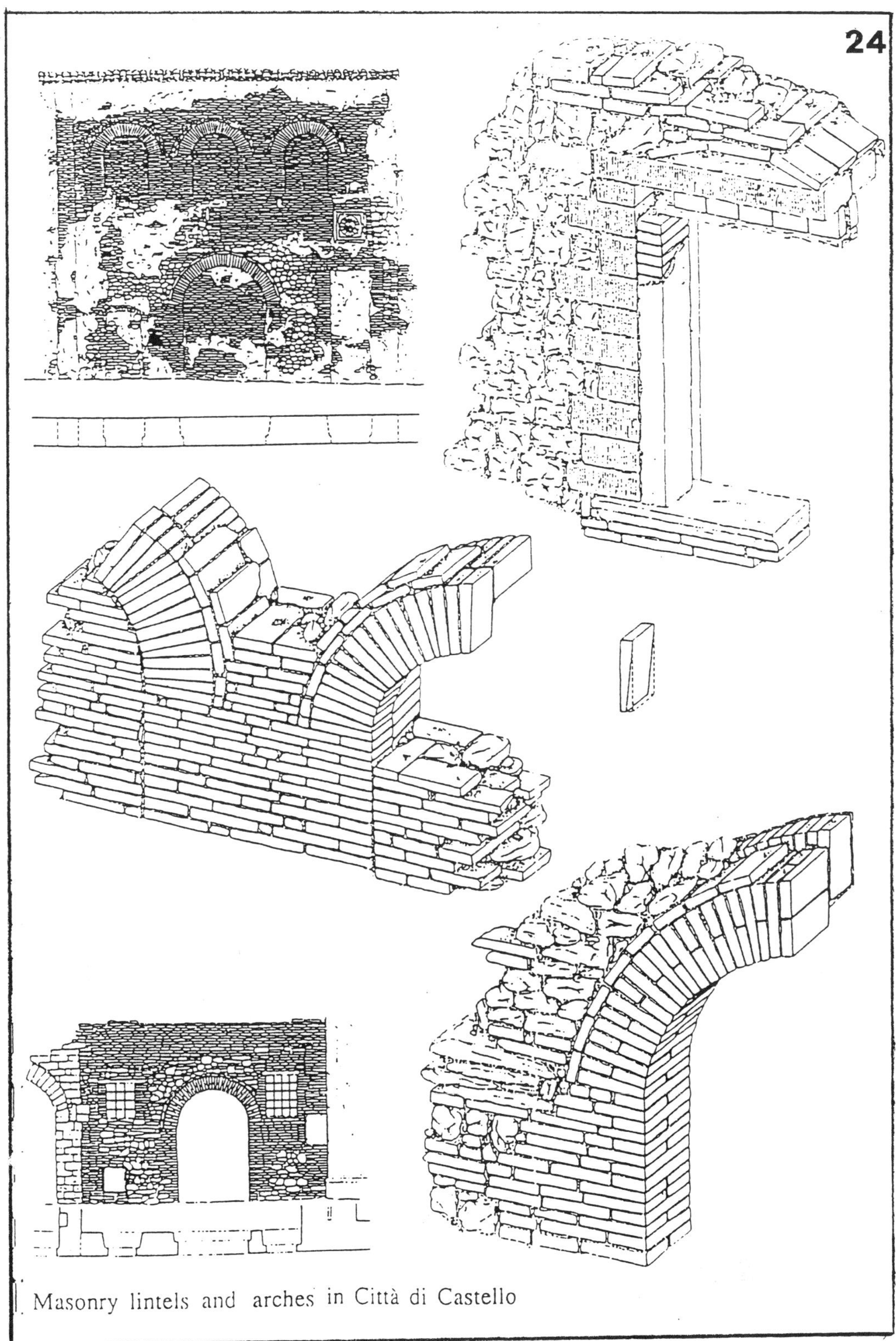

Masonry lintels and arches in Città di Castello

25

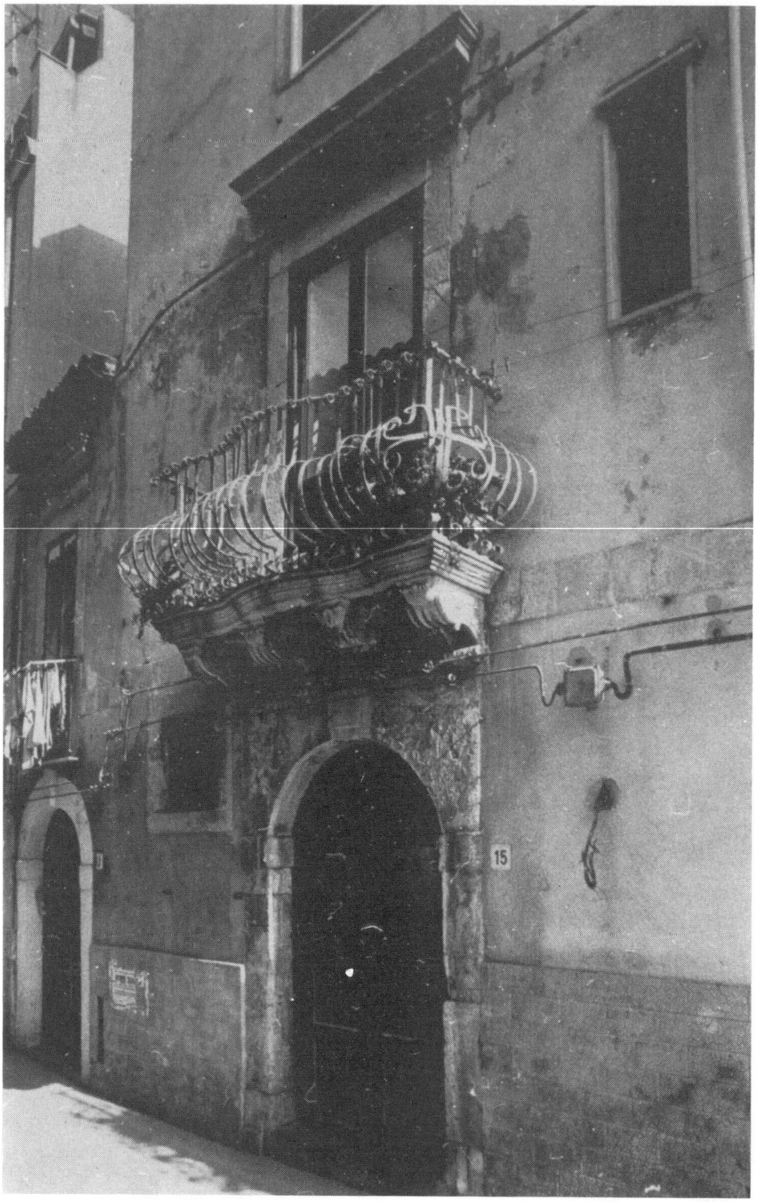

Front door and upper balcony in Ortigia

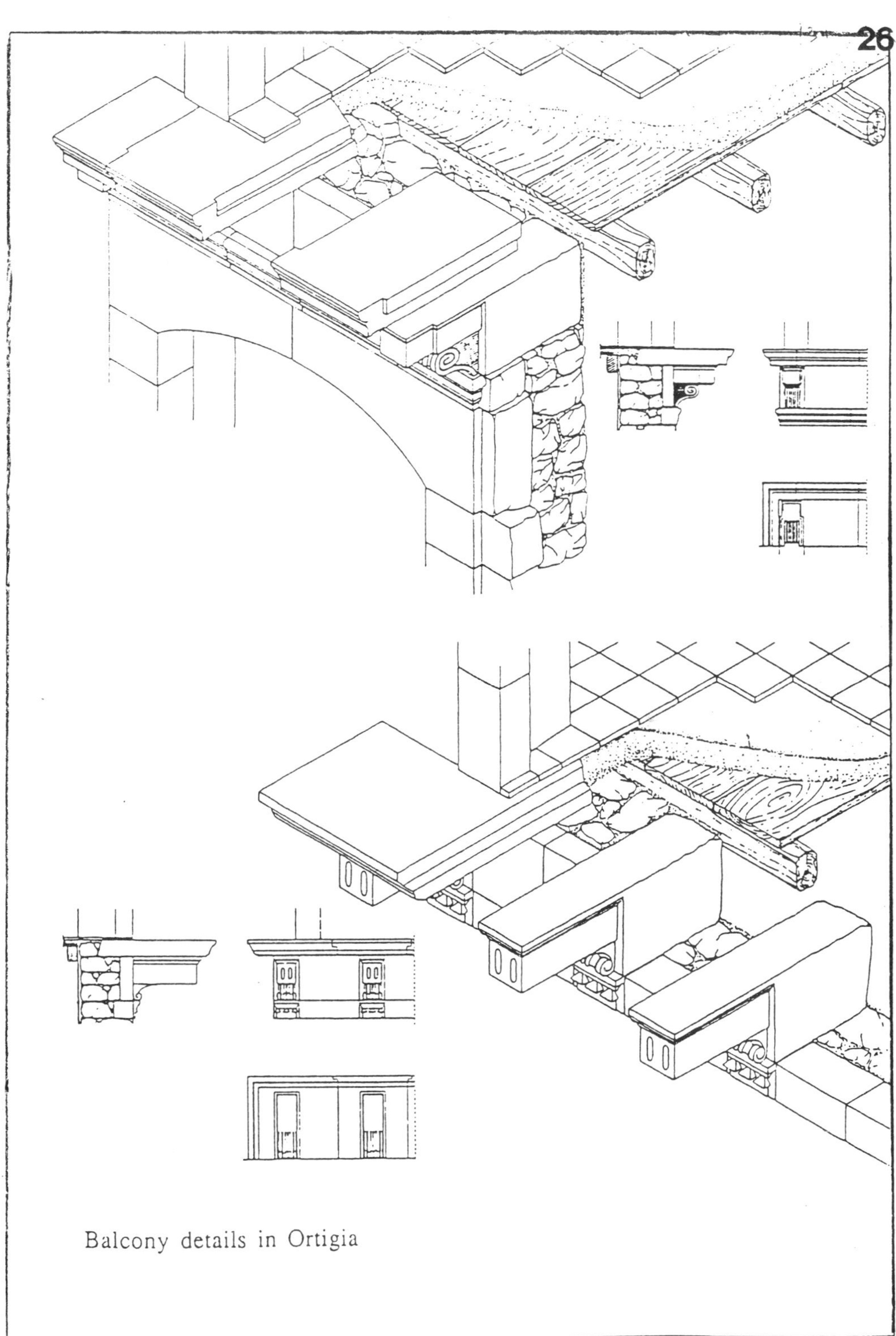

Balcony details in Ortigia

Tower transformed into house (Città di Castello)

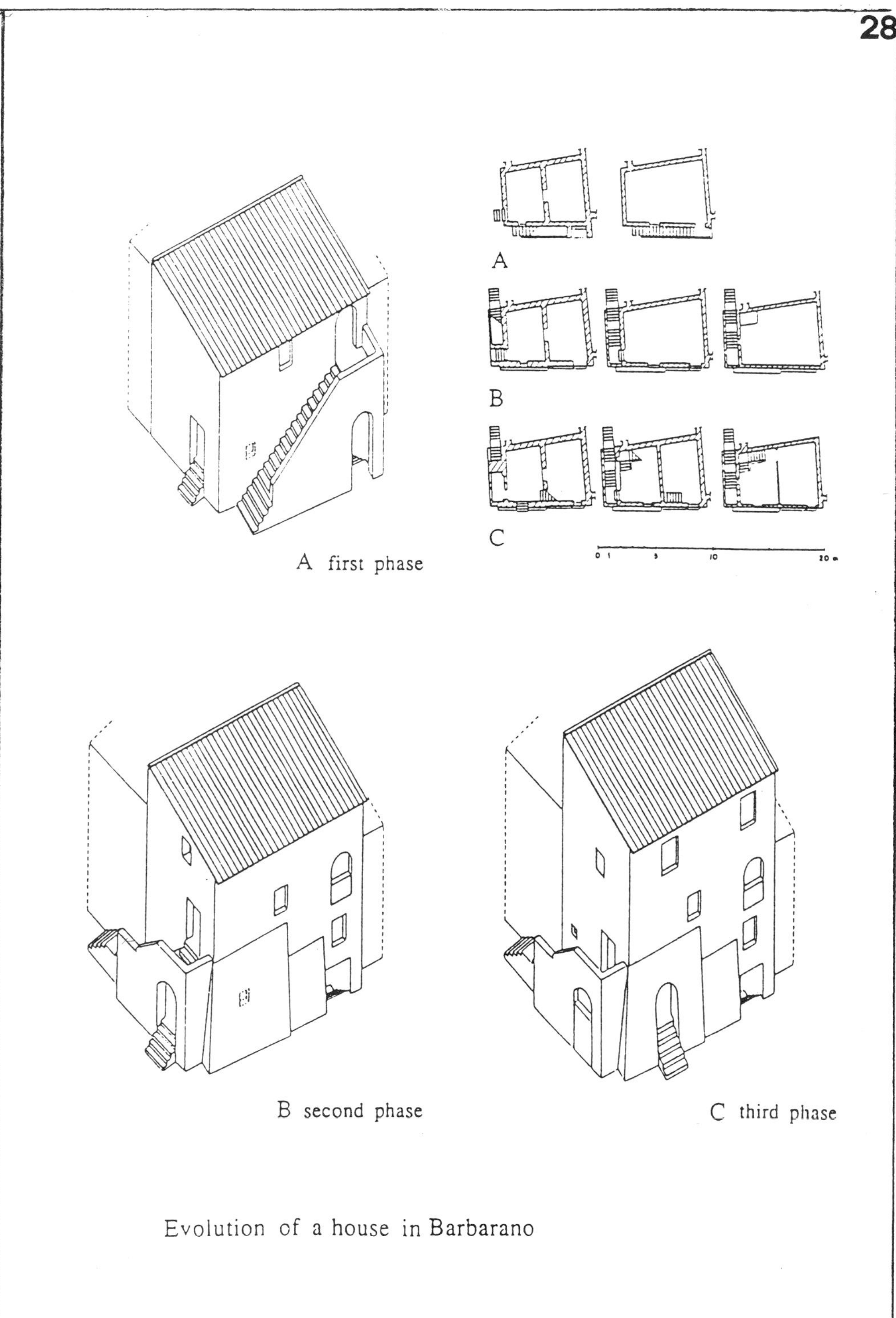

Evolution of a house in Barbarano

29

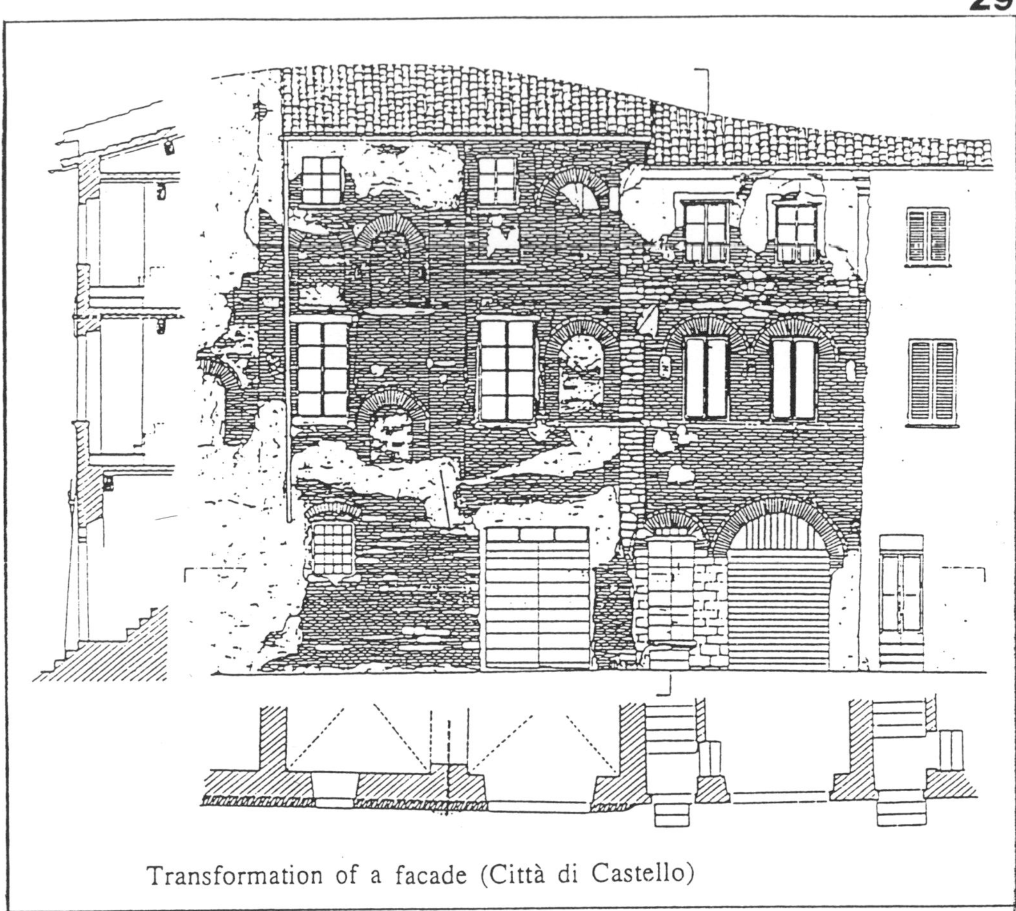

Transformation of a facade (Città di Castello)

Rocking failure of exterior wall: drawing by Rondelet and photograph of houses along a street of Messina on 1908

A. Giuffrè

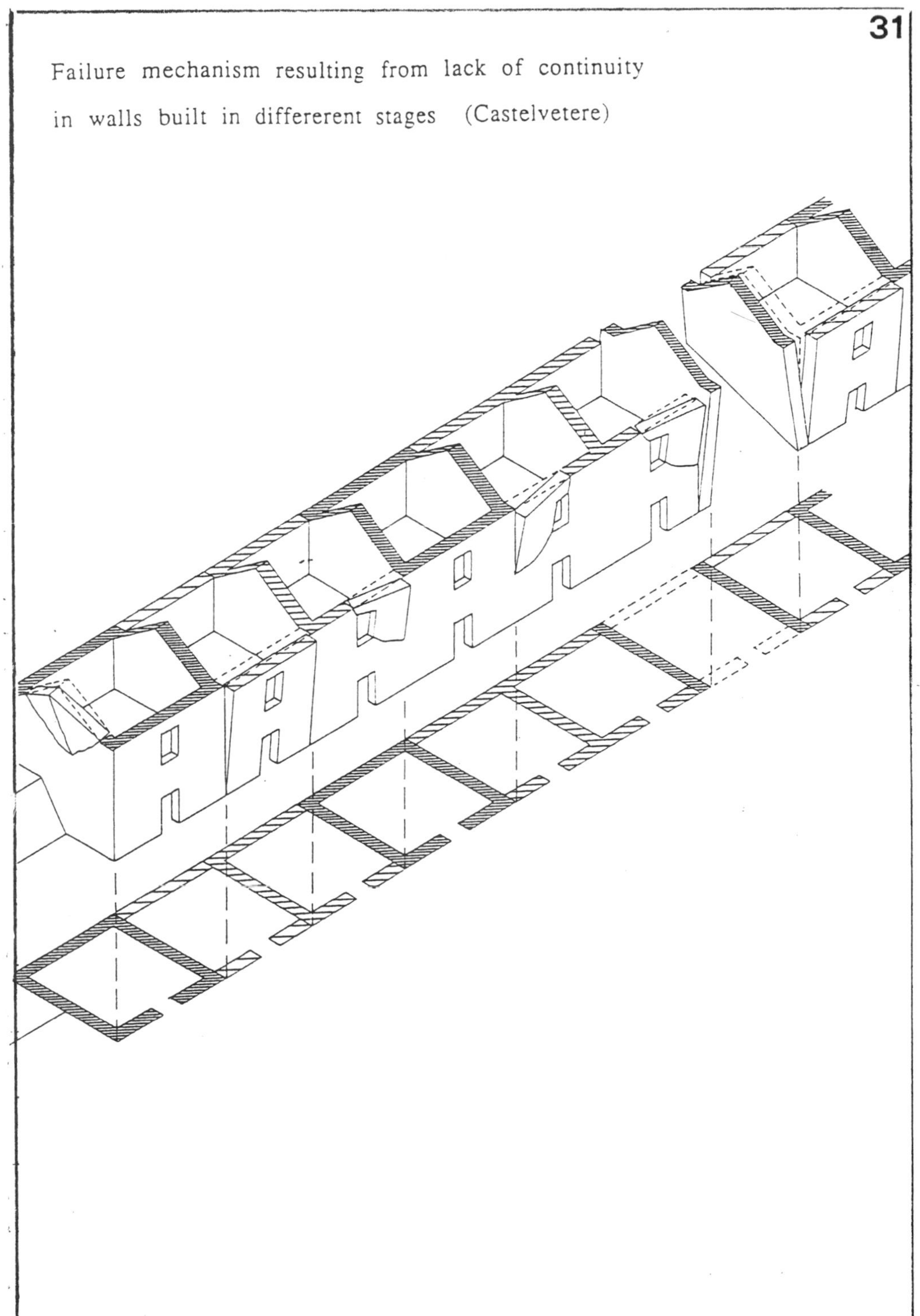

31
Failure mechanism resulting from lack of continuity
in walls built in differerent stages (Castelvetere)

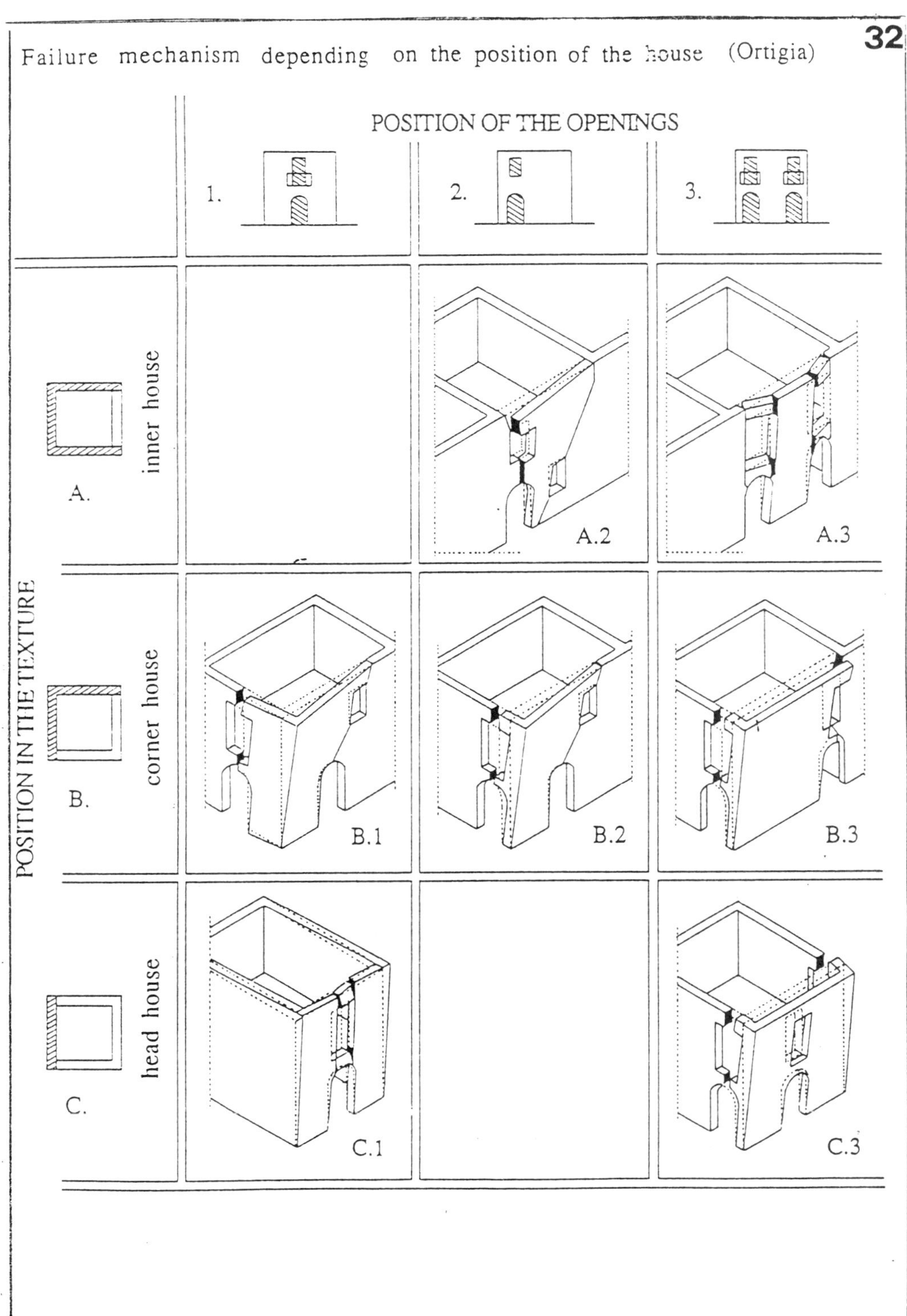

Failure mechanism depending on the position of the house (Ortigia)
32
POSITION OF THE OPENINGS
1.
2.
3.
POSITION IN THE TEXTURE
A. inner house
A.2
A.3
B. corner house
B.1
B.2
B.3
C. head house
C.1
C.3

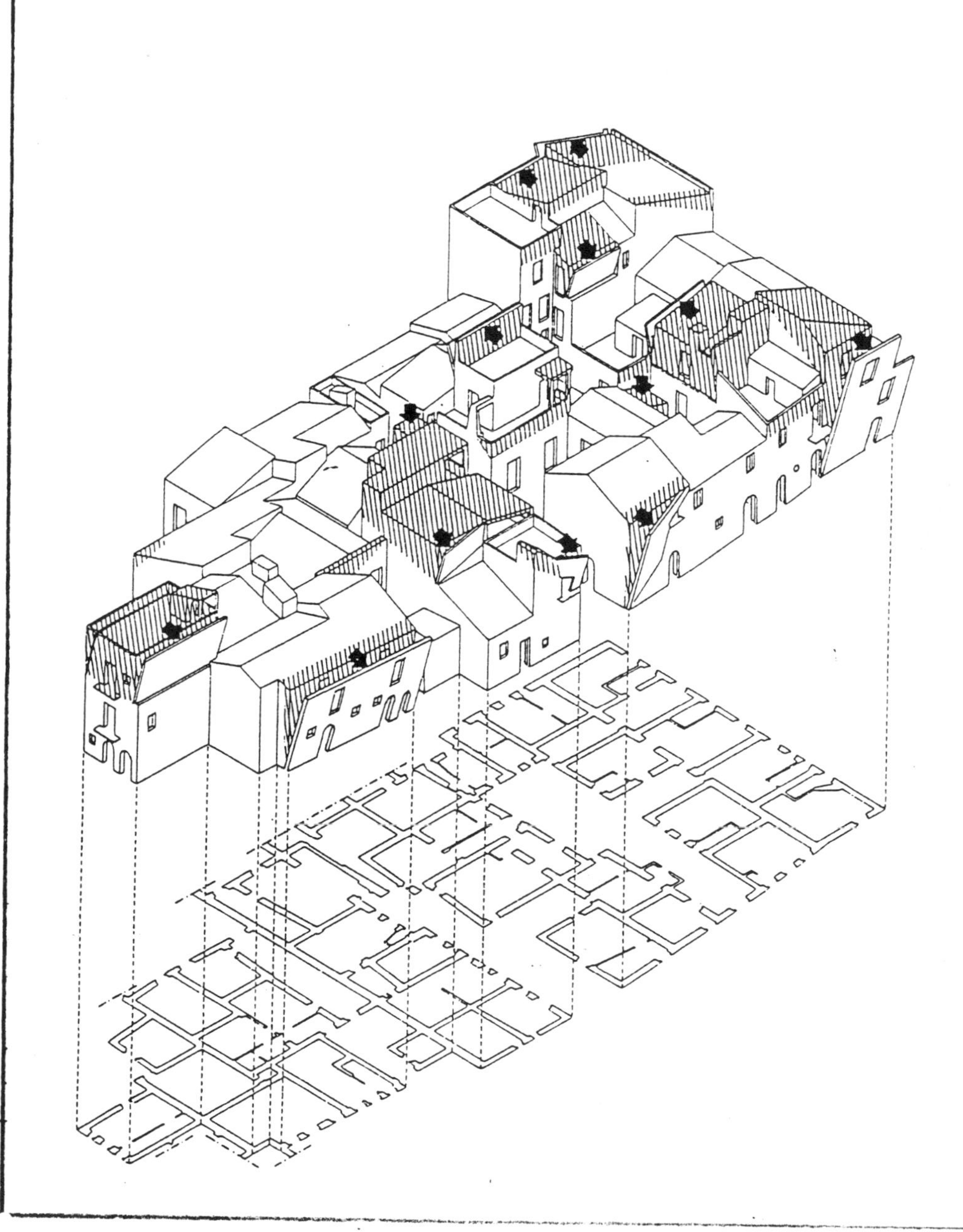

PREDICTION OF DAMAGE IN A PORTION OF URBAN FABRIC (ORTIGIA)
33

34

a
Far   ACCELERATION SPECTRA   Near
Amplification
Mean values
Mean+s.d.
T (sec)
Amplification
Mean values
Mean+s.d.
T (sec)

b
P
h/2
K
B
b/2
F
Fo
K
So
Su
S

c
Bilinear   ULTIMATE DISPLACEMENT   Takeda
Smax (cm)
Su=20cm   Su=5cm
Su=30cm
Su=10cm
Su=20cm
Su=15cm
Su=30cm
T (sec)
T (sec)

d
Bilinear   VISCOUS DAMPING   Takeda
Smax (cm)
C=.44
C=.74
C=1.11
C=1.48
C=.44
C=.74
C=1.11
C=1.48
T (sec)
T (sec)

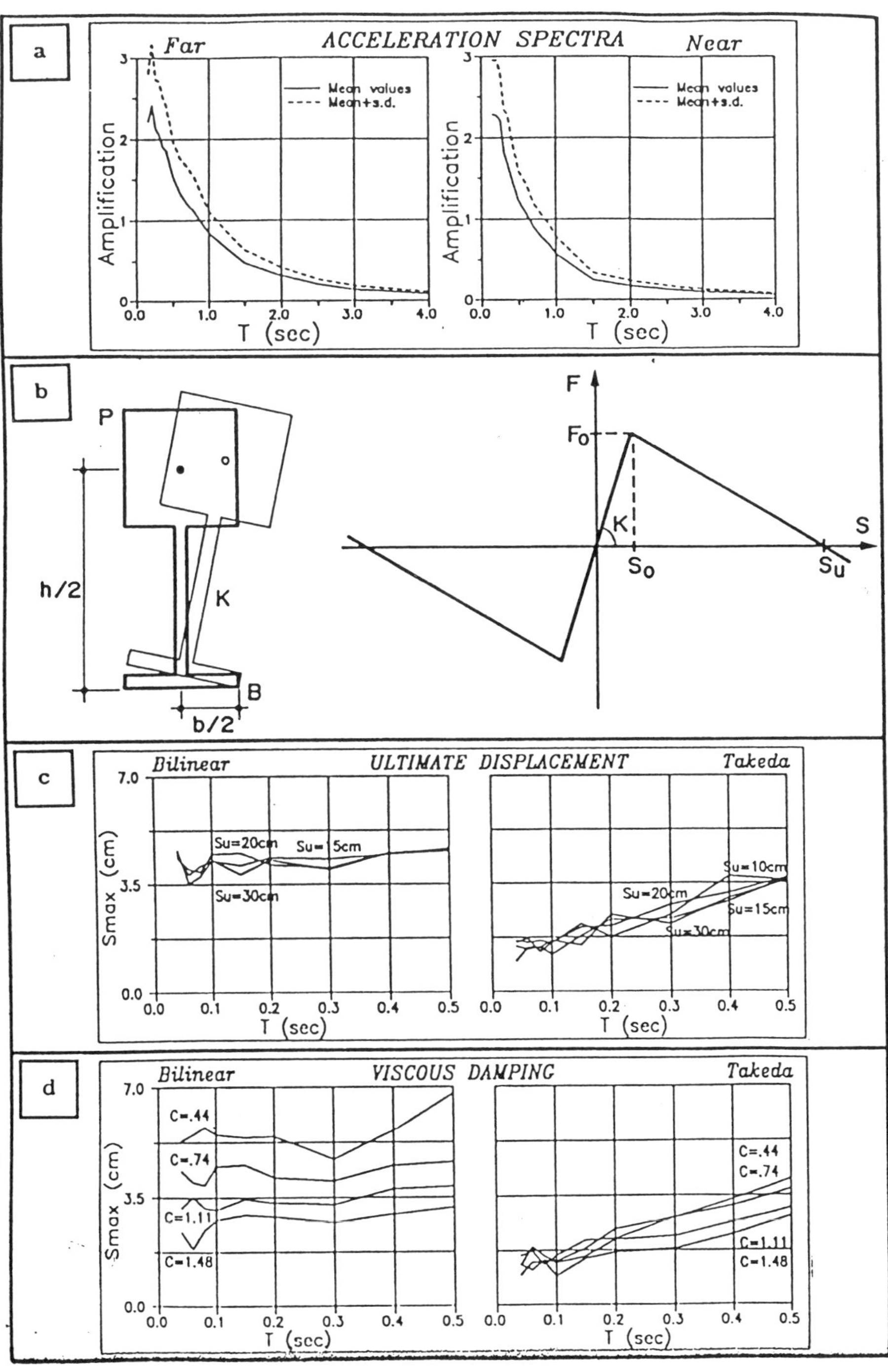

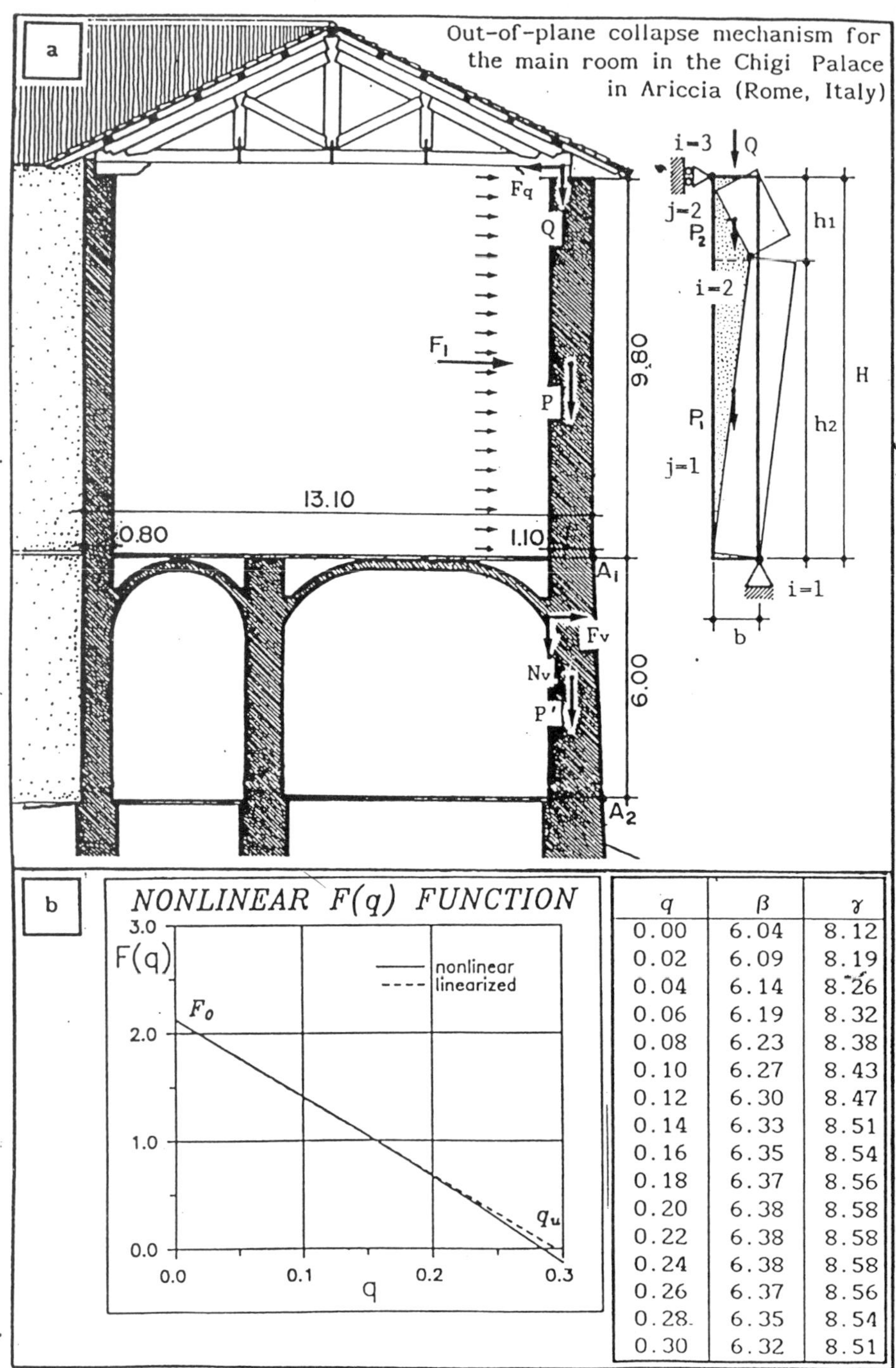

q	β	γ
0.00	6.04	8.12
0.02	6.09	8.19
0.04	6.14	8.26
0.06	6.19	8.32
0.08	6.23	8.38
0.10	6.27	8.43
0.12	6.30	8.47
0.14	6.33	8.51
0.16	6.35	8.54
0.18	6.37	8.56
0.20	6.38	8.58
0.22	6.38	8.58
0.24	6.38	8.58
0.26	6.37	8.56
0.28	6.35	8.54
0.30	6.32	8.51

36

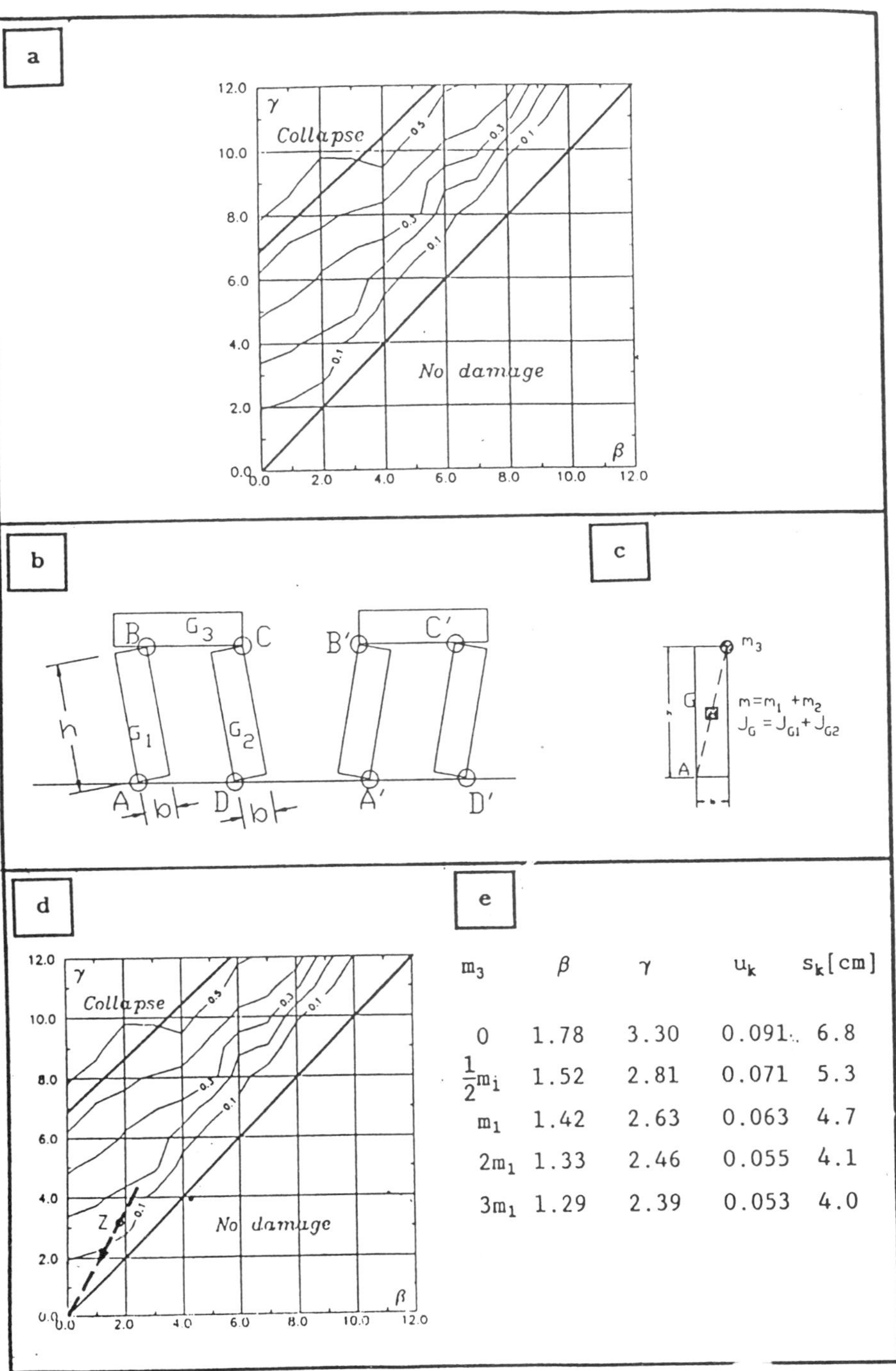

m_3	β	γ	u_k	s_k [cm]
0	1.78	3.30	0.091	6.8
$\frac{1}{2}m_i$	1.52	2.81	0.071	5.3
m_1	1.42	2.63	0.063	4.7
$2m_1$	1.33	2.46	0.055	4.1
$3m_1$	1.29	2.39	0.053	4.0

REBUILDING WALLS IN SQUARED UNITS

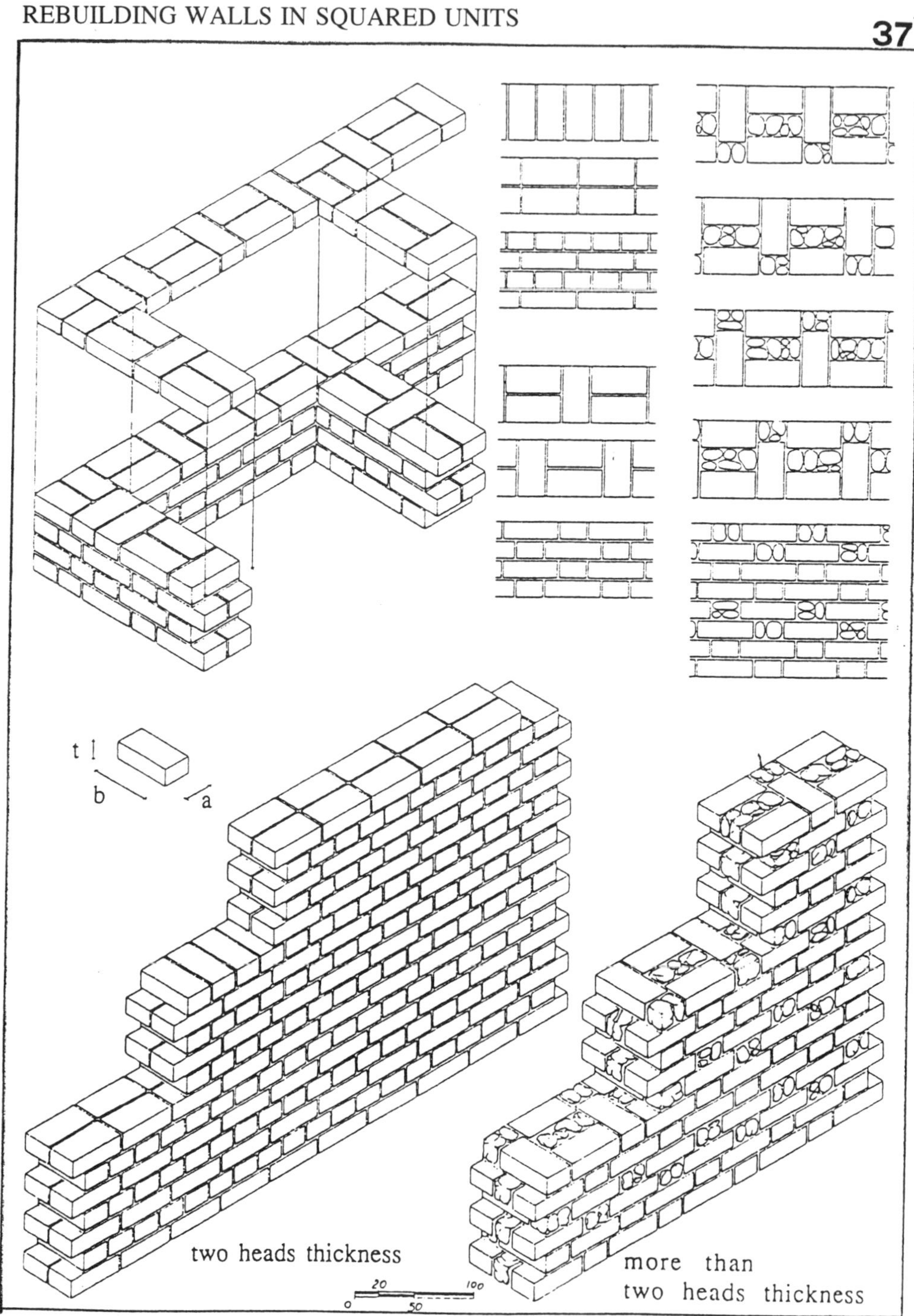

INSERTING TRANSVERSE KEYS IN MASONRIES

38

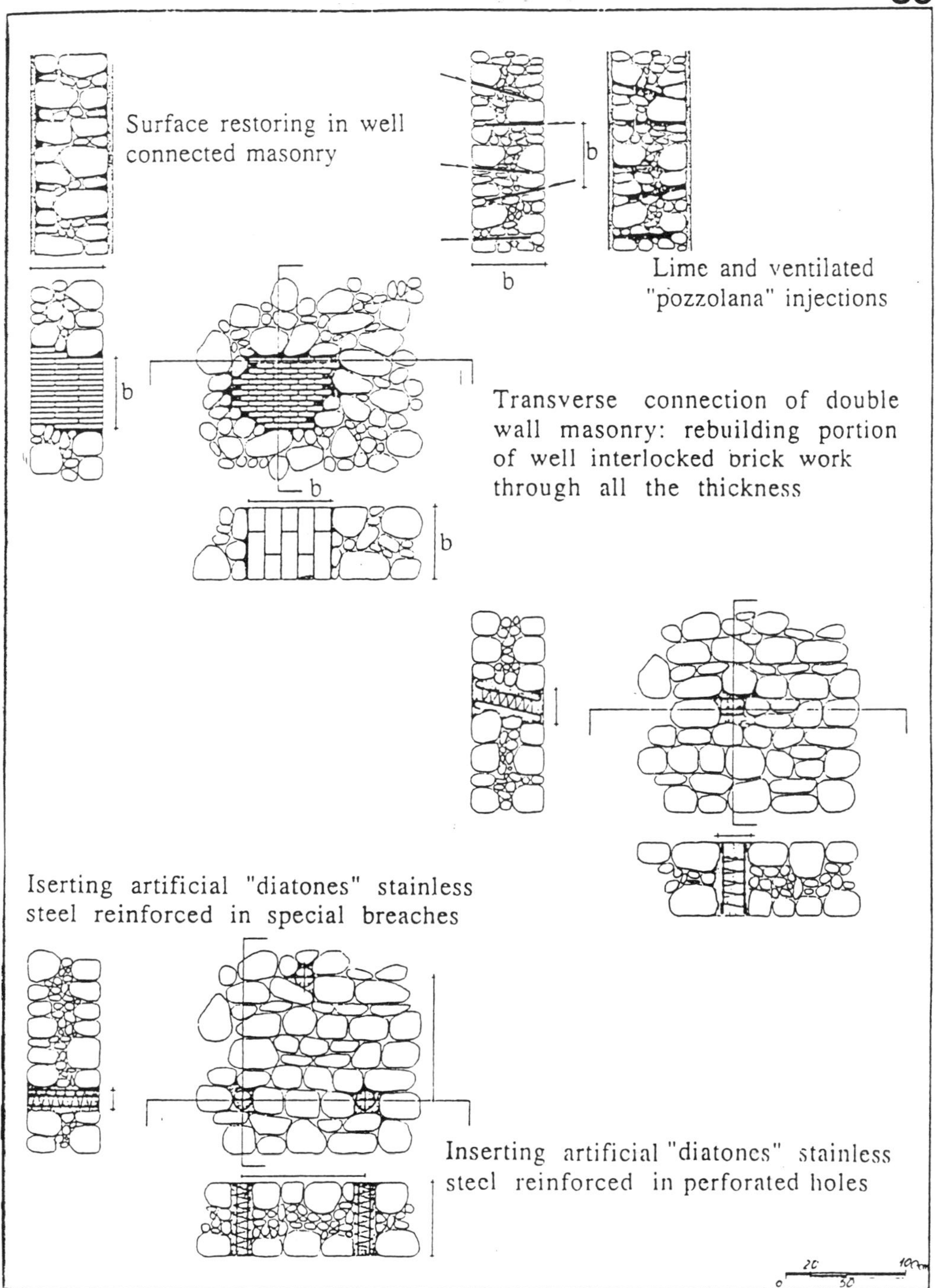

MASONRY TIE-BEAM ON TOP OF WALL   **39**

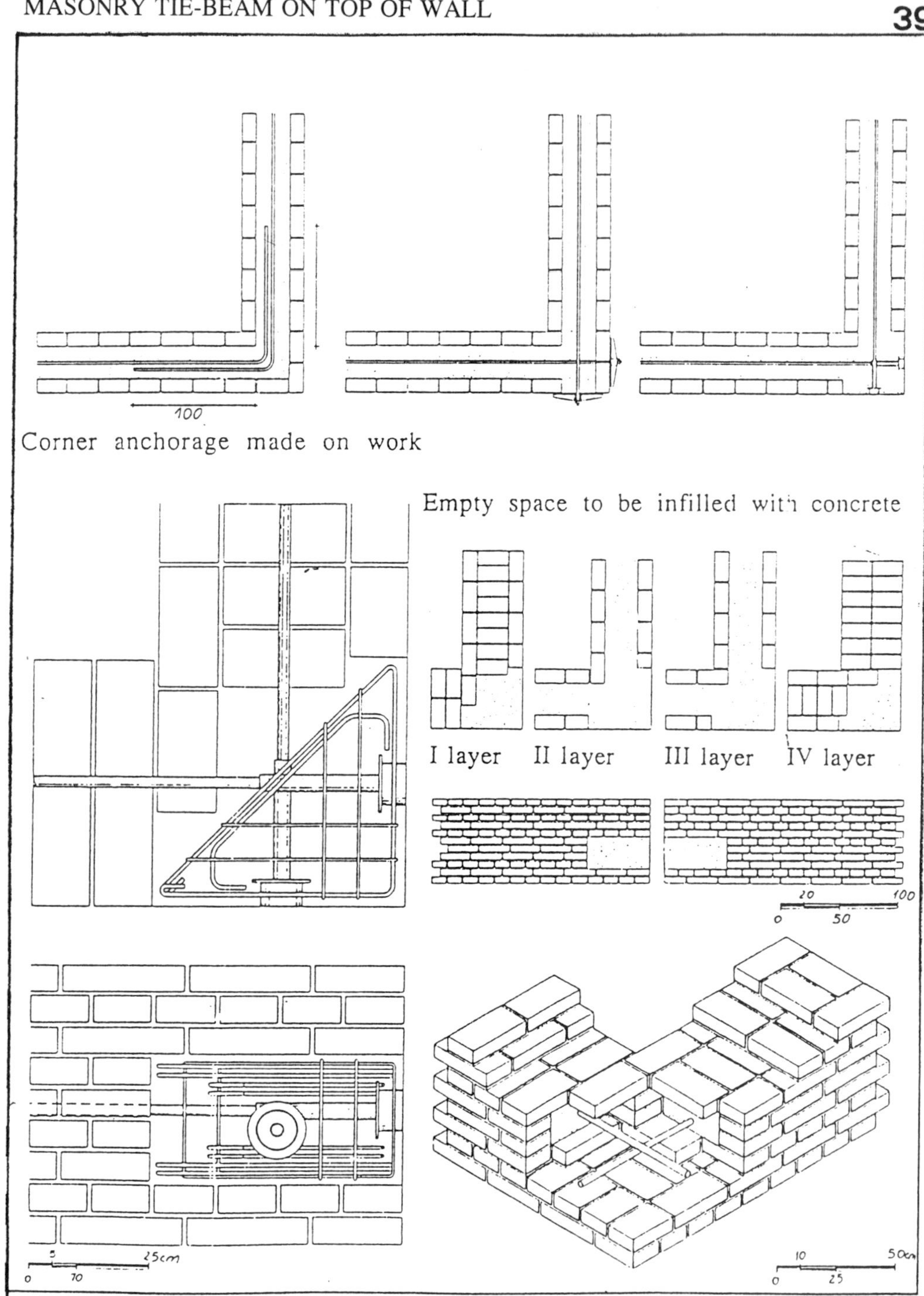

THE MASONRY TIE-BEAMS ON THE TOP OF THE WALL

40

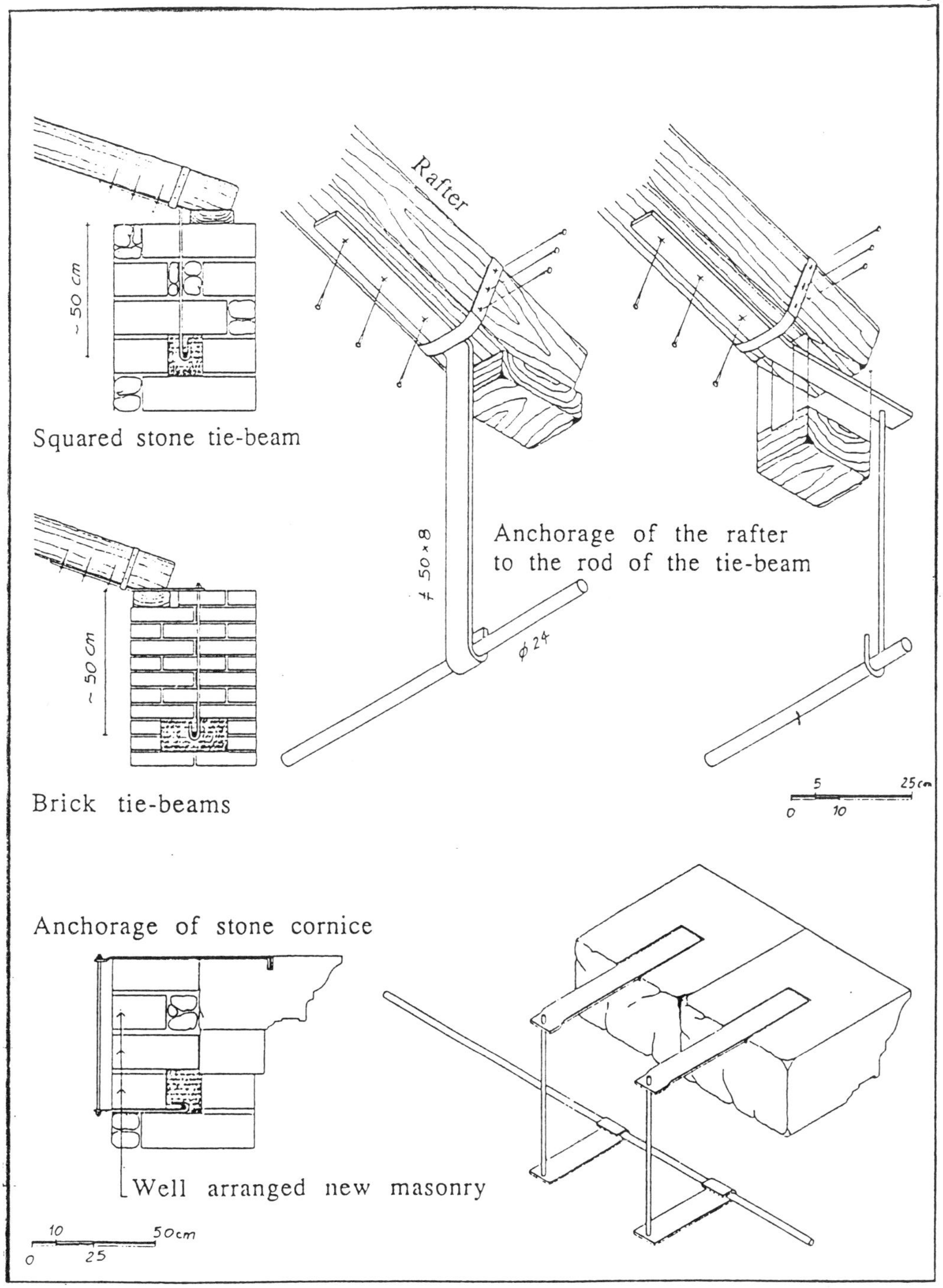

ARRANGEMENT OF TIE-RODS

41

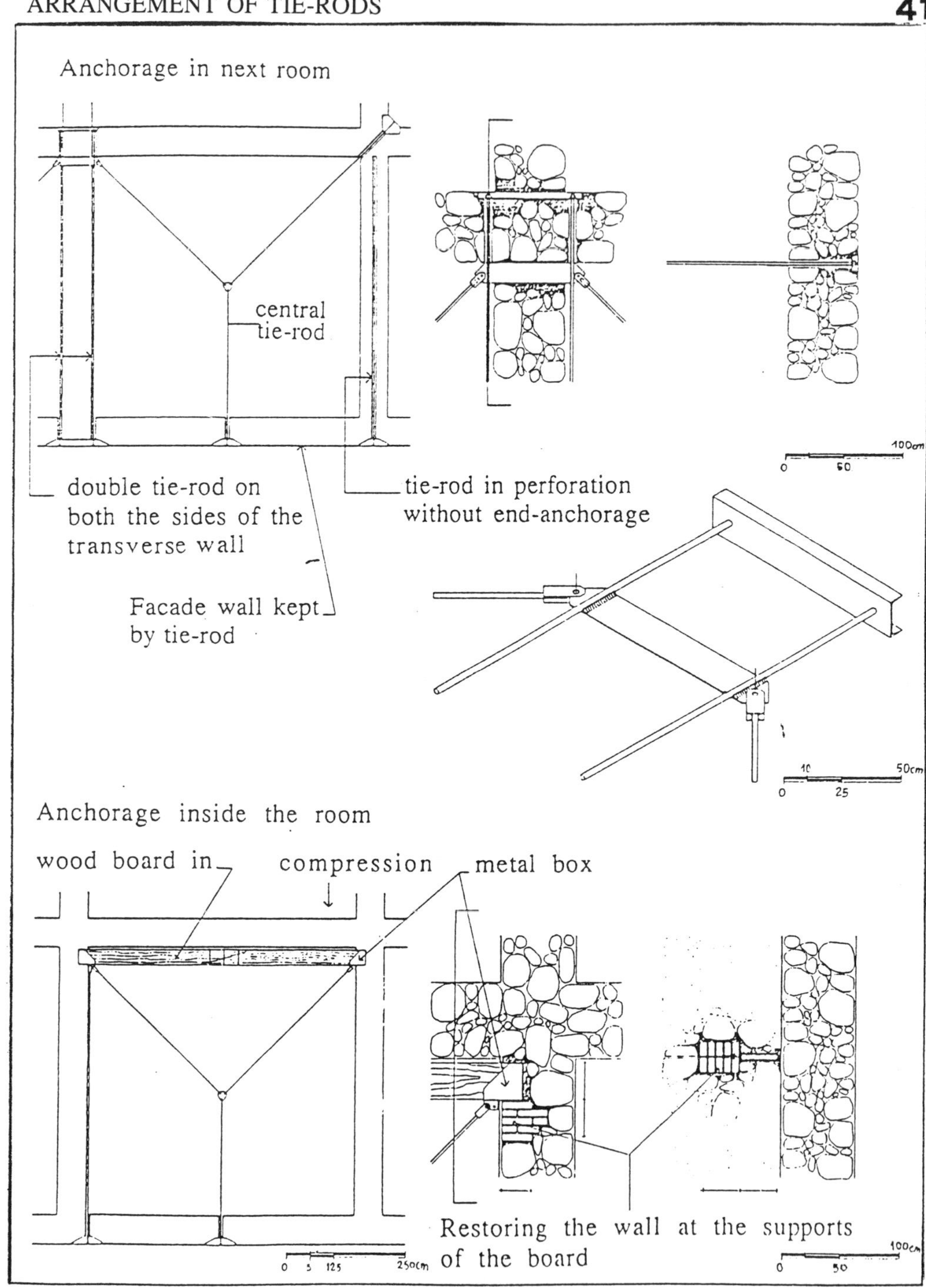

42

DETAIL OF THE WOOD BOARD WITH METAL BOX AT THE ENDS

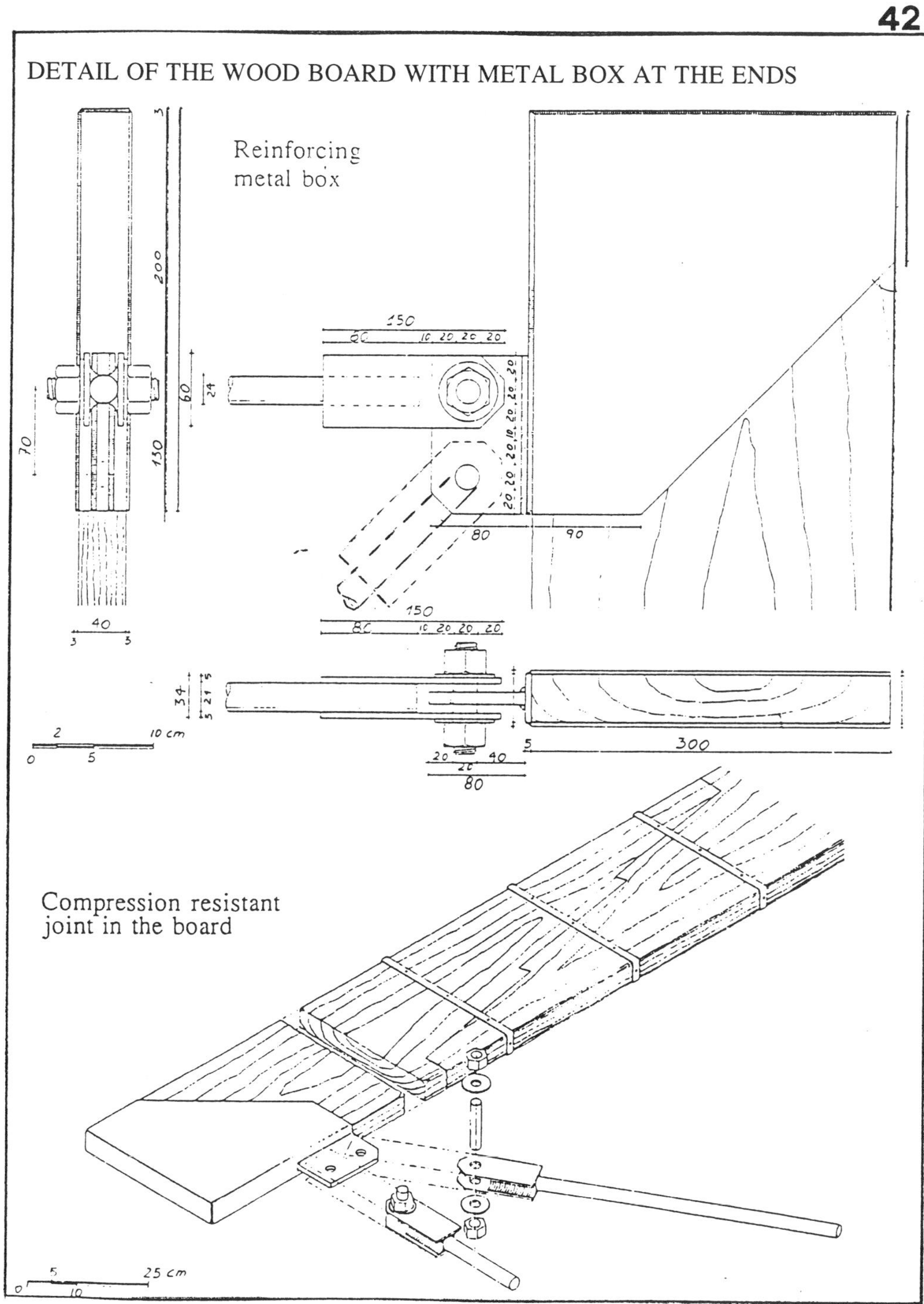

ARRANGEMENT OF TIE-RODS CONNECTED TO THE WOOD FLOOR

43

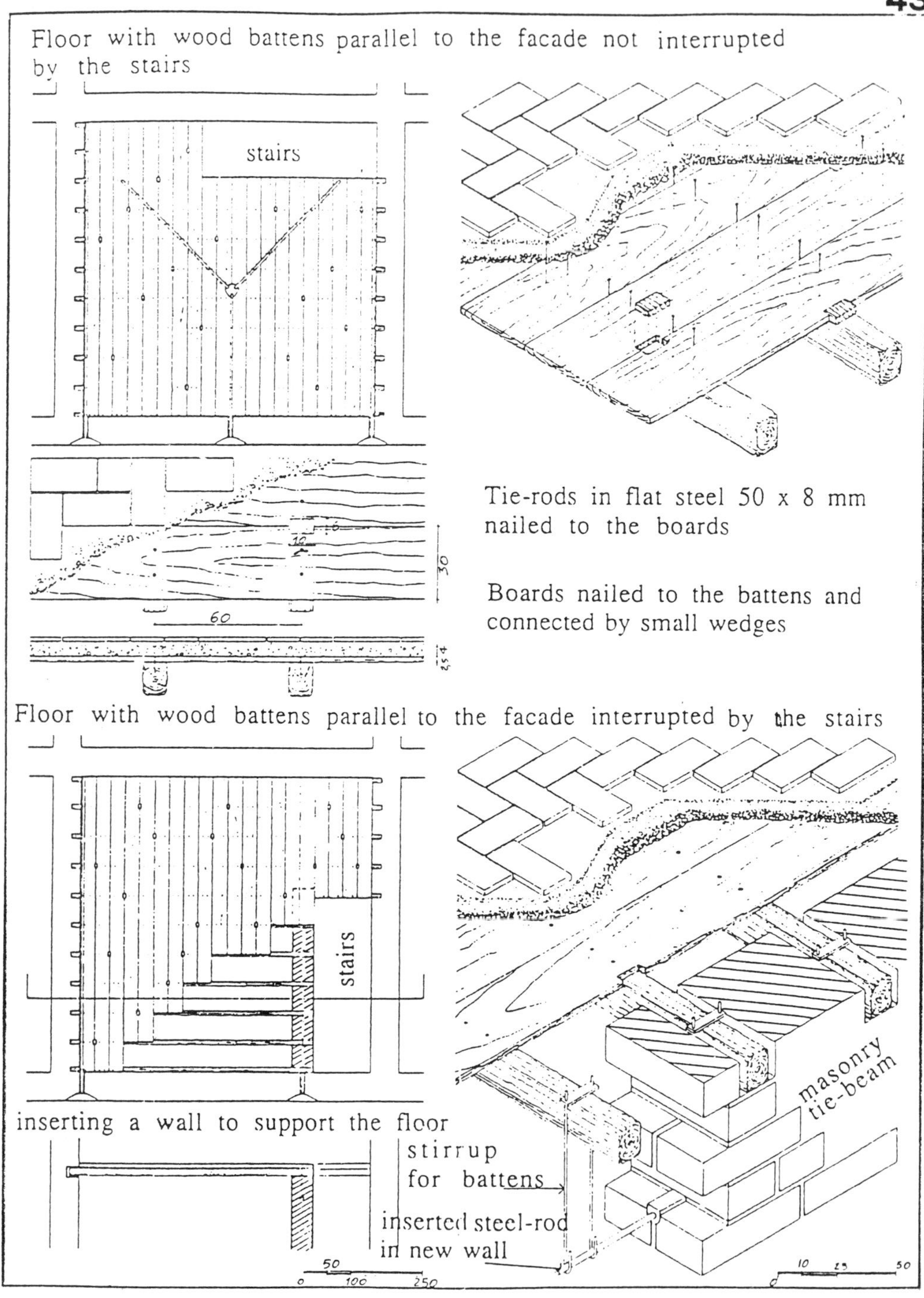

ARRANGEMENT OF TIE-RODS **44**

Floor with battens orthogonal to the facade

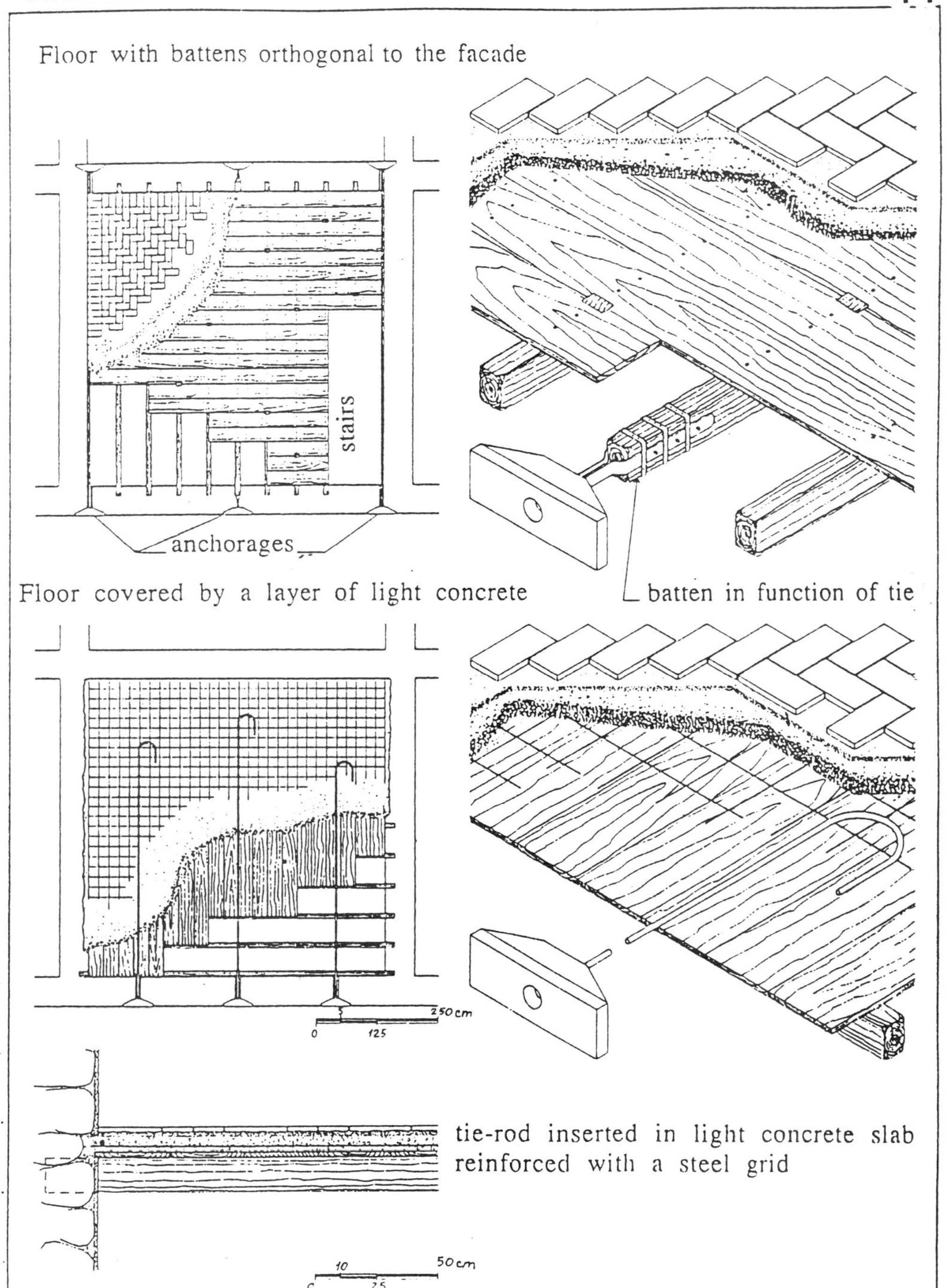

45

Traditional steel key for well layered masonry
Steel plate poorly connected walls
1/2
Restoring the
masonry by
injection round
the key
400
50
20 40
20
50
100cm

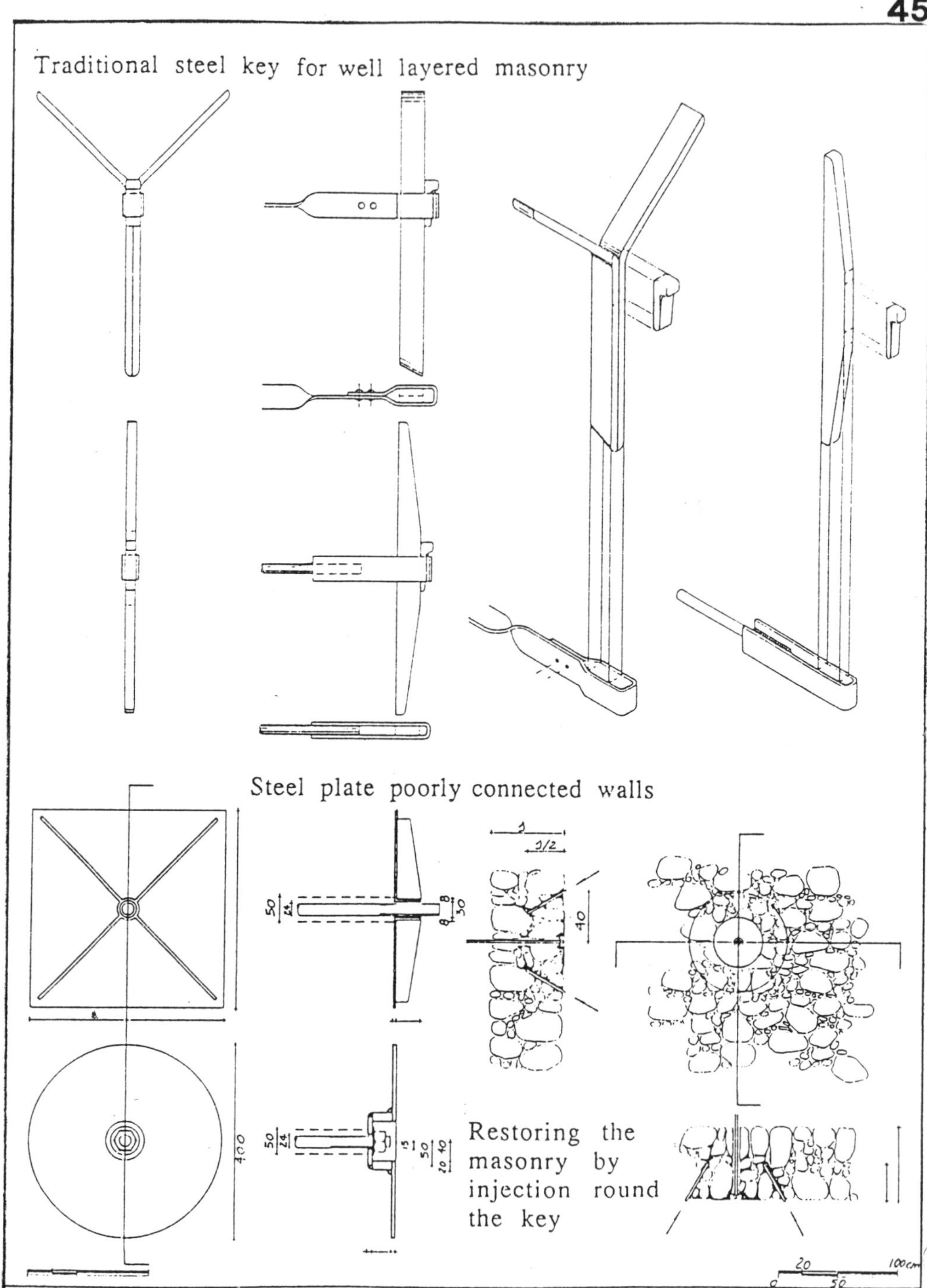

REINFORCED CONCRETE ANCHORAGE MADE IN BREACH

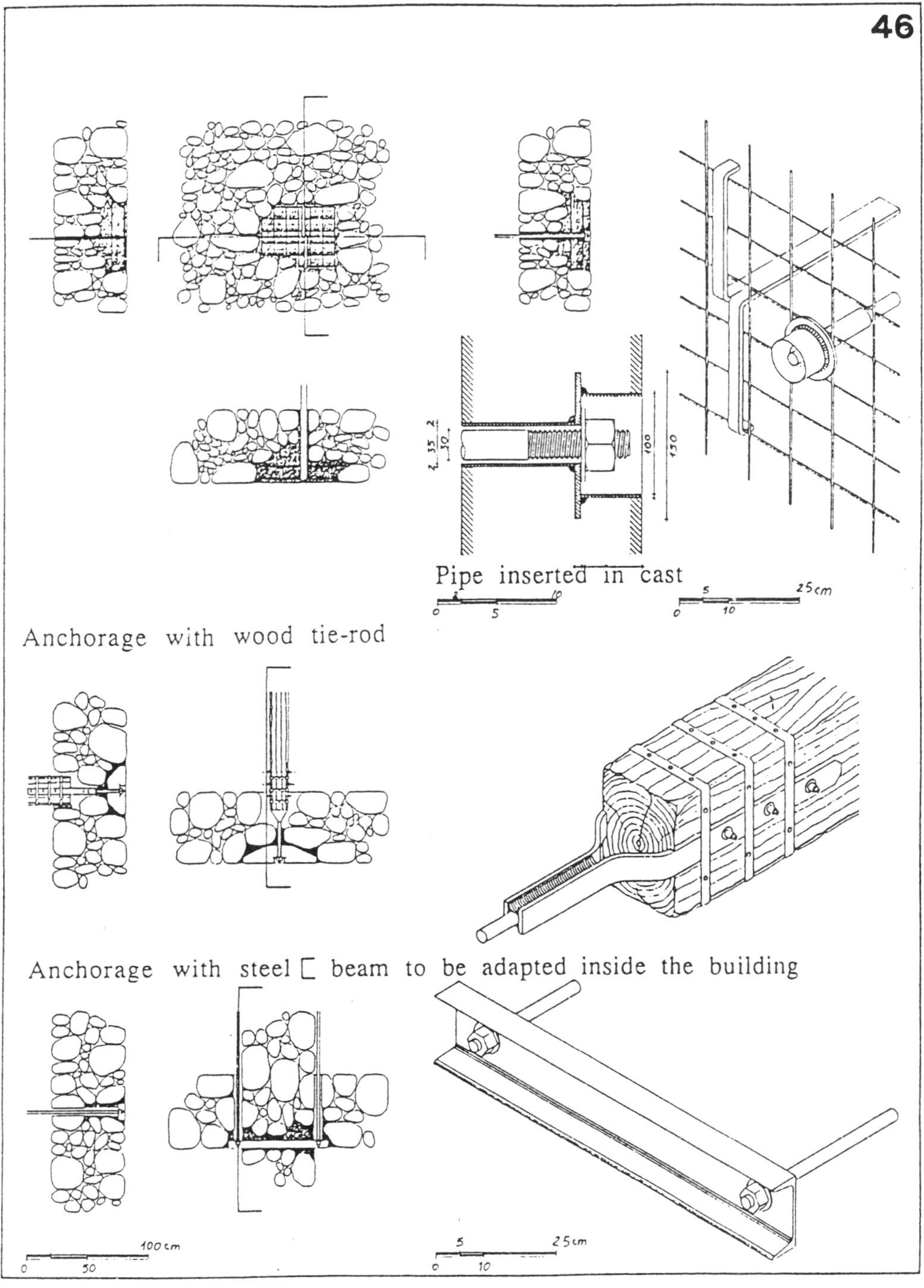

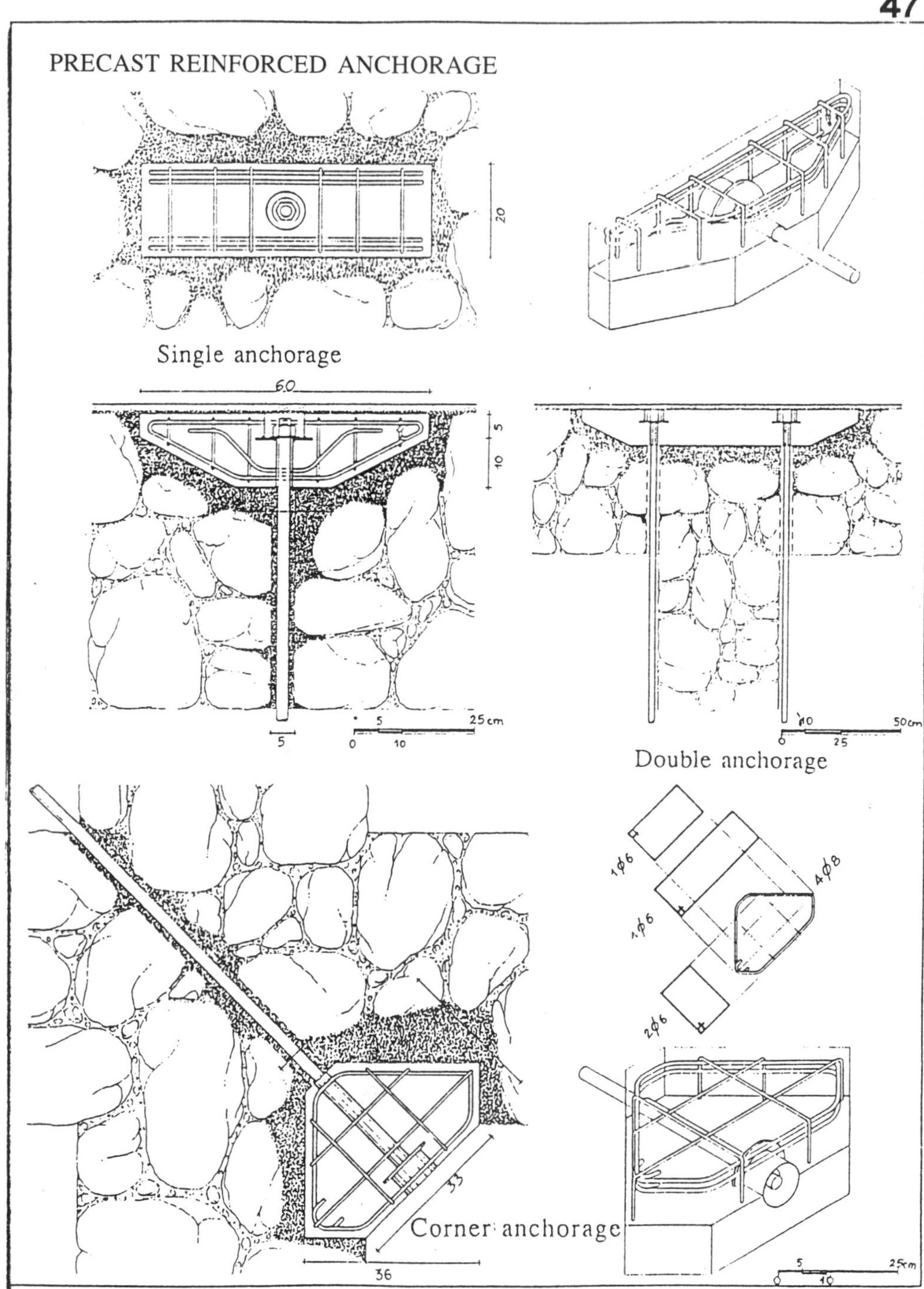

47
PRECAST REINFORCED ANCHORAGE
Single anchorage
Double anchorage
Corner anchorage

EXPERIMENTAL KNOWLEDGE OF THE TRADITIONAL BUILDING MATERIALS

S. Briccoli Bati

University of Florence, Florence, Italy

ABSTRACT

The aim of the following is to present a brief survey of the characteristic mechanical properties of the traditional materials used in buildings from the very earliest constructions up to the industrial revolution.

Until a century ago the most popular building materials were earth, stone, timber and bricks used in several ways, raw or baked, laid dry or with mortar between layers.

The use of such materials has decreased over the centuries and they have thus been replaced by modern materials such as steel, reinforced concrete and, probably in the near future, by composite materials specifically designed in order to satisfy the most diverse needs. Advances in the sciences and engineering have forced researchers in the field to devote much of their attention to the study of the chemical, physical and mechanical properties of these new building materials, while long neglecting similar research on traditional materials.

Nowadays, the need or, rather, the will to conserve and restore buildings constructed using traditional materials has compelled us to broaden our knowledge of such materials in order to assure the appropriateness of our treatment of ancient buildings, not only in attempting to restore them but also, and perhaps above all, in order to avoid adding further damage to the structures above and beyond those already occasioned by time and nature.

In fact, in the past builders had a "feel" for their materials, determined by the experience gained over centuries and used as guidelines for their choices in the art of building; today instead, the seems to be a loss of memory of these experiences, as well as a lack of thorough knowledge of the properties of those materials which have played such a prominent role in our architectural heritage, and the art of building in general.

In the hope of rectifying this lack, the results of some experimental analysis performed on materials of historical buildings are hereby presented.
Such analysis has allowed us, not only to highlight some of the materials' characteristic behaviour, but also their most valuable peculiarities, the consequence of a finely balanced combination of properties arrived at through exquisite and complex mechanisms. We can therefore say that attempts at modifying traditional building materials through artificial improvement of their properties may end up detrimental rather than beneficial.

1.   INTRODUCTION

The preservation of historical sites depends on the ability to deal with ancient buildings correctly. However, today many of them are in precarious structural state, due primarily to the scarce respect shown for their original conception during restorations and to the lack of adequate maintenance. Furthermore, recent earthquakes have also contributed to the urgent need to face the problems of restoration/preservation of ancient buildings on a scientific basis - meaning also that we must emphasise the damage caused by works, carried out with the best intentions, but which instead in the end proved to be disastrous. The main problem then was the need to furnish more accurate tools for structural reliability determinations and to conduct sound restorations in compliance with structural codes and guidelines, on the one hand, and in harmony with the original identity of the ancient buildings, on the other.

Thus, the question arises: how is it possible to provide a scientific basis for the assessment of the safety and serviceable life span of such ancient structures?

Evaluating the vulnerability and seismic risk of buildings made with traditional materials is still hard and uncertain due to the limited knowledge in the field. This is due to the fact that in the past such kinds of problems have not been confronted within a scientific framework. This of course does not mean that they should have been left their traditional empirical status. Such a controversy gives rise to the need to place the problem of ancient buildings on an objective basis determined by the actual structural conditions which vary from case to case, place to place and in both the techniques and materials used. One of the possible attitudes would be to examine, in light of the available theoretical results, whether there might be some way to provide some help and guidelines to those workers and professionals confronted daily with this kind of problem. This could be done, for instance, by providing them with a better understanding and a better way of interpreting and applying both theory and the results of experimental tests aimed at determining the behaviour, safety and reliability of age-old buildings. While a wide set of experimental results on the behaviour of new materials already exists, little is known about the mechanical properties of traditional building materials. In order to define the characteristic parameters and any other information concerning the mechanical behaviour of such materials, several researchers have devised experimental studies.

The results obtained have demonstrated, first of all, that neither the classical mechanical characteristics assayed, nor the testing procedures usually adopted are by no means the most appropriate ones. This underscores the necessity of defining new and more appropriate parameters in order to portray some vital aspects such materials' behaviour.

Such knowledge is quite critical whenever restoration work need to be carried out on ancient buildings, and also whenever their load-bearing capacity must be assessed. In fact, in some cases obsolete or damaged parts of ancient buildings must be replaced with newly manufactured similar elements, since the original materials are no longer available. In such cases, the compatibility between the original and the new materials must be investigated. There is often a significant difference between their characteristics, due mainly to their different composition, manufacturing techniques and the load history they have experienced.

Secondly, the experimental analyses already performed, allow us to make some interesting considerations of relevance, at least to this field.

In fact, although we are dealing with so-called "poor" materials (in the sense that they present a low characteristic strength-weight ratio), they show mechanical characteristics which allow them, if appropriately utilized and combined, to resist even seismic actions. This is mainly due to their surprising capacity to activate dissipative mechanisms, which give them properties similar to what is defined as "ductility" for metallic materials.

In the following paper our aim to provide only a rough outline of the essential characteristics of some traditionally utilized materials. By shedding light upon these too often neglected materials, we hope that their potential can be evaluated for structural solutions and achieve better results in restorative work as well as with regard to their capacity to withstand actions also of a seismic nature in historical buildings.

2. MECHANICAL CHARACTERISTICS OF TRADITIONAL BUILDING
    MATERIALS AND RELATED PROBLEMS

Centuries ago determinations of materials' physical and mechanical properties were performed by simple observation of the external appearance of the material to be selected for use in constructions. Results were therefore quite dependent upon the ability and experience of the builder. During the 19C, this "naked eye" assessment came to be largely replaced by testing procedures carried out in laboratories with the aid of suitable measuring equipment and thus capable of furnishing objective data. The tests to be performed then became clearly defined procedures whose operations could be replicated any number of times. As the choice of building material for a given structural application depends clearly upon its mechanical properties, basically the deformation it undergoes in response to applied forces or actions, knowledge of these properties, both qualitatively and quantitatively, is indispensable. Such data, which only controlled experimentation can furnish, form the basis of any efficient package for detailed structural analysis and assessments of a structure's consistency, static efficiency and durability. Accurate comparisons among the theoretical, numerical and experimental results coming from the analyses performed on structures made of traditional building materials shows these are not always in agreement. The discrepancies between the numerical and experimental results indicate, on one hand, that the algorithms utilized do not provide reliable measurements; and on the other, that models of material behavior are also flawed; Correct modelling of materials behavior is a difficult challenge but nonetheless an important one for practical applications.

A personal view of the authors is that further experimental and theoretical studies are necessary in order to provide detailed descriptions of the phenomena taking place during deformation processes so that the behavioral models at our disposal may mirror more closely the reality revealed by actual experience and to improve the methodology and data processing procedures aimed at resolving structural problems relative to historical buildings built with non-standard materials (i.e. heterogeneous, anisotropic or heteroresistant).

The following discussion, limited to the experimental aspects and taking a primarily qualitative point of view in describing the phenomena in question, deals with the basic behavioral characteristics of traditional construction materials. It is based upon review of the results available in the literature and will therefore be limited to the concepts and ideas contained therein and will not include the mathematical formulations.

At the phenomenological level the mechanical behavior of a body is described by the stress / strain relation and that between their derivatives. We are thus dealing with a relationship of cause and effect between actions upon a material body of finite dimensions and presumed homogenous makeup, under varying conditions, and the response of the body itself in terms of strain or deformation. The observed mechanical properties are then correlated to formulated theoretical conceptions of these properties. The resulting theoretical relations are engineering formulations, more empirical than functional, and as such, their validity, in the strict sense, is limited to the specific conditions under which the test were conducted.

The qualitative characteristics of a material's response to various states of induced stress permits it classification as rigid, elastic, viscous, plastic, and so forth. Each is described by a corresponding specific mathematical theory, which description however is limited to the material's behavior alone, giving no indication whatever as to the physical mechanisms underlying the observed phenomena.

Despite the wide differences in the nature and structure of matter, in general, and building materials, in particular, analogies among their behavior are paradoxically observable even at the macroscopic level. The same parameters which define mechanical properties, such as elasticity, viscosity, plastic strain, brittle fracture, plastic collapse, etc., are applied to all materials, albeit with widely differing orders of magnitude. On the basis of demonstrated similarities in behavior, which in turn depend upon the internal structure of each, materials are roughly divided into the following:

Metallic materials: made up of atomic assemblies bound by electromagnetic forces which develop among the electrons of nearby atoms; such an internal structure determines well-defined physical mechanisms of deformation and breakage. An example of a metallic material is steel, which demonstrates behavior conforming very closely to that predicted by the theory of elasticity and plasticity.

Polymers: made up of molecular chains whose backbone is formed mostly by carbon atoms. The class of polymers includes all those substances commonly termed plastics; the class of natural polymers, on the other hand, comprises all types of wood, whose mechanical properties are strongly dependent on the fact that the macro-molecules of which they are composed have a particular orientation and crystallinity. All these materials are undergo distinctive processes of deformation and rupture.

Granular materials: These are composed of granules which, whether bound to each other or not, determine the mechanisms of deformation and rupture peculiar to this group, which, among others, includes all those materials of particular concern to us here: stone and rock, unbaked earth and bricks, as well as the different types of mortar and concrete.

Ceramics: the most prevalent being made up of minute crystals of metal oxides and silicates bound to each other by a non-crystalline matrix. The baked bricks used in the construction of the greater part of traditional buildings belong to the sub-category of porous-matrix ceramic.

While modern materials such as steel and reinforced concrete have been the subjects of so much study that the inelastic phenomena underlying their behavior have been well defined and some aspects of their properties codified and included in new building codes, the same cannot be said of the materials used in older, more traditional constructions.

The fundamental issue, ignored until recently, is that the behavior of materials needs to be studied beyond the elastic phase if we are to define the significant parameters needed for assessment of historical buildings' capacity to withstand earthquakes. This implies clear understanding, not only of the post-peak phase, from which it is possible to deduce their degree of ductility, but also their responses to cyclic, static, impulsive and sustained loading.

In fact, only through such studies will we be able to define the characteristic mechanisms which confer upon traditional materials a sort of "ductility", that is to say, a certain ability to impede the speed of crack propagation.

Classic tests to define material properties consist basically of uniaxial monotonic tension and compression tests at constant temperature under conditions of either static or impulsive loading. These are supplemented by tests under cyclic stress, either static or impulsive, which permit determination either of a material's serviceable life-span when subjected to repeated loads or the stresses which a material sample of standard dimensions can safely bear over a given number of cycles. Even under conditions of constant load, materials continue to undergo deformation indefinitely; thus the rational for the sometimes performed fracture tests under constant load in order to define the relative creep or viscous flow.

Whichever test is carried out, the sample is subjected to a stress state (due to force or displacement), mostly along a single axis, which generates a state of strain or deformation throughout it entire volume. If the resulting strain or deformation can be considered reasonably uniform, the test sample can be treated as a volume element of continuous-media mechanics. However, it should be emphasized that the thermodynamics and mechanics of continuous media represent only the basic theoretical points of reference for formulation of the physical phenomena of deformation and breaking, and that, strictly speaking, treating the test sample as a volume element or a volume element representative of continuum mechanics is clearly valid only in the case of homogenous and isotropic materials, or even statistically homogenous ones. Dealing with strongly dishomogeneous materials, on the other hand, such treatment would raise more than a few perplexing doubts. Inevitably, however, in order to gain knowledge of phenomenological aspects, experiments upon the volume elements of materials need be performed by submitting them as much as possible to uniform fields of strain or deformation. Unfortunately, it is current practice for laboratories to perform tests on material samples that from the start exhibit localized defects. In these cases the "average behavior" of the sample depends on boundary conditions and the concept of effective properties no longer applies. Such circumstances involve resolving problems of experimental procedure: particular attention is called for in the selection and preparation of samples of to appropriate shape and dimensions, which will necessarily vary from one type of material to the next, as well as in the ever difficult process of definition and implementation of suitable homogenous boundary conditions.

The evident and proven anisotropy and heterogeneity of traditional materials - several tests are required to assess the validity of any constitutive law of anisotropic materials - demonstrate that the usual mechanical characteristics and classical testing techniques used cannot alone suffice for their treatment. It thus seems opportune to identify new parameters more suitable for representation of some aspects of inelasicity. It is more appropriate, then, to reconsider testing techniques and equipment, the size and dimensions of samples and their influence on the tests results since, as it is well known, the quantitative aspects of the material response will depend on the material properties, the sample's geometry and the nature of the imposed loading history.

Another fundamental characteristic of traditional building materials resides in the fact that they exhibit different compressive and tensile strengths; more specifically, their tensile strength is particularly poor, on the order of about 1/10 of their compressive strength (Fig. 1). Thus imparting special significance to compression tests.

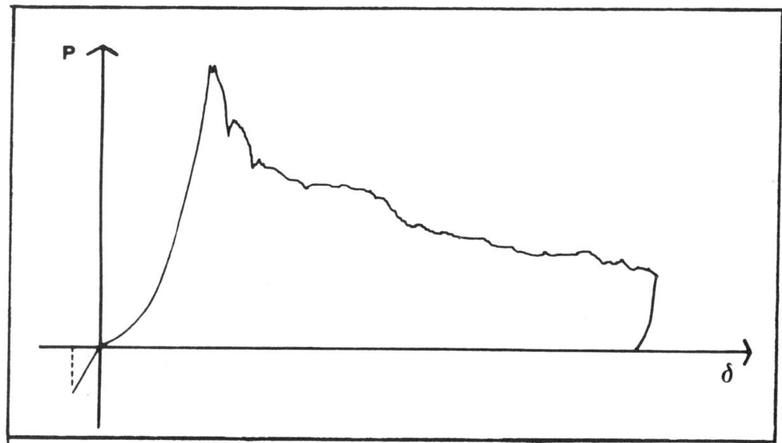

Fig. 1, Typical results of compression and tensile tests in heterogeneous material

Tests of stone materials' strength to direct tensile stress are difficult to perform, and their results are not always free of spurious influences. Therefore, recourse is usually made to tests of either indirect tensile stress or bending. Tensile strength values so obtained result 20-30% higher than those from direct tensile tests. In any case, complete tensile test measurements, either direct or indirect, provide indications of the parameters which characterize the different behavioral stages, if any, of the materials and of strength. In fact, if the test is conducted with equipment capable of proceeding with guided displacemen, it is possible to study the evolution of the separation process up to complete rupture of the body into two or more sections. This process, the object of study in fracture mechanics, a field which has gained considerable popularity in recent years, varies from one material to the next and is influenced by the nature of the applied stress, the sample's geometric conformation, the temperature and the deformation speed. Depending on the combination of mechanical constraints and material properties, fracture can either occur suddenly or be preceded by stable crack growth. Normally fracture is termed ductile if it occurs after extensive plastic deformation and through slow crack propagation, while it is considered brittle, if subsequent to a small or no plastic deformation with rapid and uncontrollable crack propagation even if the loading process is deflection-controlled. On these bases, the behavior of materials is roughly divided into ductile, brittle, semi-brittle and softening (Fig. 2).

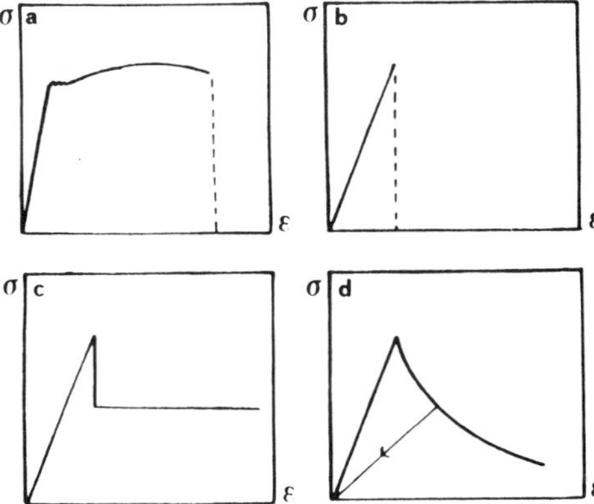

Fig. 2, Mechanical behavior: ductile (a); brittle (b); semi-brittle (c); softening (d)

In mechanical testing of heterogeneous materials, the test sample neither yields nor fails when maximum stress is reached, but undergoes a gradual decrease of stress at increasing strain, called strain softening. The gradual decrease of load with increasing displacement is the softening branch of the load displacement curve, and reflects a non-brittle fracture mode of the heterogeneous sample. A softening branch has been measured for several traditional materials under uniaxial tension as well as compression. This sort of behavior can be explained by recourse to fracture mechanics, since the area under the load-displacement curve may be associated to the amount of fracture energy necessary to cause the sample to fail. The shape of the load-deflection curve can be accurately predicted if the shape of the stress-displacement curve for the strain-softened fracture process zone is known.

Although fracture mechanics is a field of study which has only recently attracted the attention of researchers [1], [2], [3], [4], [5], [6], [7], [8], [9], [10], a number of models for the fracture process zone have already emerged. For rock-like materials, the "cohesive cracks model" proposed by Hillerborg [6] has without doubt achieved the most success, because of both its physical significance and its simplicity. As first pointed out by Hillerborg, the description of the behavior of concrete-like material in traction must include both a stress-strain law for undamaged material and a stress-separation law for localized fracture. The cohesive crack model is based on the following assumptions:

– the cohesive fracture zone - plastic or process zone - begins to develop when the maximum principal stress achieves the ultimate tensile strength σ_u;

– the material in the process zone is partially damaged but still able to transfer stress that is dependent on the crack opening displacement w.

The basic element is represented by the softening curve, which relates the stress which acts through the crack faces and opening. When the traction-separation softening relationship is assumed to be linear, it becomes a linear fictitious crack model (Fig. 3), and is the simplest cohesive crack model that preserves all the deformation-dependent and softening features. Adoption of bilinear or exponential stress-separation laws provides only slightly better results than the simpler linear law.

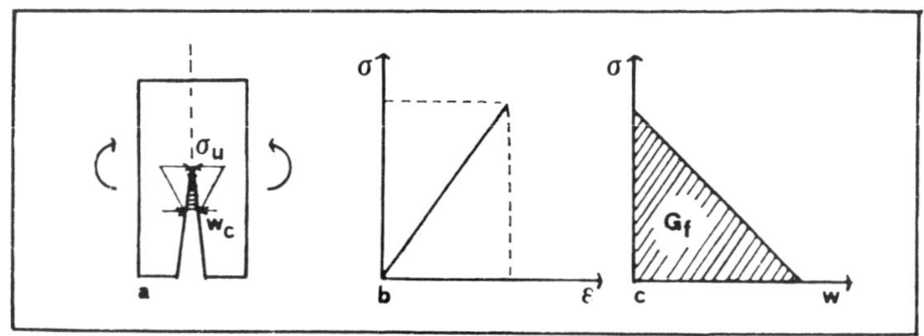

Fig. 3, Linear stress distribution across the cohesive zone (a); stress-strain constitutive law for the undamaged material (b); linear traction-separation law (c)

By adopting the cohesive crack model it becomes possible to account for the softening-type descending path of a structural element such as, for example, a three point-bending beam (Fig. 4).

For structural design, the relevance of the tensile fracture behavior of quasi-brittle materials is well established and can be quantified by means of the fracture energy G_f. This parameter is useful for design purposes, as well as for modelling the fracture behavior of cohesive materials. The most direct way of determining G_f is by means of a stable uniaxial tensile test. Unfortunately, it is very difficult to perform stable and representative tensile tests; so a simpler procedure, based on determination of the total work of fracturing three-point bend notched beams was proposed by the RILEM TC-50, though dependence of G_f on sample size was detected. This apparent size-dependence casts doubts upon the possibility of considering G_f as a material parameter.

However, as already stated, when dealing with traditional materials, given the exceedingly low value of their limit tensile strength, it is compressive, rather than tensile, strength which is most interesting. In fact, for this reason, in carrying out structural analyses of elements constructed with these materials, tensile strength is completely ignored and the behavioral model of a so-called "no-tension material" is assumed.

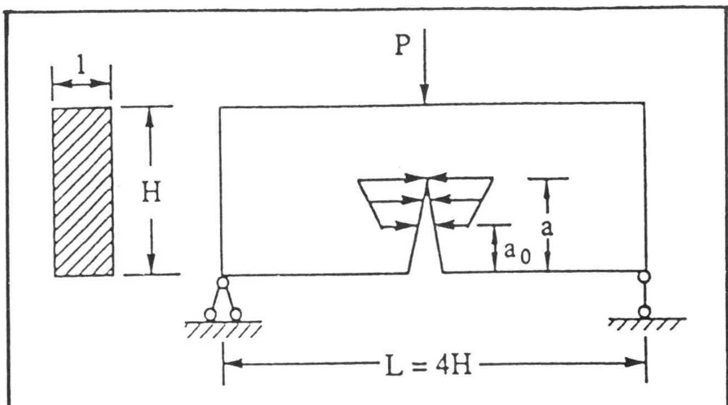

Fig. 4, Cohesive crack configuration of a three-point beam

Uniaxial compression tests are generally carried out on cubic or prism-shaped samples with a height to depth ratio of 2:1. The tests performed relative to the present discussion confirmed that the cubic compressive strength always yields a higher value than that obtained through testing a prismatic sample. On the other hand, the choice of a prism-shaped test offers the advantage of being able to simultaneously determine the strength value, the Young's modulus and the Poisson ratio for a given stress direction by means of a single trial on only one sample. The aforementioned procedure allows us to reduce the number of samples used and therefore the quantity of material needed for testing.

According to current rules, the uniaxial compression test always results in a compressive bending state of stress. Such a circumstance is due to several reasons, impossible to avoid when using such materials. First of all, it is impossible obtain two sides of the test sample perfectly parallel and level. Secondly, the heterogeneity of the material brings about complex internal stress states due to either the internal geometry of the material or local differences in elastic or plastic behavior. Moreover, the presence of more or less visible voids never allows the testing board to lie horizontally during the test. The rotations revealed on the testing plate in contact with the top side of the sample corroborate this. The difference in the vertical displacements - measured in at least three different points of the testing plate - when increasing the load provides precise information on the state of strain to which the testing sample is subjected to, as well as indications of its heterogeneity.

In an attempt to approximate as close as possible a uniaxial state compression, it is necessary to place a layer of material capable of reducing the friction between the bottom side of the test sample and the testing plate. This is done in order to eliminate, as much as possible, shear stress in the nearby regions of the bottom sides of the sample. The anti-friction layer should not be a Teflon sheet or any similar material, as often suggested, as these materials have their own Poisson ratios which are very different from that of the sample. As is well known, the difference in Poisson ratios of the two materials induces a pluriaxial state of stress, at least in the areas close to the interface, and consequently may increase or decrease the measured strength values. Relevant tests showed that the most suitable anti-friction device would be a very thin layer of talcum powder introduced between the testing plate and the bottom side of the sample. Such a material, having no cohesive properties, is free to flow and consequently to prevent any kind of friction. Using this device, it is always possible to obtain, in the uniaxial compression test, a cracking pattern parallel to the load direction, which clearly shows that the friction has been eliminated. Obviously, in terms of strength, compression tests performed using this device, give results which are about 20% lower than those obtained with no anti-friction provisions.

Another aspect of the problem is measurement of deformation and displacements. These measures are considered useful in order to provide reliable information concerning the strain state experienced by the sample during the loading process. From deformation of the sample, the inappropriately termed "modulus of elasticity" can be obtained. Such a term is to be considered inappropriate because here we are dealing with a heterogeneous material. Generally, in order to determine this parameter, strain gauges are applied, which implies the use of glue spread on both sides of the sample. If the material is highly heterogeneous and porous, such as in our case, the results given by the strain gauge are related to the length of the application area. Not only does the information obtained have a local character, but it is also altered by the presence of the glue. Dealing with a heterogeneous material - with more or less macroscopic voids distributed throughout the sample volume - it is appropriate to consider an overall variation of the length related to the whole sample. Similar considerations can also be made for measurements of the transverse deformations. The overall deformation can then be linked - when referred to the measurement length - to average unitary deformation. This should not be confused with what it is intended in the strict sense by Young's modulus or the Poisson ratio.

In order to better illustrate some peculiarities of the mechanical response of traditional building materials, it is useful to consider uniaxial compression tests which, when dealing with a heteroresistant material with weak tensile strength, turns out to be the most significant (Fig. 5).

Thus, the test curves can be divided schematically into four more or less extensive regions, related to the different behavioral stages of the sample volume element. In the first stage behavior is non linear, in the second, linear, while the third - the stage of damage initiation or microcracking - is non linear and is characterized by a diffuse microcracking developing on the macroscopic scale throughout the entire volume. This process continues until cracks propagate within a very localized area. The spread leads to more or less marked discontinuities in the test curve. In stage four the cracking process expands on the front of the main crack, leading to decrease in the load and failure. While pre-peak stages present roughly the similar characteristics in all materials, the post-peak varies from material to material (Fig. 6).

When the test is performed with an applied displacement, the collapse phase generates one or more unstable branches showing a rather steep slope until they join in a final linear branch which is almost parallel to the displacement axis.

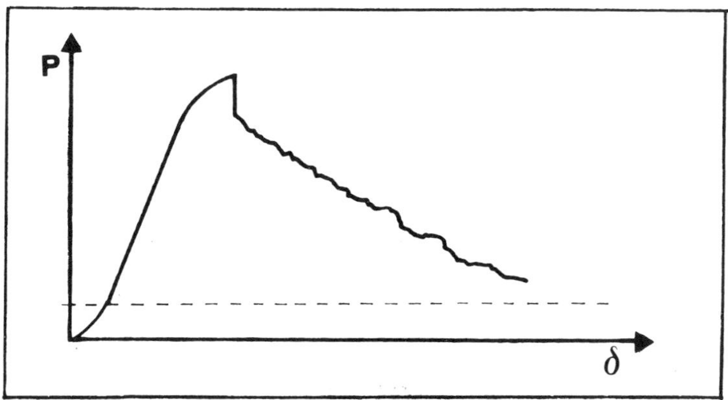

Fig. 5, General mechanical behavior of heterogeneous material under compression

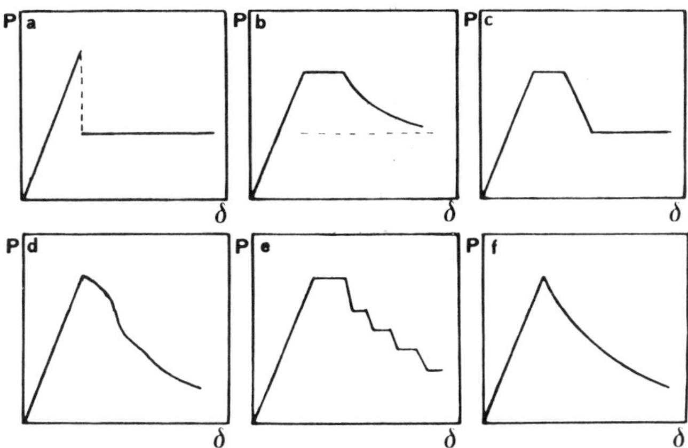

Fig. 6, Post-peak behavior

In particular, a number of equilibrium points can be found in the descending branch. If there are two points of equilibrium (Fig. 6a), one just before and one following fracture, the phase of transition between the two stages is not observed, and the phenomenon thus demonstrates internal instability. This signifies that once a fracture has occurred the sample looses part of its load-bearing ability and keeps a lower load level which can be sustained indefinitely. In all other cases (Figs. 6b,..f) an additional displacement is required to cause the diagram to move along and down its post-peak response.
Several small-scale mechanisms govern the large-scale response: compaction,

microcrack initiation, growth, coalescence and frictional movement of the microcrack interface. These mechanism are complicated and not independent of each other. It is however clear that during compression, the process of brittle fracture is linked essentially to the material's internal structure and defects thereof. Materials which possess heterogeneous texture, variable grain and pore sizes or microcracks within their structure exhibit the strain-softening type of behavior at fracture, while highly homogenous ones with very small grain and pores demonstrate more unstable behavior. It can be seen that in general, limit strength decreases with increasing pore size, while the softening phase is prolonged.

The application of an increasing compressive load or displacement to any sample also results in:

–   compaction with increasing stiffness;
–   microcrack initiation and growth: the microcracks are parallel to the applied stress or strain;
–   the microcracks coalesce into a large crack and progressive failure ensues.

It should be noted that for heterogeneous materials, some permanent deformation will remain upon unloading.

The **P-δ** curves, recorded during the whole loading process (Fig. 7a), are characterized by common features which can be schematically represented in a typical path (Fig. 7b).

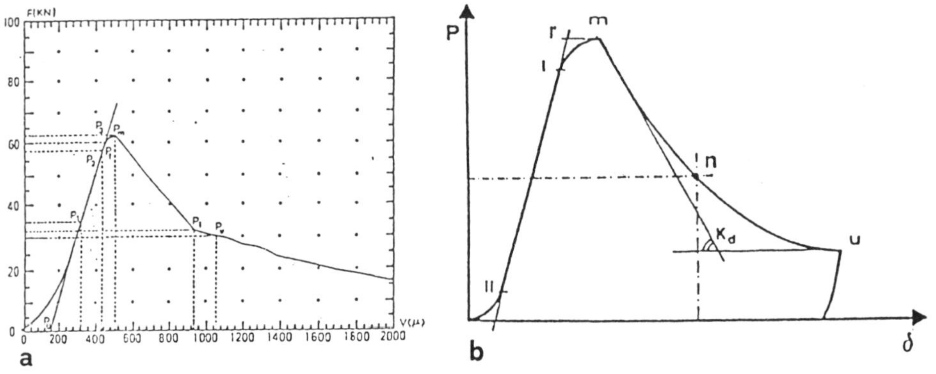

Fig. 7, P-δ curve: recorded (a), schematic equilibrium path (b)

Delving into further detail, it is possible to trace a non-linear initial line - represented by an upwardly concave curve - which extends for about 20% of the whole ascending branch of the equilibrium path. The non-linearity is due both to modification of the contact area between the testing plate and the bottom side of the sample, and also to the so-called compaction of the material which depends upon its micro and macro-porosity. The specific behavior observed under compression is explained by the reclosing of inherent microcracks or voids, thus bringing about stiffening of the material; the curve then presents a line, which can be considered linear to a good degree of approximation, which goes up to 85% of the peak load; this part of the curve is suitable for calculation of the tangent stiffness. Before the appearance of the first visible cracking, the curve turns into a second non-linear line, whose concavity, this time, points downward toward the δ axis.

This deviation from linearity is due to microcracking. The behavior can not be regarded as reversible, in the sense that unloading does not follow the same path of loading. This stage signals the beginnings of damage to the material, a phenomenon which reveals the creation and evolution of microcracking within the material's interior; such microcracking affects its elastic properties and is generally manifested by a sharp decrease in stiffness. This phenomenon, of particular interest to the field of damage mechanics, has also been the subject of a number of proposals and theories which have emerged in recent years [11], [12], [13], [14], [15], [16], [17], [18], [19], [20], [21].

Since the pioneering works by Kachanov [11] and Rabotnov [12], the theory of damage mechanics has quickly developed and evolved into a practical means for modeling the damage processes of engineering materials at a macroscopic continuum level. However, it seems that a physically and geometrically consistent definition of material damage has yet to appear.

The phenomena which these materials undergo are strongly influenced by alterations at the microstructural and mesostructural levels. At the level of microstructure two inelastic phenomena dominate [19]:

plastic slip, which is the macroscopic manifestation of the process of dislocation, as well as particle movement and rearrangement;

nucleation and growth of microdefects in the form of microscopic cracking and void formation.

The separate contributions of microcracking and plastic deformation can be estimated by observing the unloading slope. If the unloading slope is about the same as the initial loading slope, the inelastic strain is essentially plastic and is given by the offset of the unloading diagram at $\sigma=0$. If the unloading slope is lower with respect to the loading path, which is typical throughout the strain-softening branch, the inelastic strain, due to the slope decline, must be attributed essentially to microcracking or fracturing.

Based on microscopic observations, various mathematical models have been introduced in an effort to analyze the failure process of brittle solids under compression [22], [23], [24], [25], [26], [27], [28], [29], [30].

The overall force-displacement response typically shows an ascending part which is more or less linear up to the peak load; the post-peak response is instead typically non linear.

It is worth noting that, although the ascending branch of the first equilibrium path is linear over a long segment, the behavior of traditional building materials cannot be considered perfectly elastic, in that residual strain is always encountered after unloading. This is indicative of the fact that not all the strain or deformation exhibited up to a given level of stress is reversible. Furthermore, if a second cycle of loading is executed, the slope of the ascending segment is steeper than that of the first cycle, signifying that the test sample has undergone an increase in stiffness. In addition, after the first loading, the loading-unloading loops are also very closed (Fig. 8).

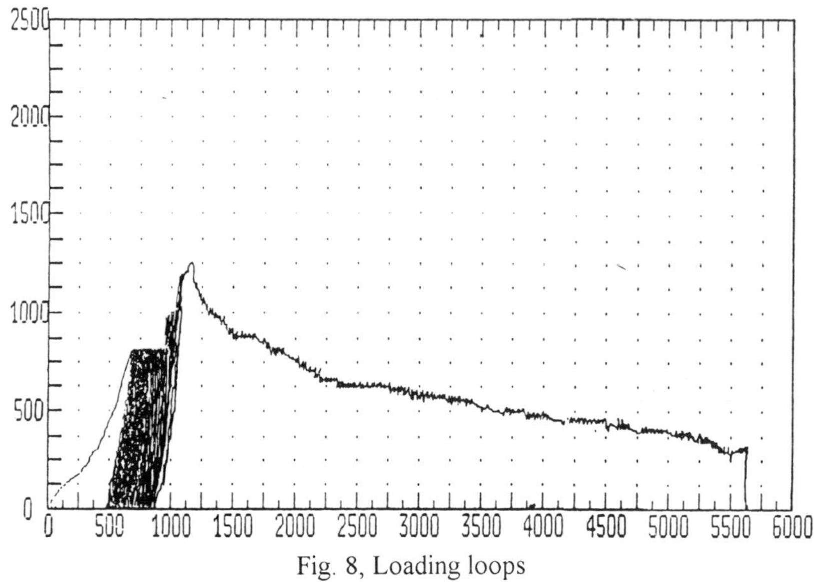

Fig. 8, Loading loops

In order to quantify the parameters considered essential for representation of the individual branches of the equilibrium path, the following characteristic points have been identified:

l_i: represents the beginning of the line which can be considered linear to a good degree of approximation;

l: represents the linearity limit, after which the curve can no longer be considered linear. When computing the ductility indexes, this point is assigned the meaning of elastic limit with a good degree of approximation;

l': represents the point whose δ coordinate intercepts the elastic displacement at the peak load;

m: represents the point where the curve reaches the maximum P coordinate, that is the peak-load;

n: represents (when it exists) the so called "stabilization" limit, (i.e., the point where the post-peak equilibrium path stops after the first load falloff);

u: represents the so-called "ultimate point", (i.e., the point corresponding to the end of the sample's load-bearing capacity); it is conventionally assumed to be a pre-established percentage of the maximum load value reached during the test.

Points n and u characterize the post-peak branch of the equilibrium path since their presence provides information on the strain-softening behavior of the analyzed material.

According to the above mentioned characteristic points, some associated parameters can be calculated. They permit a sufficiently reliable evaluation of the mechanical properties of the material. Furthermore, it has been considered useful to characterize mechanical behavior through the following parameters:

K_l: stiffness computed on the line which can be considered linear to a good degree of approximation; it can be taken as a measure of the tangent stiffness;

K_s: secant stiffness up to the linear limit;

E_a: absorbed energy up to the peak load; it is represented by the area under the load-displacement curve and the δ axis in the interval $(0, \delta_m)$;

E_u: total absorbed energy; it is represented by the area under the load-displacement curve and the δ axis in the interval $(0, \delta_u)$;

μ_k: kinematic ductility; it measures the ability of the sample to show deformation in the non-linear range before reaching the peak load; it is assumed equal to one when the load-displacement curve is linear up to the peak-load, and greater than one in other cases;

μ_{kd}: available kinematic ductility; it measures the tendency of the sample to show strain-softening behavior after the peak-load.

Lastly, it should be pointed out that the materials used in traditional construction exhibit the macroscopic properties common to most brittle materials, namely:

stiffening due to closure of microcrack or pores;

anisotropic degradation of elastic properties;

development of a dishomogeneous deformation field in the proximity of the peak load;

softening-type behavior, which is more or less unstable depending on the loading direction, in the post-peak phase under compression.

In any case:

– the post-peak tensile behavior can be considered of the brittle type independent of the loading direction;

– the ratio between the tensile strength in one direction and the compressive strength in the perpendicular direction can be considered nearly constant and equal to the Poisson ratio;

– it is possible to identify a proportionality between the peak compressive strength and the more or less stable behavior in the post-peak region.

3.   THE MECHANICAL BEHAVIOR OF ANCIENT MANUFACTURED BRICKS

The baked brick manufactured in ancient times constitutes one of the two major components of the greater part of masonry structures, the other being mortar. Ancient brick possesses certain characteristics which distinguishes it markedly from brick produced in modern times.

Bricks - either of new or ancient manufacture (Figs. 9 and 10) - belong to the class of ceramic materials of porous quality, and although the raw, unbaked material, clay, is nearly the same in composition, its mechanical and morphological characteristics are quite different.

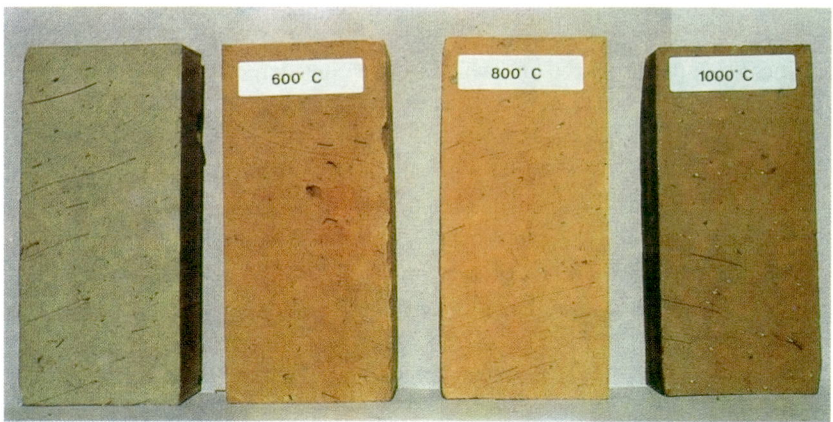

Fig. 9, Brick of modern manufacture

Fig. 10, Brick of ancient manufacture

First of all, the bricks of current production appear regular in both size and shape, while their ancient counterparts are irregular and variable in size and colour. These latter show areas of marked unevenness over the broader surfaces, a macroscopic porosity made up of voids varying in size from microns to millimeters, as well as colour variation from one part of the same brick to the next. These superficial peculiarities, which vary according to the period and place of production, can be ascribed to the lack of precise control of mixture composition and the rudimentary techniques used for shaping and baking.

Furthermore, one of the two broad sides of the brick is usually rough but nearly flat, while the other is smooth but slightly curved. This characteristic - present in all samples - corresponds perfectly to historical accounts found in the literature on the matter. In fact, during shaping of the brick, one of the broad sides would lie on the working table covered with a layer of sand, while the other side was smoothed out by hand after removal with a wooden ruler of the excess clay mixture. Very often during the drying process, performed in the open air, regular rain drops and sleet left their impressions on the brick's upper surface (Fig. 11).

Fig. 11, Bricks removed from Bishop's Palace in Pienza (XVI C.)

As is well known, ancient manufacturing techniques consisted of manual mixing and moulding of the brick, while the current practice takes advantage of wire-draw procedures. The disparity existing between the two techniques is the basis for the relevant differences between the two types of brick, both from the qualitative and quantitative point of view, as well as their different mechanical attributes. With the aim of elucidating these differences in behavior, numerous tests have been performed on true ancient bricks, simulated old-style bricks (newly manufactured using techniques similar to the traditional ones) and, finally, on wire-drawn bricks [31], [32], [33]. Here we will not detail the various equipment, sample size or geometry employed, or the testing procedure adopted [34], [35], as this would, in any case, add very little to the overall qualitative description which is the aim of this paper. The following considerations are the result of analyses of the experimental data recorded during tests.

–   The different procedures of manufacturing, mixing and moulding, determines differences in the composition and internal structure of the material, as well. Therefore, such material may, as a result of the technique used, possess different characteristics in different parts of the same brick.

–   Manual pressing during moulding allows the material to lie in layers parallel to the broad side of the brick. This process determines a characteristic anisotropy above and beyond that due to the drying process.

–   The baking temperature of each individual brick has a great influence upon its behavior.

–   The characteristics revealed by a test sample removed from the central part of a brick are very different from those of a like sample drawn from the outer regions (rims).

–   Compressive strength measured in test samples whose sides belonged to a brick rim is usually 20% higher than that of test samples drawn from the interior; this means that brick is not homogeneous.

–   Compressive strength varies according to the stress direction; the material thus shows anisotropy toward the strength parameter.

–   Direct tensile strength varies according to the stress direction.

–   The elastic coefficients are different for the three mutually orthogonal stress directions, parallel and perpendicular to the layering direction, respectively; this indicates that the material is not isotropic.

–   Post-peak tensile behavior can be considered to be of the brittle type, independent of the loading direction.

–   Post-peak compressive behavior can be considered to be of the strain-softening type, more or less unstable, depending on the loading direction.

From what has been briefly outlined above, it is evident that the tests required by the current rules are not by no means extensive enough to provide a full description and characterization of the behaviors of brick. Furthermore, certain aspects of primary importance when dealing with materials having been subjected to chemical, physical and mechanical actions influencing the durability, damage and deterioration of their mechanical characteristics have not been given due consideration.

It has already been stated that the main difference between a wire-drawn brick and one manufactured through a manual pressing process lies in the post-peak branch of the equilibrium path (Figs. 12 and 13).

In order to define the descending branch, it is necessary to perform the test using a load apparatus with displacement controls. Then, using a prismatic sample with a height to depth ratio of 2:1 and applying at least three displacement transducers on the testing plate and two Ω displacement transducers on the lateral sides of the sample, a complete load-deflection and load-lateral deformation curve is obtained which is valid up to the peak load. The data recorded during such a test, furnished all the information deemed necessary in order to define and quantify the parameters underlying the characteristics of the uniaxial compression behavior of a brickunit. The P-δ curves, recorded over the entire loading process, are characterized by common features which are represented in the typical path seen, as depicted in figure 7b.

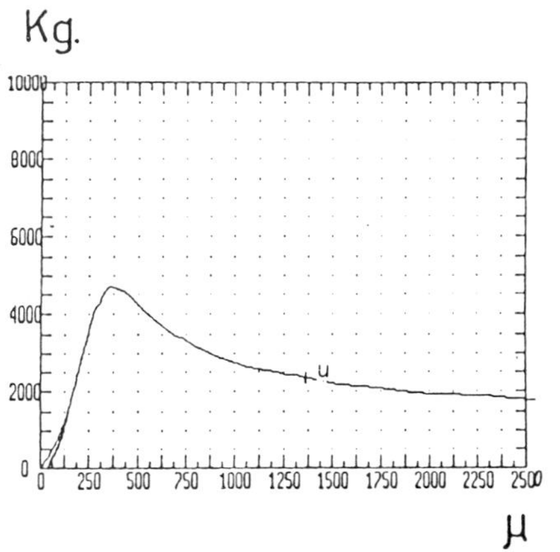

Fig. 12, Equilibrium path of an ancient brick

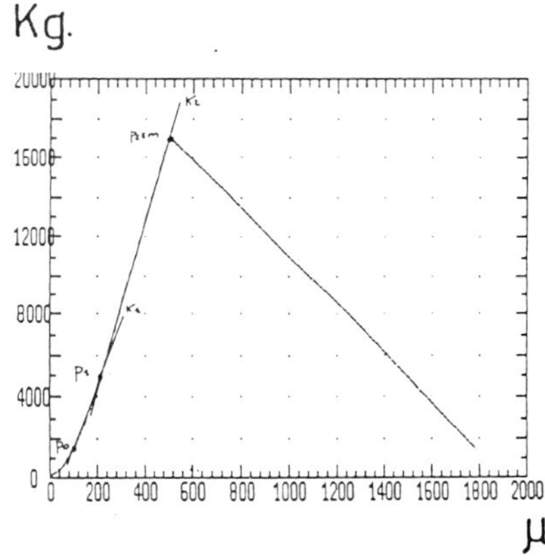

Fig. 13, Equilibrium path of a newly manufactured brick

Figures 14, 15 and 16 show the **P-δ** curve which is characteristic up to limit load and is representative of the uniaxial compression tests - along the three perpendicular directions, parallel and perpendicular to the compaction layers - performed on samples removed from bricks drawn from a wall of the Bishop's Palace in Pienza.

The parameters considered essential to represent the individual branches of the

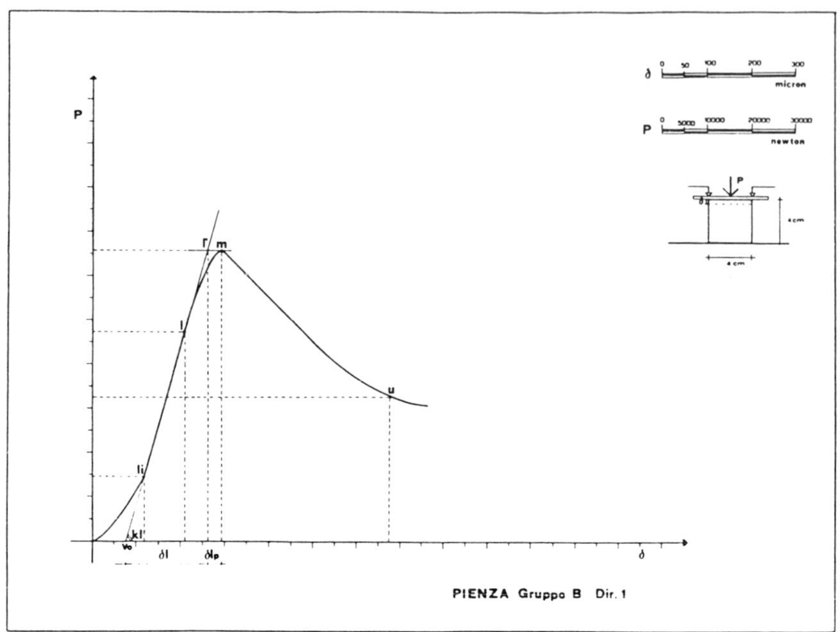

Fig. 14, Mean P-δ diagram under compression along direction 1

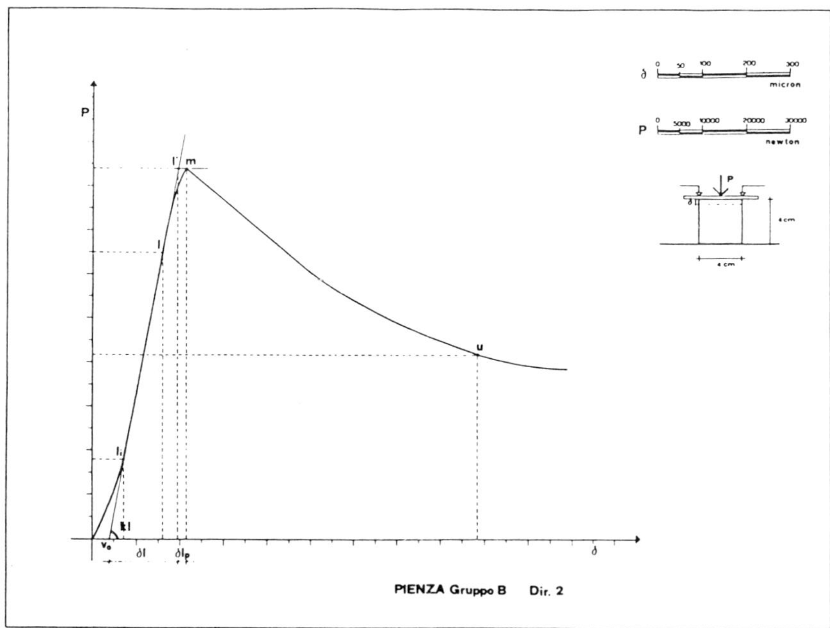

Fig. 15, Mean P-δ diagram under compression along direction 2

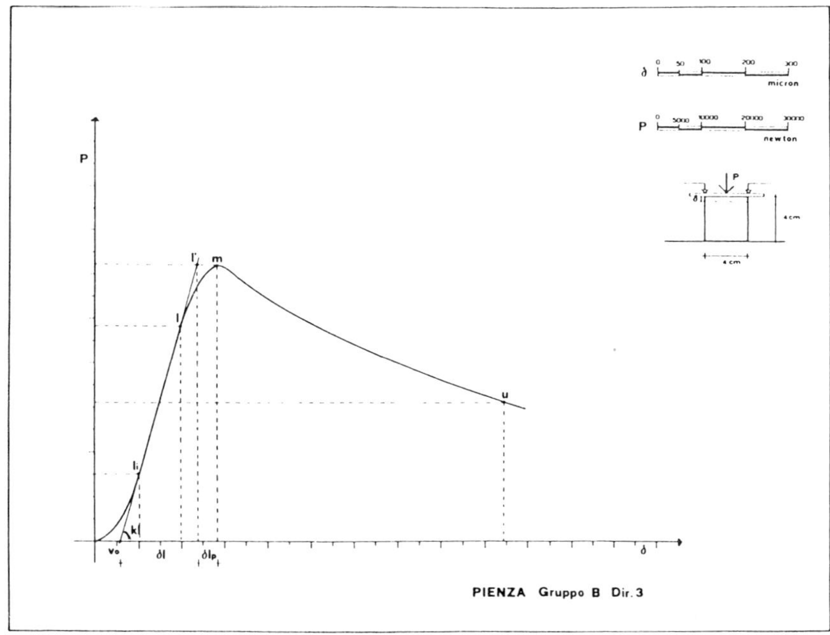

Fig. 16, Mean P-δ diagram under compression along direction 3

equilibrium path have already been stated, while the related mechanical characteristics are reported in Table I.

group	γ (N/m³)	σ_c (MPa)	σ_c (MPa)	σ_c (MPa)	σ_c (MPa)	$1/s_{22}$ (MPa)	$1/s_{21}$ (MPa)	$1/s_{23}$ (MPa)	σ_{tt} (MPa)	σ_{tf} (MPa)	μ_{k3}	μ_{kd3}
A	15557	33,0	34,9	35,6	28,3	11787	80635	70333	3,4	5,7	1,1	4,0
B	16077	43,6	40,4	51,4	39,1	17984	115040	78255	4,3	9,4	1,2	4,8
C	15329	44,8	49,3	48,5	36,7	18641	111140	96628	3,0	8,5	1,1	6,0
D	14700	27,8	31,0	30,7	21,7	-	-	-	3,2	7,5	1,2	8,5

γ	bulk density
σ_c	compressive strength (mean of the three directions)
σ_{c1}-σ_{c2}-σ_{c3}	mean compressive strength for each direction, respectively 1, 2, 3
$1/s_{22}$	elastic modulus under compression in direction 2
$1/s_{21}$-$1/s_{23}$	elastic coefficients, respectively for directions 1 and 3
σ_{tt}	tensile strength by shear test
σ_{tf}	tensile strength by flexural test
μ_{k3}	kinematic ductility assessed in compression test along direction 3
μ_{kd3}	available kinematic ductility assessed in compression test along direction 3

Tab. I, Mechanical properties of brick according to a subdivision in homogenous groups.

The differences in the behavior of samples, in terms of strength according to load direction, is evident. In particular, the values obtained in direction 3 - perpendicular to the brick layers - are always lower than those resulting in the other directions, with the ratio σ_{C3}/σ_{C1}, equal to 0,67 and σ_{C3}/σ_{C2}, equal to 0,71.

We would like to point out that such ratios remain almost unaltered even after samples have been subject to artificial degradation cycles.

All tests performed on ancient bricks have inevitably shown that load direction 3 was that exhibiting the lowest values of uniaxial compressive strength. This peculiarity may be due to two different reasons: the first is linked to the manufacturing process; the second is that, since load direction 3 is perpendicular to the brick layering plane, the material has deteriorated due to the sustained loading. Both hypothesis seem valid.

Unfortunately, the question cannot be settled, as there is no ancient virgin material available to test. It should be pointed out though, that similar tests performed on samples from newly-manufactured simulated "ancient" bricks (moulded by procedures resembling those once performed) have demonstrated similar behavior along the same three loading directions.

The direct and indirect elastic coefficients determined also show some differences. Consequently, the hypothesis of isotropic behavior can not be considered appropriate, as experimental data contradicts it.

It is remarkable that the values of available ductility - relative to load direction 3, measured at a point corresponding to residual load bearing capacity equal to 0,25 of that at the peak-load - are always higher than those obtained in the other two load directions. Instead, along these directions the material shows a higher strength value. Indeed, the material manufactured according to traditional procedures shows a rather high value of this parameter when compared to the results recorded in similar tests performed on newly manufactured bricks, especially in the direction perpendicular to the brick layers.

The trials performed so far, of which only some significant results have been presented here, allowed us to point out the dependence of many mechanical characteristics on the shaping of the raw material composing brick of new or ancient manufacture.

Of all the results obtained, we would most like to highlight the characteristic parameters of post-peak behavior. Such parameters indicate, in fact, the remarkably peculiar ductility demonstrated by bricks of ancient manufacture. Such behavior was not revealed in modern brick.

Some trials also indicated a certain relationship between the available kinematic ductility and the dimensions of the pores in the sample. An increase in pore size results in an increase in the value of this parameter.

Furthermore, the mechanical testing done after artificial degradation cycles (Tab. II), also demonstrated the greater durability of material of ancient manufacture as compared to that of newly manufactured brick (Fig. 17).

In fact, variations in bulk density and mmechanical characteristic are modest; more specifically, bulk density is generally 0,2% lower, while the values regarding the mechanical characteristics are subject to an average decrease of 8%.

Mineralogical analysis also prove useful in order to define the probable relationships existing among mixture-texture, mechanical behavior and durability factors [36], [37], [38].

The attempts made in this direction have encountered certain difficulties due to the fact that the porosity and quantity of amorphous material - both of which may influence mechanical behavior - show very similar values. Defining the relationships between mechanical characteristics and composition is to be considered paramount because, by accounting for the manufacturing technique as well, it would allow replacement of elements of ancient manufacture with brand new ones, providing their characteristics are similar to the original material.

group	$\Delta\gamma$ %	σ_c* MPa	$\Delta\sigma_c$ %	σ_{c1}* MPa	σ_{c2}* MPa	σ_{c3}* MPa	$1/s_{22}$ MPa	$\Delta(1/s_{22})$ %	$1/s_{21}$* MPa	$1/s_{23}$* MPa	s_{tt} MPa	Δs_{tt} %	m_{k3}*	m_{kd3}*
A	-0,2	30,2	8,4	31,5	33,1	25,9	11510	-0,7	69455	56842	2,6	-17,5	1,1	3,7
B	-0,2	41,7	4,4	38,9	47,6	38,7	17542	-2,5	112960	77102	3,9	-9,4	1,1	4,0
C	-0,2	41,5	7,4	48,2	41,0	35,2	16491	-11,5	97699	74713	3,0	-1,0	1,1	5,o
D	-0,2	23,6	15,2	29,1	20,9	20,8	14512	-5,5	82844	75653	2,1	-3,7	1,2	9,3

γ	bulk density
σ_c	compressive strength (mean of the three directions)
σ_{c1}-σ_{c2}-σ_{c3}	mean compressive strength for each direction, respectively 1, 2, 3
$1/s_{22}$	elastic modulus under compression in direction 2
$1/s_{21}$-$1/s_{23}$	elastic coefficients, respectively for directions 1 and 3
σ_{tt}	tensile strength by shear test
σ_{tf}	tensile strength by flexural test
μ_{k3}	kinematic ductility assessed in compression test along direction 3
μ_{kd3}	available kinematic ductility assessed in compression test along direction 3

Tab. II, Mechanical properties of brick after artificial degradation cycles according to subdivision in homogenous groups

In fact, variations in bulk density and mechanical characteristics are modest; more specifically, bulk density is generally 0,2% lower, while the values regarding the mechanical characteristics are subject to an average decrease of 8%.

Mineralogical analysis might also Comparisons among results obtained from tests on samples removed from ancient and new brick provide the opportunity to note that the solid wire-drawn brick generally exhibits strength values comparable to those of ancient brick [39]. On the other hand - unlike the latter, it is characterized by an unstable equilibrium path. This behavior decreases as the baking temperature is lowered, as does the strength value as well (Figs. 18, 19 and 20).

It would be seem appropriate to ascribe the above-mentioned effect to the structure imparted to the material during the wire-draw process. Instead, the general nature of the phenomenon leads us to attribute it to the characteristics of the mixture or the baking techniques used in modern kilns. In contrast, the strength value of newly manufacturing but hand-shaped brick, although showing an initially descending line similar to that of the ancient bricks, decreases very quickly toward zero; in fact the long almost-horizontal post-peak line with stabilization of the peak value equal to about 0,25 is lacking. By considering the resulting cracking pattern, moreover, the peculiar behavior can be attributed to a material which, since it is mixed by hand and therefore consists of a heterogeneous mixture, cracks according to meandering lines which allow consequent equilibrium states also in the post-peak phase.

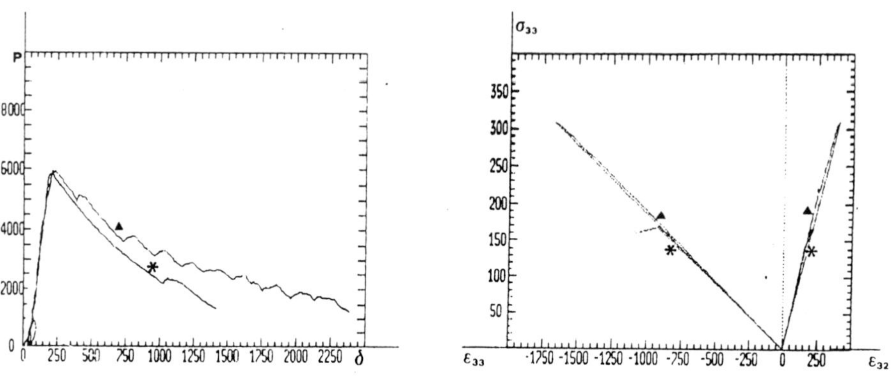

Fig. 17, Characteristic diagrams of P-δ and P*-δ* (a) and σ-ε e σ*-ε* (b) in samples before and after artificial degradation cycles

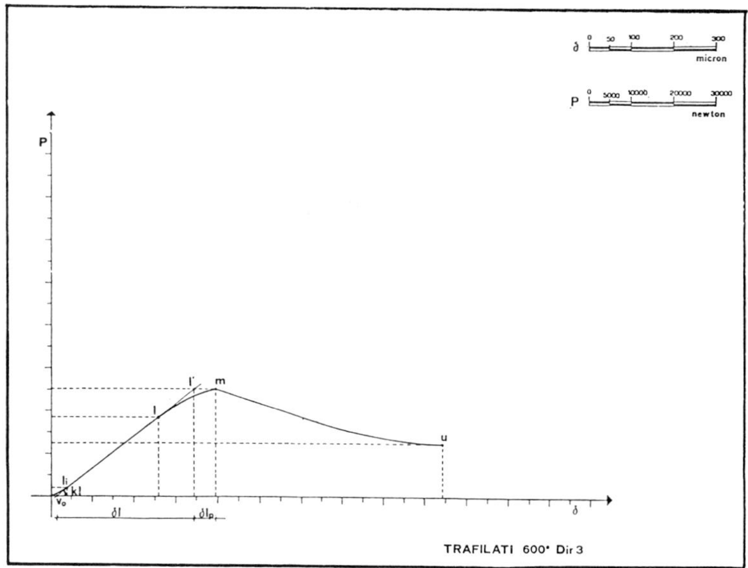

Fig. 18, Mean P-δ diagram under compression in direction 3 of new brick samples baked at temperature 600°C.

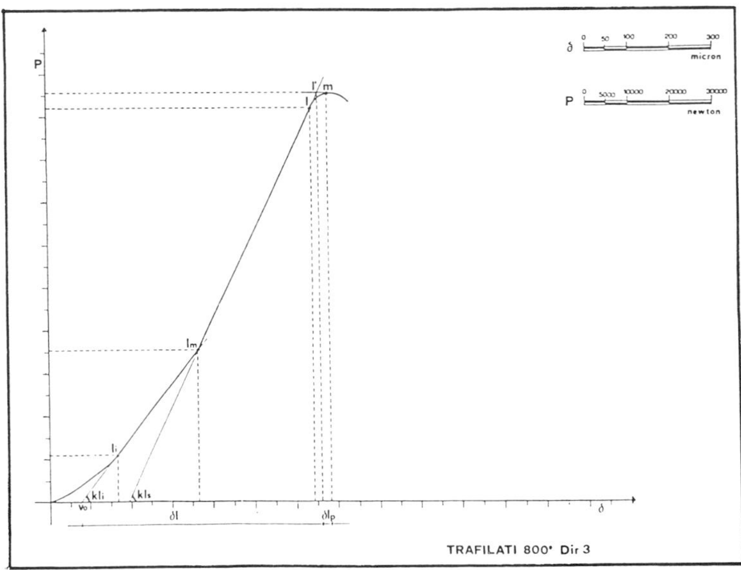

Fig. 19, Mean P-δ diagram under compression in direction 3 of new brick samples baked at
temperature 800°C

Finally, in conclusion, we wish to consider some aspects which have as of yet not
been sufficiently investigated and would therefore require further analysis.

First of all, since no set of relevant experimental results is available, it is impossible to
provide formulation of the damage laws operating in brick under sustained and cycling
loading histories, despite the fact that these seem to be a common cause of disastrous
collapse in masonry structures.

Secondly, specific and thorough research on the problem of durability should be
undertaken.

Such research, together with the damage laws, would allow an up-to-date assessment
of how long ancient structures made of materials already subjected to actions of varying
nature might be expected to last.

For the sake of brevity, we have only outlined the essential characteristics of the
behavior of brick of ancient manufacture.

We are of the opinion that the influence of this wide-spread material needs be
considered more carefully in order to achieve better results when working on ancient
masonry buildings.

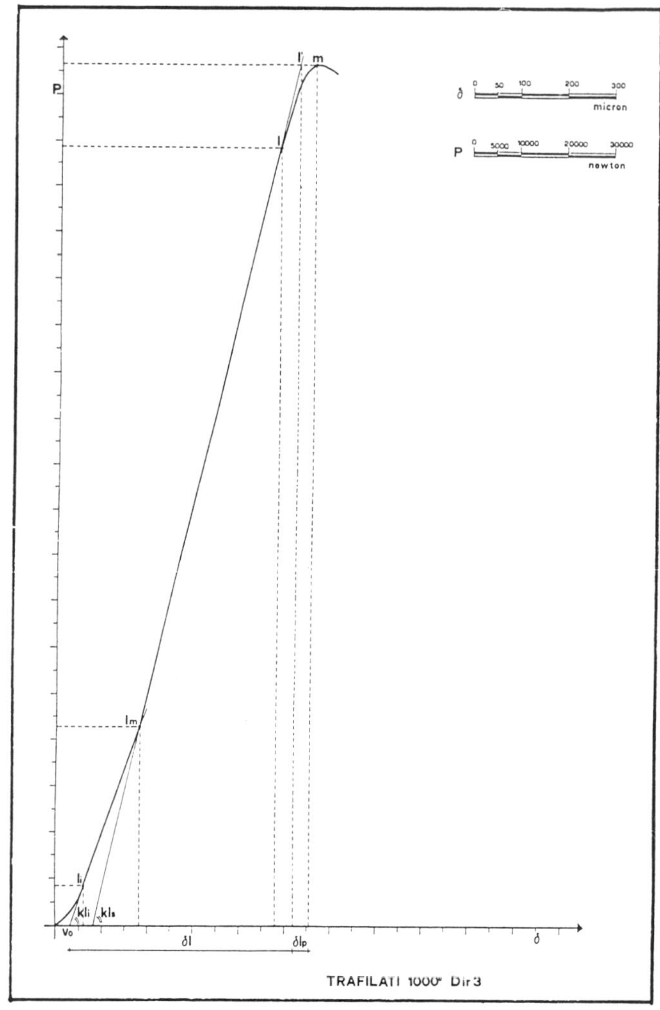

Fig. 20, Mean P-δ diagram under compression in direction 3 of new brick samples baked at
temperature 1000°C

4.   THE MECHANICAL BEHAVIOR OF EARTH AND GREEN BRICK

The use of earth in its various forms as a building material, both in times past and at
present, is much more common than one might imagine. Existing earthen architecture
testifies to an ancient technology which in the past found wide-spread use all over the world
and still persist in many regions. Earth is used in architecture with various techniques,
depending on the clay available in the soil, on the climate and on the environment.

The building techniques used in earthen structures can be divided into two basic types, called pisé and adobe. Both are currently in use under certain circumstances in some countries due to their low cost, the speed with which they can be erected and because they require no specialized labor in order to carry out. Obviously, their application is strongly limited by the prevailing environmental conditions, and above all, by the types of natural resources present (Figs. 21 and 22).

Fig. 21, Pisé building: home in Monte S. Savino - Arezzo

Knowledge of the mechanical properties of these materials is important for the conservation of the historical patrimony of many countries. In the following, the results of some experiments are reported that deal with either ancient material or new material produced according to ancient building rules. The technique called pisé is based on the compaction of a clay slurry in mobile formworks during the various stages of construction. The percentage of clay mixed with sand and water varies from 10 to 20% so as to produce a dense easily moulded slurry, humid, but not wet - a consistency able to avoid the formation of cracks during the stage of drying.

Under ideal conditions the clay and sand mixture is found ready-formed in natural deposits; in this case the material is used as is, after the sole operation of extracting it from the ground. Generally, raw earthen walls of the pisé type are about 50 cm. thick and are built up in layers: the slurry for each layer is cast in the formworks and then beaten and compacted with specifically designed instruments. Clearly, this sort of earthen material represents the most rudimentary form of building material, as such. Nonetheless, its mechanical characteristics earn it a place among this category of substances.

Fig. 22, An adobe building: Sardinia

The laboratory tests performed on suitably prepared samples have brought to light many of its basic properties, previously uninvestigated, which allow several considerations to be made.

Based upon the results obtained, it can be asserted that, contrary to what might be expected, this earthen substance possesses fair compressive strength and considerable ductility, though its tensile strength is very low. Figures 23 and 24 present the full mean equilibrium paths of earthen samples tested with compression in two directions, one parallel to that of compaction layers, the other perpendicular to it.

Analysis of the behavior observed in such tests permitted reaching the conclusions that samples whose pressed layers are oriented in a direction other than that of the loading direction, exhibit the same strength, similar stiffness, but distinct post-peak behavior and thus, a different value of the ductility parameter.

Sample were compacted in wooden moulds; the percentage of water was taken to be 6% of the earth dry weight, the samples were aged for at least 45 days at room temperature. The main physical properties of earth used for the pisé technique are reported in Table V and the main results of the tests performed are presented in Tables III and IV.

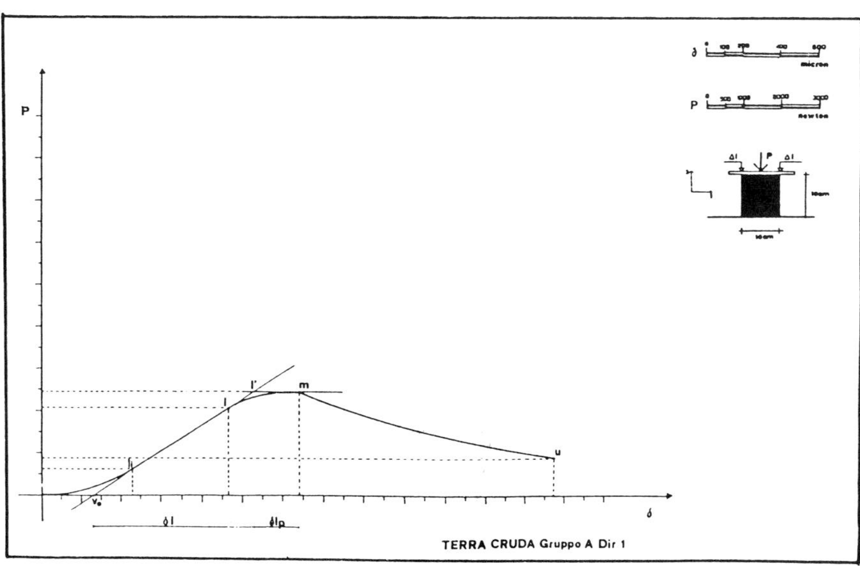

Fig. 23, Mean P-δ diagram in samples subjected to compressive stress parallel to the compaction layers of pisé

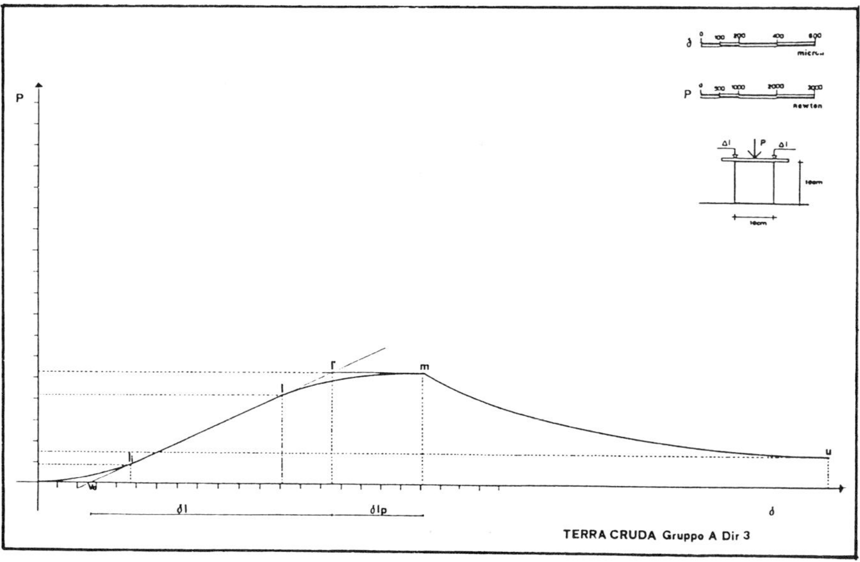

Fig. 24, Mean P-δ diagram in samples subjected to compressive stress perpendicular to the compaction layers of pisé

	γ (g/cm³)	σ (MPa)	μ_c	μ_{cd}
sample mean	1,84	2,68	1,26	4,47
stand. deviation	0,041	0,600	0,060	1,52
coeff. of variat.	3%	21%	4%	15%

γ = bulk density; σ = compression strength; μ_c = kinematic ductility; μ_{cd} = available kinematic ductility.

Tab. III, Main mechanical parameters for samples tested with load acting orthogonal to the compaction layers

	γ (g/cmc)	σ (MPa)	μ_c	μ_{cd}
sample mean	1,86	2,69	1,20	3,76
stand. deviation	0,061	0,410	0,110	1,77
coeff. of variat.	3%	15%	9%	47%

γ = bulk density; σ = compression strength; μ_c = kinematic ductility; μ_{cd} = available kinematic ductility.

Tab. IV, Main mechanical parameters for pisé samples tested with load acting parallel to the compaction layers

	W	γ (g/cm³)	W_l	W_p	I_p	W_s
earth	14,70%	2,46	19%	14%	5%	21%

W = water content; γ = bulk density; W_l = liquid limit; W_p = plastic limit; I_p = plastic index; W_s = shrinkage limit.

Tab. V, Main physical properties of earth

Earth material exhibits a relatively high strength, good for curtain walls and adequate for very short buildings as well.

The building procedure called adobe, which utilizes unbaked earth, is the most ancient of masonry techniques and is still used in many countries for erecting homes and rural structures. Adobe is a technique that employs sun-dried bricks using a mixture of the same mud as mortar. The size and shape of the adobe bricks vary from region to region and according to the date of their fabrication, though the manner in which they are laid is the same throughout the countries in which they are found.

Uniaxial monotonic compression tests were performed along two mutually perpendicular directions on cube-shaped specimens extracted from adobe brick of a building in Valdarno and having dimensions 3.25 by 3.25 by 3.25 cm. The results indicate that, as in the case of pisé, this material's reaction to compression depends on the orientation of the stress with respect to that of the compaction layers. Figures 25a and b present the load - displacement diagrams of the two samples subjected to loads perpendicular and parallel to the compaction layers, respectively, while Tables VI and VII contain the mean values of the mechanical parameters relative to each. From the graphs of test results it can be seen that adobe's compressive strength is lower when the sample is subjected to stress oriented perpendicular to the brick's laying plane; its stiffness value in this direction is also lower. The presence of a long segment in the post-peak phase, on the other hand, indicates greater ductility and thus greater regularity of the material's behavior in response to such stress.

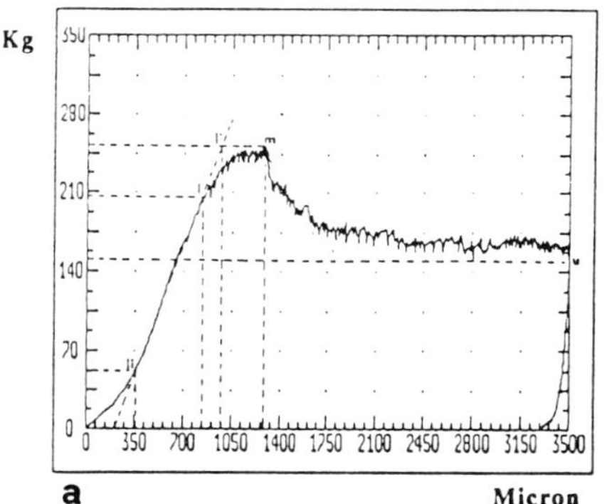

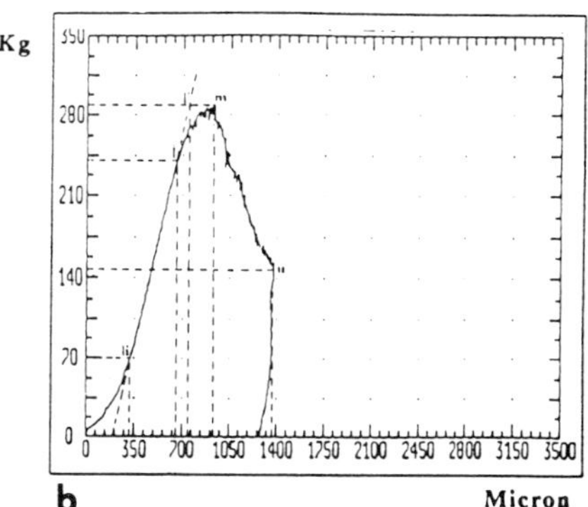

Fig. 25, Mean P-δ diagram relative to compression tests on cube-shaped adobe specimens in direction perpendicular (a) and parallel (b) to compaction layers.

	γ (g/cm^3)	σ (MPa)	μ_c	μ_{cd}
sample mean	1,78	2,37	1,22	2,53
stand. deviation	0,015	0,460	0,120	0,39
coeff. of variat.	0,85%	19,34%	10,09%	15,42%

γ = bulk density; σ = compression strength; μ_c = kinematic ductility; μ_{cd} = kinematic available ductility

Table VI. Main mechanical parameters for adobe specimens with load acting along the (long) side of the brick (perpendicular to compaction layers)

	γ (g/cm^3)	σ (MPa)	μ_c	μ_{cd}
sample mean	1,78	2,63	1,09	2,28
stand. deviation	0,036	0,330	0,030	0,54
coeff. of variat.	2,02%	12,45%	2,74%	23,77%

γ = bulk density; σ = compression strength; μ_c = kinematic ductility; μ_{cd} = kinematic available ductility.

Tab. VII, Main mechanical parameters for adobe specimens with load acting parallel to compaction layers

Analysis of existing adobe constructions has revealed the existence of a mixed or composite building system with green and fired bricks laid in the same structure. With the aim of evaluating the mechanical properties of such heterogeneous masonry material, simple compression tests were carried out on 1:4 scale models in the attempt to uncover the effects of insertion of fired bricks within a green brickwork panel. The models were built by combining green bricks (made following traditional casting techniques and a mixture used in manual brick production) with fired elements in different percentages and positions so as to reproduce typical conditions observed in existing constructions. The results of these experiments are reported and illustrate the effect of such insertion on both mechanical characteristics and fracture pattern. Specifically, fired bricks seem to have be employed systematically in construction of the ground floor, as well as in corners and other highly stressed sections, such as the wall plates of ceiling beams; other insertions seem to be due to restoration or maintenance intervention such as additions, occlusion of existing openings, or remakes of damaged areas. Observations of the behavior of model under simple-compression conditions highlighted the effect of the insertion of cooked elements with regard to fracture pattern and mechanical characteristics as well.

Two distinct series of 20 by 20-cm wall models simulating brickwork at a 1:4 scale were constructed: The first series is made up of 10 single-leaf and 10 double-leaf adobe walls. The second series is comprised of 26 single and double-leaf walls, to which a composite structure was imparted with both green and fired bricks. The two types of bricks were combined according to different assembly patterns, so as to reproduce typical existing structures. Specifically, seven different assembly patterns were considered:
    fired-brick corners;
    incomplete layers of fired bricks;
    full fired-brick layers;
    occlusion of windows with toothing;
    occlusion of windows without toothing;
    raising;
    scattered placement.
The joint bed between successive layers of green bricks was made up of a thin layer of the same material used in constructing the brick; this type of mortar was also used in laying the fired bricks. This compromise, which was dictated basically by the difficulty of

producing layers of lime mortar to a 1:4 scale, does not significantly affect the resulting properties of the brickwork model, as these are relatively independent of mortar type.

The materials employed in the models' construction were subjected to simple compression tests and the most important results analyzed statistically; the relative mechanical characteristics and statistical parameters are reported in Tables VIII through X.

Static compression tests were performed with two different loading patterns:

1) distributed load over the upper face of the model

2) distributed load over part of the upper face of the model including its thickness for a length of 5 centimeters. During compression tests, the lowering of the loading surface and the vertical and horizontal strain in the middle of the brickwork were measured by means of a set of electrical displacement transducers. The results of the experiments are shown in Tables XI through XIV.

	γ (g/cm^3)	σ (MPa)	K (N/μm)	μ_c	μ_{cd}
sample mean	1,88	15,22	17,86	1,28	1,49
stand. deviation	0,012	2,749	0,430	0,171	0,301
coeff. of variat.%	3,28	18,06	24,30	13,40	20,20

γ = bulk density; σ = compressive strength; K = stiffness; μ_c = kinematic ductility (taken as the ratio of peak to linear displacement); μ_{cd} = available kinematic ductility (taken as the ratio of limit to peak displacement); limit displacement was presumed to correspond to peak load/2.

Tab. VIII, The main mechanical properties of 30 fired-brick specimens tested for compressive strength.

	γ (g/cm^3)	σ (MPa)	K (N/μm)	μ_c	μ_{cd}
sample mean	1,98	1,75	1,68	1,05	1,79
stand. deviation		0,210	0,320	0,070	0,300
coeff. of variat.%		12,00	19,05	6,67	16,76

Tab. IX, The main mechanical properties of 30 green-brick specimens tested for compressive strength with load perpendicular to compaction layers

	γ (g/cm^3)	σ (MPa)	K (N/μm)	μ_c	μ_{cd}
sample mean	1,99	1,97	1,69	1,06	1,90
stand. deviation		0,290	0,300	0,070	0,520
coeff. of variat.%		14,72	17,75	6,60	27,37

Tab. X, The main mechanical properties of 30 green-brick specimens tested for compressive strength with load parallel to compaction layers

The insertion of fired elements within green brickwork has an evident effect on its mechanical properties: even at low insertion rates, both the strength and toughness exhibit are greatly increased [40]. This effect is largely conditioned by the positioning of fired elements within the overall structure. In particular, a spread or scattered distribution achieves a relative increase in strength without concomitant concentrations of stress.

tipe of specimen	b/a ratio	σ (MPa)	K (N/μm)	μc
A n. 3 spec.	0.00%	1.84	6.57	1.110
B n.2 spec	40.91%	3.87	13.700	1.063
C n.2 spec.	15.15%	2.15	8.730	1.250
D n.2 spec	18.18%	2.13	8.735	1.098
E n.2 spec.	10.61%	2.37	9.840	1.170
F n.2 spec.	21.97%	2.16	0.943	1.251

Tab. XI, The main mechanical properties of double-leaf walls with load distributed over the entire upper boundary

The presence of stiffer elements causes the fractures to divert to their boundaries; a spread distribution generates a diffuse fracture pattern. Regarding the example of the occluded window, the main fracture appears at the interface areas between fired and green bricks; for this reason the toothed occlusion exhibits more or less the same strength, but higher ductility with respect to the non-toothed one. Experimental results indicate that an even scattering of fired elements avoids instability within the overall post-peak path; this is confirmed by the fact that crack distribution preserves a spread pattern (Fig. 26).

tipe of specimen	b/a ratio	σ (MPa)	K (N/μm)	μc
A n. 3 spec	0,00%	2,17	2,46	1,71
B n. 3 spec.	12,12%	3,42	3,07	1,25
C n. 2 spec.	15,15%	3,26	2,04	1,39
D n. 2 spec.	21,21%	3,45	2,41	1,28
E n. 2 spec.	33,33%	3,81	3,65	1,20

Tab. XII, The main mechanical properties of single-leaf walls with load distributed over part of the upper boundary

tipe of specimen	b/a ratio	σ (MPa)	K (N/μm)	μc
A n. 3 spec.	0,00%	2,68	5,95	1,21
B n.3 spec	30,30%	3,90	6,44	1,25
C n.2 spec.	10,61%	3,29	6,11	1,38

Tab. XIII, The main mechanical properties of double-leaf walls with load distributed over part of the upper boundary

tipe of specimen	b/a ratio	σ (MPa)	K (N/μm)	μc
A n. 3 spec.	0,00%	1,96	3,690	0,930
B n.3 spec	12,12%	2,38	0,232	1,206

Tab. XIV, The main mechanical properties of single-leaf walls with load distributed over entire upper boundary

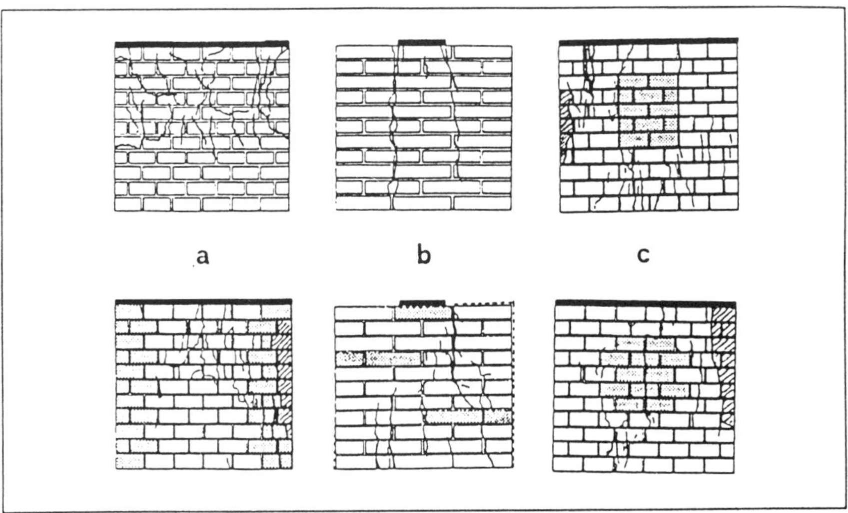

Fig. 26, Fracture patterns of some wall models: corners of fired bricks (a); scattered placement of fired bricks (b); occlusion with and without toothing (c).

5.   THE MECHANICAL BEHAVIOR OF STONE

The stone materials used in building are represented by rock fragments of variable size and more or less irregular shape. Their nature is of fundamental importance, in that both the style and, above all, the construction technique adopted in regional architecture, at least for minor cities and towns, are very strongly conditioned by the types and characteristics of the stone available in nearby areas. The great number of stone varieties making up the earth's crust, a determining factor in the historical manifestation of architecture, prohibits review of all the behavioral characteristics observable in stone materials. Therefore, the present discussion will be limited to a sole type of stone, tuff, a material which merits particular attention, because of both its extremely particular properties and its historical significance - it was already widely used during the time of ancient Rome and is thus found in many present-day historic sites.

Still today this stone is frequently used in many areas of Italy, even areas classified as seismic risk zones. However, despite its wide-spread use as a building material, tuff's mechanical behavior has never been adequately described in terms of modern theories, and theoretical treatment and comprehensive experimental data of its behavior are still lacking. A well-known fact of earthquakes' actions on buildings is that the ability of a masonry structure to withstand seismic actions, even very strong ones, depends mainly on its mechanical behavior in the post-elastic range. An important area of practical concern is the response to cyclic disturbances arising from earthquakes. Therefore, laboratory tests of both cyclic static and dynamic load are performed on tuff samples and simple masonry elements made of the same material in order to define typical post-peak behavior and damage laws and to collect the data necessary for predicting life-span and eventual failure of such materials and structures. In what follows, we report the results of experimental tests performed on tuff samples drawn from the quarries of Sorano in the province of Grosseto.

Tuff is a soft sedimentary volcanic rock that appears very heterogeneous because of the non-homogeneity of its grains and the simultaneous presence of micro and macroporosity with voids diameters varying from a few microns up to a centimeter; it can therefore be classified as an anisotropic porous compressible material (Fig. 27).

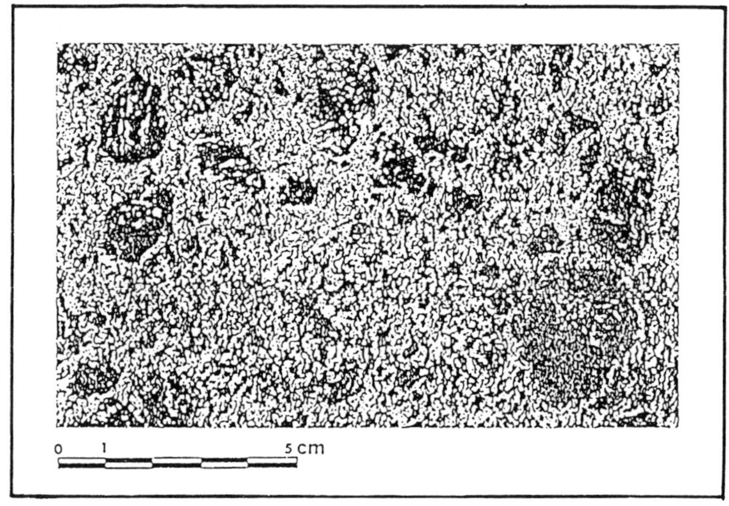

Fig. 27, Superficial porosity in a tuff sample

The initial void/volume ratio is about 60%; the in-plane mean "nearest-neighbor" pore spacing is approximately 3-4 cm; the mean void area fraction was about 17% for surfaces parallel to sedimentation layers, and 23% for perpendicular surfaces. The range of pore sizes is very large and the geometrical properties of the void system are quite difficult to define. This material is a porous rock material that is heterogeneous, fails progressively by breakdown of its structure during deformation and exhibits strain-softening behavior. Although tuff presents rather low strength values, it possesses several qualities which, contrary to expectations, render it quite capable of resisting repeated actions of both a static and dynamic nature.

The local properties within a material of such a structure may differ considerably from on region to the next. This heterogeneity is at the basis of one of the basic problems in arriving at a rigorous mathematical description of the local variables such as strain tensor. Given these circumstances, the problem becomes even more difficult to approach if a precise macrodescription of the material's behavior is needed.

Interpretation of test results in terms of stress-strain relationships is not attempted here, as testing samples are so dishomogeneous and the irregularities so randomly distributed that results are rendered unreliable. In fact, the presence of large dishomogeneities causes local fields which, under an external homogeneous field like stress or strain, vary from one point to the next. Localization processes develop within the specimen and prevent direct definition of the relation between the load-displacement curve and the assumed stress-strain curve.

On the basis of the above considerations, it is meaningless to speak in terms of characteristic strength etc., thus, we will only deal with average mechanical properties referred to some given size. Such properties, particularly strength and elastic coefficient, are strongly affected by the anisotropy in the void area fraction - they yield higher values and a better distribution curve in experimental results when corresponding to an orthotropy axis roughly perpendicular to the sedimentation layers.

Therefore emphasis is placed on data collected concerning the phenomenological aspects of behavior during cyclic loading and displacement histories (Figs. 28 and 29).

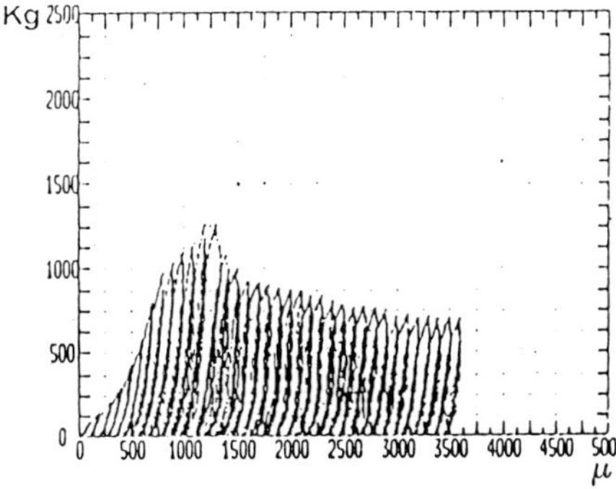

Fig. 28, Typical cyclic compression response

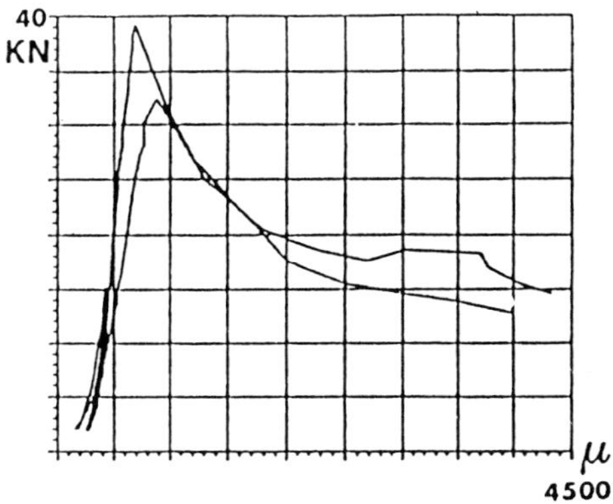

Fig. 29, Cyclic and monotonic compression response

Both diagrams demonstrate strongly ductile behavior and in no case is there evidence of degradation of stiffness along the direction of applied stress. In fact a sufficient amount of experimental evidence is already available concerning the shape of the hysteresis loop and the evolution in response to cyclic loading, as well as that of the unloading and reloading curve shapes. For the sake of brevity, we present only some conclusions drawn from the vast amount of experimental work carried out. Over the range of cyclic tests with load control, on both bare material and structural elements, some global characteristics may be summarized as follows: two significant features of this materials' mechanical properties are irreversible compaction and hysteresis during cyclic loading. The permanent volume decrease or progressive compaction results from rearrangement of the material's granular structure.

The difficulty of modeling this property is due to the impossibility of macroscopically distinguishing the two different co-existing effects induced by loading: porosity reduction i. e. densification and progressive damage. The densification rate is determined by the sum of contributions from all the mechanisms operating; the effects of creep are small at low load levels, and plastic flow becomes the dominant mechanism by which densification occurs in the early stages of compaction during loading. Also the detailed distribution of porosity has a strong influence on the material response.

As the number of cycles increases, the loops approach a closed stable limit loop. A common idealization - confirmed by experimentation - is to suppose that the limit loop is reached after one or at most a few cycles (Fig. 30).
For low load levels, the material is soft, but becomes stiffer as the load increases. An analysis of the response to a single sequence of loading, unloading and reloading indicates that the development of compaction depends on the advance in load level. Further load cycling produces wider and wider hysteresis loops, and usually the material can no longer be returned to its initial state because permanent damage has been caused. Cycle evolution through repeated unloading and reloading at constant load amplitudes leads to a gradual transformation of the loop; however, the evolution rate is remarkably higher at the beginning and end of cycling.

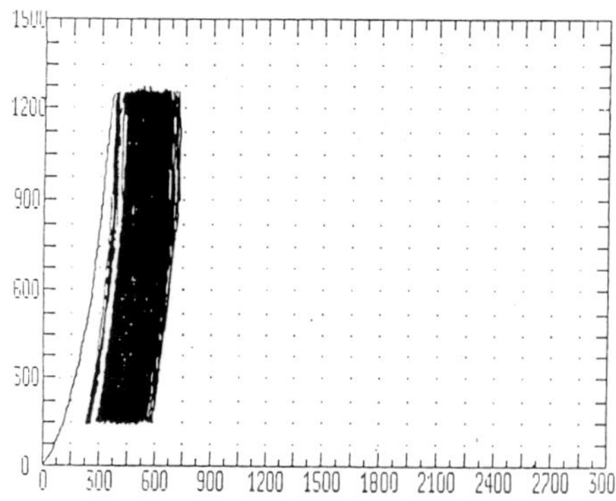

Fig. 30, Cyclic static compression test

The occurrence of cyclic failure, or fatigue, depends on the ratio of the load range to the monotonic limit load; for constant load amplitude, cyclic failure occurs after a number of cycles which increases with decreasing load amplitude.

When the load is repeated within a range of 0.5 of the monotonic strength F_m, both maximum and residual displacements within the prescribed number of cycles seem to fit an asymptotic curve (Figs. 31 and 32).

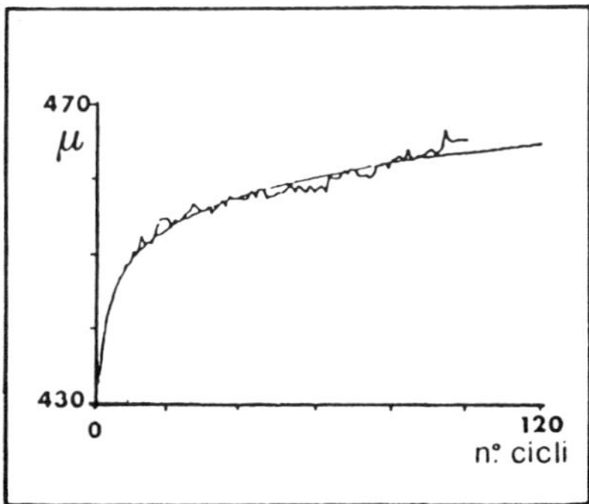

Fig. 31, Maximum displacement versus cycle number and best fit curve

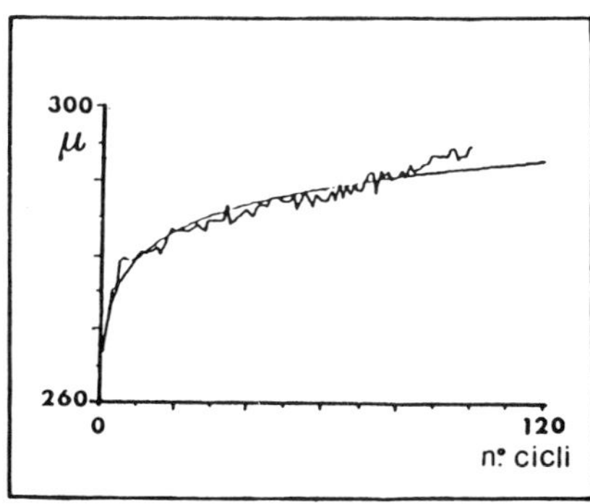

Fig. 32, Residual displacement versus cycle number and best fit curve

Failure is not reached within a technically relevant number of cycles, but the peak strength in subsequent monotonic test results higher than prior to the cyclic test. If the load is in the range 0,7-0,8 F_m, both maximum and residual displacements fit a typical "s" trend before failure occurs within the prescribed number of cycles (Figs. 33 and 34).

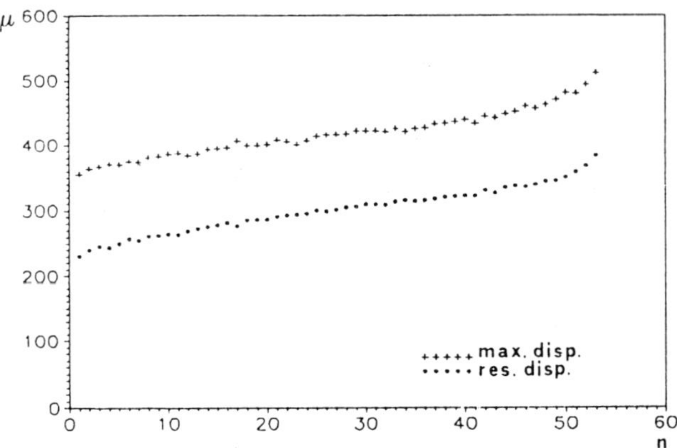

Fig. 33, Maximum and residual displacement versus cycle number for a tuff sample

The inflection point is found roughly at the middle of the load history. For this reason damage has been related to the dimensionless equation of residual strain [41]. In our opinion, this function, a cubic polynomial, does not fully represent damage growth: until the inflection point is reached, it represents the prevailing of compaction, which causes the strength to increase (Fig. 35).

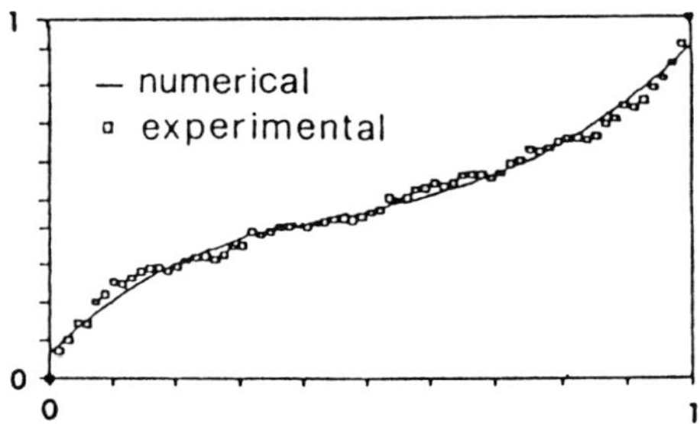

Fig. 34, Maximum displacement versus cycle number for a tuff masonry panel

This behavior, together with the lack of stiffness decay with increasing number of cycles, is yet another indication of the presence of densification taking place within the substance's interior. For comparison's sake, figure 35 presents the load-displacement diagrams obtained in both the cyclic (envelope curve) and monotonic tests. It is evident that the sample subjected to cyclic loading revealed a nearly 30%-higher fracture load than the monotonically loaded one. Compaction thus seems to act as a sort of compensating mechanism, averting the decay of stiffness which would otherwise be evident as the loading history proceeded; the loading increase makes the sample stiffer and more brittle, up to a value of about one third of the peak load.

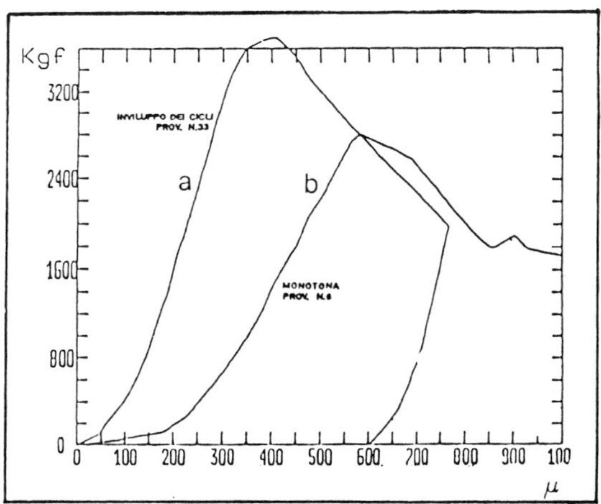

Fig. 35, Recorded cyclic (a) and monotonic (b) compression P-δ curves

A model accounting for both the compaction and damage processes may be developed on the basis of measured changes in material volume as a parameter for predicting this phenomena. In our opinion, the lateral expansion coefficient can be taken as representative of the variation in volume. When compaction occurs, volume decreases and, for short specimens (slenderness ratio of about 1), lateral expansion often assumes a negative value. The lateral expansion coefficient becomes smaller and smaller when the loops stabilize; if damage takes place, the coefficient changes sign with an asymptote, and when fracturing becomes visible it approaches 2 (Fig. 36), [42], [43], [44].

As all laboratory experiments on the uniaxial compressive response of tuff samples show that both damage and compaction occur during the loading process, it is very difficult to describe the progressive material deterioration, and the growth and pattern of individual cracks. Crack initiation is closely related to the localized strain produced by cycling, and depends on the structure of the material as well.

Finally, while performing mechanical tests on this material, we were able to observe a series of anomalies and size effects, which, when dealing with a highly heterogeneous material, might be explained by taking into account the ratio between the largest defect and the size of the laboratory sample, the presence of more than one phase, or the topology of the defects. Material of such a structure very often has a fractal character, but a fractal dimension is not the only cause of size effects. The mechanical characteristics of tuff depend on relative air humidity because tuff behaves like a sponge and moisture affects time-dependent behavior as well as post-peak properties. However, that which seems most distinctive about the behavior of this material is above all the cyclic behavior which passes through both compaction and damage stages, thus, we can observe strong dishomogeneity of the strain field in cyclic tests, due perhaps to movement of a loose phase within the pores of the sample.

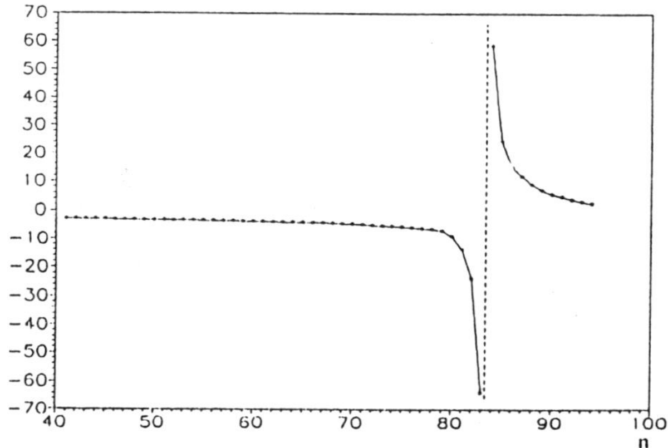

Fig. 36, Average lateral expansion coefficient for a tuff sample

6.   CONCLUSIONS

The studies carried out so far - of which only some significant results have been covered in this paper - allow us to underscore that many of the mechanical properties of traditional building materials depend heavily upon their internal structure.

Though we have only outlined the essential characteristics of such materials' mechanical behavior, we maintain that these data call for careful consideration in order to obtain better results in managing ancient masonry buildings whose constituting elements are represented by brick or stone, bound by layers of mortar .

Among the results mentioned, we would like to call special attention to the characteristic parameters of post-peak behavior which serve to highlight the peculiar and remarkable quality of ductility under compression, and a certain relationship between void dimensions and available kinematic ductility - an increase in the former results in an increase in the latter.

From what has been briefly referred to above, it is evident that the tests required by current regulations are far from sufficient to provide full description and characterization of the behavior of traditional building materials. Moreover, many aspects of primary importance when dealing with materials subject to chemical, physical and mechanical actions and thus strongly related to the durability, damage and deterioration of their mechanical characteristics have not been given due consideration.

The aim of this limited review is to call the attention of workers in the field of conservation of historical buildings to some important aspects of the behavior of the materials making them up. Knowledge of these will surely affect the structural solutions adopted for ancient buildings' and the assessments made of their capacity to resist external forces, including seismic actions, and therefore promote the attainment of better results in maintaining and restoring them. Indeed, it is essential to intervene with a guarantee of compatibility between the mechanical as well as the chemical-physical behavior of traditional materials and the newly manufactured ones; this, in order to avoid combining materials of widely different characteristics which, as is well known, can wreak serious damage to masonry structures.

REFERENCES

1.   Griffith, A. A.: The phenomena of rupture and flow in solids, Phil. Trans. R. S. A221 (1921), 163-198.
2.   Irwin, G. R.: Fracture, in: Handb. der Physik Vol. VI, Springer, Berlin 1958, 551-590.
3.   Barenblatt, G. I.: The formation of equilibrium cracks during brittle fracture: general ideas and hypotheses. Axially symmetric cracks, J. Appl. Math. Mech., 23 (1959), 622-636.
4.   Dugdale, D. S.: Yelding of steele sheets containing slits, J. Mech. Phys Solids, 8 (1960), 100-104.
5.   Barenblatt, G. I.: The mathematical theory of equilibrium cracks in brittle fracture, Ad. in Appl.. Mech., 7 (1962), 55-129.
6.   Hillerborg, A., Modeer, M. and Peterson, P. E.: Analysis of crack formation and crack growth in concrete by means of fracture mechanics and finite elements, Cem. Concr. Res., 6 (1976), 773-782.
7.   Hillerborg, A.: Results of three comparative test series for determining the fracture energy Gf of concrete, Mater. Struct., 18 (107) (1985), 407-413.
8.   Sha, S. P. and Gopalaratnam V. S.: Softening response of plain concrete in direct tension, J. Am. Conc. Inst., 82 (1985), 310-323.
9.   Sih, G. C.: Mechanics of Material Damage in Concrete, in Fracture Mechanics (Ed. A. Carpinteri and A. R. Ingraffea), Martinus Nijhoff Publishers, Dordrecht 1984, 1-29.
10. Carpinteri, A.: Scale effect in fracture of plain and reinforced concrete structures, in Fracture Mechanics (Ed. G. C. Sih and A.Di Tommaso), Martinus Nijhoff Publishers, Dordrecht 1985, 95-140.
11. Kachanov, L. M.: On the creep rupture time, Izv. AN SSSR, Otd. Tekn. Nauk, 8 (1958), 26-31.
12. Rabotnov, Yu. N.: On the equations of state for creep, Progr. Appl. Mech., (1963), 307-315.
13. Lemaitre, J. and J. L. Chaboche: Aspect phénoménologique de la rupture par endommagement, J. Méc. Appl., 2 (1978), 167-189.
14. Kachanov, L. M.: A continuum model of medium with cracks, J. Eng. Mech. Div: A.S.C.E., 106 (1980), 1039-1051.
15. Lemaitre, J. and J. Mazars: Application de la théorie de l'endommagement au comportement non linéaire à la rupture du béton de structure, Annales I.T.B.T.P., 401

(1982), 115-136.

16. Krajcinovic, D.: Constitutive theory of damaging materials, J. Appl. Mech., 50 (1983), 355-360.

17. Krajcinovic, D.: Continuous damage mechanics, Appl. Mech. Rev., 37 (1984), 1-6.

18. Krajcinovic, D.: Continuous damage mechanics revisited: basic concepts and definitions, ASME J. Appl. Mech., 52 (1985), 829-834.

19. Krajcinovic, D.: Micromechanical Basis of Phenomelogical Models, in: Continuum Damage Mechanics Theory and Applications, International Center for Mechanical Sciences Courses and Lecture (Ed. D. Krajcinovic and J. Lemaitre), CISM, Udine 1987, 195-206.

20. Lemaitre, J. and J. L. Chaboche: Méchanique des Matériaux Solides, Bordas, Paris 1988

20. Krajcinovic, D.: Damage mechanics, Mech. Mater., 8 (1989), 117-197.

21. Lemaitre, J.: A Course on damage Mechanics, Springer, Berlin 1992.

22. Dougill, J. W.: On stable progressively fracturing solids, J. Appl. Math. Phys.,27 (1976), 423-437.

23. Burt, W. J. and Dougill, J. W.: Progressive failure in a model of heterogeneous medium, J. Eng. Mech. Div: A.S.C.E., 103 (1977), 365-376.

24. Dragon, A. and Mroz, Z.: A continuum model for plastic brittle behaviour of rock and concrete, Int. J. Eng. Sci., 17 (1979), 121-137.

25. Bazant, Z. P. and Kine S. S.: Plastic fracturing theory for concrete, J. Eng. Mech. Div.: A.S.C.E., 105 (1979), 407-428.

26. Bazant, Z. P.: Work inequalities for plastic fracturing materials, Int. J. Sol. Struct., 16 (1980), 837-901.

27. Dougill, J. W. and Rida, M. A. M.: Further consideration of progressively fracturing solids, J. Eng. Mech. Div: A.S.C.E., 106 (1980), 1021-1038.

28. Krajcinovic, D. and Sumarac, D.: A mesomechanical model for brittle deformation processes: partI, J. Appl. Mech., 56 (1989), 51-56.

29. Krajcinovic, D. and Sumarac, D.: A mesomechanical model for brittle deformation processes: part II, J. Appl. Mech., 56 (1989), 57-62

30. Lubarda, V. A. and Krajcinovic, D.: Damage tensors and the crack density distribution, Int. J. Sol. Struct., 20 (1993), 2859-2877.

31. Binda, L., Anti, L and Valsani, L.: Indagine sperimentale sul comportamento meccanico dei materiali recuperati dalla torre civica, Tema, 4 (1993), 27-41.

32. Barbi, L., Briccoli Bati, S. and Ranocchiai, G.: Problemi inerenti il comportamento meccanico di laterizi di nuova ed antica produzione, Costruire in Laterizio, 36 (1993),544-549.

33. Barbi, L., Briccoli Bati, S. and Ranocchiai, G.: Problemi inerenti il comportamento meccanico di laterizi di antica produzione, Costruire in Laterizio, 39 (1994), 244-250.

34. Morton, J., Kotsovos, M. D., Pavlovic, M. N. and Seraj, S. M.: Initial investigation of the shape factor platen effects when testing masonry units to determine the material compressive strength, Proc. 9th Int. Brick/Block Masonry Conf., Berlin, (1991), 653-657.

35. Briccoli Bati, S. and Ranocchiai, G.: A critical review of experimental techniques for

brick material, Proc. 10th Int. Brick/Block Masonry Conf., Calgary, (1994), in press.

36. Tagliani, A., Binda, L. and Molina, C.: Interpretazioni statistiche di alcune proprietà fisiche e meccaniche dei laterizi, Industria Italiana dei Laterizi, 3-4 (1989), 126-139.

37. Baronio, G. and Binda, L.:Physico-mechanical characteristics and durability of bricks of some monuments in Milan, Edilizia, 7-8 (1990), 503-509.

38. Briccoli Bati, S. e. al.: Analisi comparativa delle proprietà composizionali, fisiche e meccaniche di mattoni usati a Pienza nel XVI secolo, Atti Convegno di Studi Le suprfici dell'architettura in cotto - caratterizzazione e trattamenti, Bressanone, (1992), 429-444.

39. Briccoli Bati, S.: The current experience of the mechanical properties of ancient brick, (Invited Lecture) Proc. 8th. CIMTEC-World Ceramic Cong., (Ed. P. Vincenzini), (1994), Firenze, in press.

40. Ranocchiai, G. and Rovero, L.: Influence of insertion of fired bricks on the mechanical characyeristics of adobe brickwork, Proc. 8th. CIMTEC-World Ceramic Cong., (Ed. P. Vincenzini), (1994), Firenze, in press.

41. Hwan Oh, B.: Cumulative damage theory of concrete under variable amplitude fatigue loadings, ACI Mater. J., 1 (1991), 41-48.

42. Schatz, J. F.: Models of inelastic volume deformation for porous geologic media, in: The Effects of Voids on Material Deformation (Eds. S. C. Cowin and M. M. Carrol), ASME, AMD, 16 (1976), 141-170.

43. Briccoli Bati, S. and Ranocchiai, G. : Comportamento di elementi strutturali in muratura di tufo sotto azioni cicliche, Atti V Conv. Naz. L'ingegneria sismica in Italia, (1991), Palermo, 395-404.

44. Briccoli Bati, S. and Ranocchiai, G.: Development of damage and progressive fracture of an heterogeneous rock material, Proc. Int. Conf. on Fracture and Damage of Concrete and Rock Material-2, (1992), Wien, in press.

RECENT RESEARCHES ON MASONRY CONSTRUCTION AND CRITICAL DISCUSSION OF EXAMPLES OF REHABILITATION

S. Di Pasquale

University of Florence, Florence, Italy

ABSTRACT

We expose the last results of the researches concerning the structural analysis of some monuments localized in seismic areas. We put forward some considerations on the behaviour of masonry materials and also on the foundation of the statics of masonry construction. In the present state of the art there are many doubts about the dynamical behaviour of these materials , expecially due to the characteristic elastic-brittle behaviour under tension. We give two examples to focalize the extrem difference of the subjects of study: cases of study in which Pozzuoli amphitheatre, and Modica cathedral are analysed.

1. HISTORICAL INTRODUCTION

C.A. Castigliano, as it is known, was the first to wonder about the equilibrium problem of a masonry arch [1]; at the end of the nineteenth century, we find the common idea concerning the masonry materials, that is the small resistance to tensile stress compared with the one to compression: see, for example, the Résal's Treatise on masonry constructions [2], expecially about arches and bridges. But similar ideas were expounded, intuitively, by E. Viollet Le Duc, in two famous works of his, the celebrated *Dictionnaire* [3] and the *Entretiens* [4], written between 1850 and 1870: in every masonry construction the structure, that is the part devoted to the transmission of loads to the ground, is only an "hidden" part of the construction.

The reasonning of E. Viollet Le Duc is summarized in the description of the Constantin basilique: "Que l'on veuille bien me suivre: voici, Fig.1, une coupe transversale et un fragment de plan d'une salle romaine, la basilique de Constantin, par exemple. Cette salle est entièrement construite en blocage avec parements de brique, revêtus de stucs, sauf les colonnes et leurs entablements qui étaient en marbre, mais qui, par le fait, ne sont qu'une décoration, car le monument peut se tenir debout sans ce grand ordre intérieur. La nef centrale A se compose d'une suite de voûtes d'arête, construites suivant la méthode romaine, c'est-à-dire au moyen de demi-cylindres se pénétrant. Ces voûtes sont en blocages, forment ainsi une masse concrète sans élasticité, passive, comme une immense carapace taillée dans un seul bloc. Il fallait cependant maintenir ces voûtes, car leur poids énorme, si elles n'eussent pas été enserrées entre des masses inébranlables, aurait causé des lézardes, et la voûte ainsi brisée fût tombée en morceaux. On a donc élevé les contre-forts B au droit de la retombée des arêtes des voûtes, percés d'une arcade C dans leur partie inférieure, et, entre ces contre-forts, qui s'élèvent jusqu'en D, on a bandé des berceaux également en blocages, couvrant les espaces B B, B' B', portant les terrasses F, et laissant la facilité d'ouvrir des jours G sous la voûte centrale. Le mur I, percé lui-meme de baies, n'est qu'une fermeture ne portant rien. Si nous dépouillons cette construction de tout ce qui est inutile à sa stabilité parfaite, nous pouvons, comme le font voir le plan et la coupe de gauche, réduire les piles intérieures au support vertical H, ouvrir davantage les baies K, supprimer le grand ordre, et bander un arc-boutant L au droit de la poussée des hautes voûtes et reportant cette poussée sur les contre-forts M. C'est la structure vraie."

* * *

In order to provide a reference model, independent from employed materials and building techniques, we suppose masonry materials are not able to resist tensile stress at all. We can imagine a plane region R delimited by a closed regular boundary and filled with this material; the resultant body is a disk of constant thickness *t* supported, in order to eliminate the effect of his weight, on a plane without friction. A generic section divides R into two parts R' and R"; if external forces are applied in an attempt to separate R' and R" in R, they will meet no resistance independently of the dimension of R: the equations of statics are necessary but not sufficient for the equilibrium of this material, exactly symmetric to the behaviour of woven clothes and strings. In other words, the microscopic model of such an ideal no-tension material is similar to a set of particles that react only through contact, and, so the two particles, when they are separated, show no mutual influences of any type. The material considered here is granular, incoherent and incohesive.

Such materials were investigated at the end of eighteenth century, and they are defined half-fluid in order to characterize their capability of having a solid-like, or fluid-like mechanical behaviour, under the effect of external forces [5].

Therefore they belong to a particular class of materials whose behaviour must be described by means of appropriate constitutive equations. Before considering a mathematical formulation of this problem, it is necessary to describe some qualitative and quantitative aspects that can be deduced very easily, and that will be very useful. It is helpful if we do the obvious analogy with non-reacting-to-compression materials that are commonly available and tested: for example wires and clothes are excellent examples of this material. The origin of this suggestion can be found in J. Heyman's research on the History of Mechanics; in particular, in the study of Poleni's problem regarding the stability of the St Peter's dome in Rome [6], [7].

It is appropriate to define first a fundamental aspect of these materials: an ideal test consists of impressing a homogeneous and uniform deformation defined by a shearing deformation γ in the region R, that has no longer external constraint no longer: the boundary B is free. If the material displays standard characteristic, i. e. isotropic, in order to maintain this deformation, it would be necessary to apply a system of boundary actions equivalent to a tangential stress $\tau = G\gamma$, or, in principal directions, $\sigma_{1,2} = \pm \tau$. But such stress are incompatible with the nature of our material, which sustains the strain γ without any internal stresses; in this case, its behaviour is characteristic of ideal fluids, while it can also behave as a solid under appropriate external actions.

Moreover: one of the classical problems of the Theory of Elasticity consist of a circular disk subject to the action of two equal and opposite forces acting on the boundary, along a diameter of the circle. The solution, provided by H. Hertz, has been validated in an excellent way through experiments carried out with a polarized light, and this demonstrates the capacity of the standard material to distribute the stress state inside, due to the external actions. From a mathematical point of view, this aspect is attributed to the ellipticity of the differential equations of the elastic equilibrium; and everybody knows very well that ellipticity produces many advantages in the research of solutions.

But, if we apply two tensile forces to a piece of a woven cloth, for example, when opposite edges of a handkerchief are pulled along a line of fibres, we can observe a completely unusual phenomenon: the only reacting parts are those constituted by the fibres of the cloth to which the external forces are applied; there is no stress in the remainder of the cloth.

Other, more complex experiments can be executed to demonstrate a fundamental principle, intuitively proposed by E.V. Le Duc, and which states that not all the area R is in a state of stress: some of its parts are stress-free.

If, speaking about masonry, we define as "construction" the region R and as "structure" the part of it under stress, we can see that these do not generally coincide; we can also agree that a change in the external actions on the same "construction" will also bring about changes in the reacting "structure".

This is not a surprising fact; it is well known in the statics of masonry columns subjected to eccentric compression: the reacting area, that is the "structure", does not coincide with the whole section of the column, if the pressure centre is external to the central core, but it is defined by a width that depends on the ratio between the bending moment M and the normal force N. Some fractures occur in the non-reacting part; the fact, that is surprising, is that these fractures are disposed almost in parallel to the axis of the column, contrary to what one might expect; so that the remedy consist of the installation of metal rings.

2. THE COMPRESSION TEST OF MASONRY MATERIALS AND THE THEORETICAL FORMULATION FOR NO-TENSION MATERIALS.

In the previous article (see prof. Silvia Briccoli Bati) on the complete description of the mechanical behaviour of masonry materials we saw that in the load-displacement diagram, there is first a multi-linear branch with some discontinuities. Our scope is to give an elementary approach to these discontinuities; they depend on many complex problems as i.e. fracture and instability; this question rises from the impossibility of manifacturing a material or a specimen made of a material rigorously with no strength in tension, because all materials exhibit some resistance to these actions. We thus imagine, according to Résal's recomandation an ideal material whose tension resistance $\overline{F}$ is one-twentieth of the absolute value of the compression resistence, with an elastic-brittle behaviour so that, when this limit is attained, resistance becomes zero. We also suppose an ideal experiment in which we impose assigned displacement, and the corresponding values of loads are determined.

For simulating a sufficiently general situation we analyse a classic problem of Boussinesq, that is an indefinite medium with a concentrated load P; the analysis is limited to a central part of the medium, obtained by a circular section, and reduced to a pin-jointed structure composed by eight bars with the same extentional rigidity K. In the elastic range the relation F,w is linear, F=Rw, (Fig.2a) where R is the rigidity of the complete structure. An elementary calculation give us the equation 4Kw=F represented by a linear branch O,a in Fig.2.

When the value F of the limit resistance is attained in the bar CA, the load F diminishes because the structure transforms itself in the one represented in the Fig.2b; in this new situation the load carried out by the structure are determined by the new equation

$$3Kw=F$$

whose representation in the space F,w is the linear branch O,b.

The fall of the load, from the point 1 to the point 1' is due to a degradation of the rigidity of the initial structure.

For a new increment of the displacement w the point 1' move along the branch b to the point 2, that is the point where the limit value of the resistance are attained in the bars CB; in this situation we reproduce the same effect: the bars CB are destroyed and the new structure is generated.

For this structure the corresponding equilibrium equation of the point C is

$$2Kw=F$$

and it is represented by the linear branch O,c in the Fig.2; at the increment of the desplacement w an increment of the load corresponds, after the fall of the load from the point 2 to the one 2'; this is the reaction of the structure to the imposed displacement (Fig.2c).

If, for example, we reach the point n, and after we reduce the displacement w, the corresponding point of the equilibrium path lies in the branch O,c; this implies that the shaded area in the figure gives the measure of the energy dissipated by the fracture.

The continuous increment of the displacement w is represented by the point n that covers the branch O,c; the limiting point 3 is attained when the absolute value of the axial force in the bar CD becomes $20\,\overline{F}$; for this value the corresponding load is $F=40\overline{F}$.

This value needs a logarithmic scale for a convenient representation; after this point we have a strong modification of the equilibrium path produced by a strong modification of mechanical behaviour (for example from elastic to elasto-plastic).

But the actual knowledge of the relative phenomena does not permit a true representation; to conclude, it is only sufficient observing the percentage of shaded area with respect to the total recoverable energy. In this sense we believe that a masonry theory founded on the hypotheses of no-tension resistance is acceptable.

3. ELEMENTARY PROBLEMS

3.1 THE MASONRY COLUMN

Let us consider a two-dimensional column subjected to two concentrated and opposite forces F (Fig.3) generating in the section a-a a normal force N, a shear T and a bending moment M.

N and M generate the normal stress σ that we suppose to be compressive and trapezoidal over S; the shear force T generates the tangential stress τ over the section S, so that we have the principal stresses

$$\sigma_{1,2} = \frac{\sigma}{2} \pm \left(\frac{\sigma^2}{4} + \tau^2\right)^{\frac{1}{2}}$$

Throughout this paper we shall regard the tensile stress as positive.

Here, since $\sigma < 0$, we have $\sigma_1 > 0$, $\sigma_2 < 0$.

This result is incompatible with the nature of our hypothetical material because it is incapable of withstanding the tensile (positive) stress. No one, however, can deny that the structure is in equilibrium. Therefore we have a paradox that requires an explanation.

In fact, our reasoning is based on the accepted idea that a masonry column is a structure whose axis is determined by its geometric form; this is true for standard materials, but it is completely wrong for masonry materials.

The paradox disappears if we consider that the action line of force F is the axis of the structure "hidden" in the column, and we take the section S normal to this axis: the tangential stress disappears and only normal stresses are generated by the external forces.

This elementary paradox can explain that the fundamental unknown in the statics of masonry is the "resistant structure" that generally does not coincide, with the "construction", as Viollet Le Duc note; but it also suggest some critical observations on the possibility of defining a Dynamic Theory of these materials as a rational foundation of their seismic behaviour: see, for example, the recent paper of J. Moreau [8] on the Dynamic of granular materials; naturally the presence of a weak resistance to tension stresses makes the problem even more complex.

Therefore the conclusion of the column paradox gives an imput idea, valid at least in the static equilibrium, for similar problems as, for example, the masonry arch; other suggestions, to which the author attributes a great importance, result from the analogy whit the materials without resistance to compression stress.

3.2 THE MASONRY ARCH

We suppose that the arch has terminal constraints sliding over fixed lines so that the reactions are ortogonal to that lines (Fig.4).

If "e", "a", "i" are respectively the extrados, the geometrical axis and the intrados of the arch, the problem will consist of determining the strut-line "p", i.e. the funicular polygon connecting reactions Ra, Rb, and the loads P1, P2...; according to this determination, we have only one unknow parameter represented by the position of the strut-line, whose form moreover is totally fixed.

If we examine the generic portion of the arch between two successive loads, we can conclude that this problem is similar of that the column; one of the possible solutions is determined by the condition that the square of the distance between the strut-line and the geometrical axis is the minimum:

$$d^2 = \int_A^B \left[a(x) - p(x) \right]^2 dx = \min$$

over all the funicular polygon entirely contained in the shape of the arch; the resistant structure is determined by the parallel displacements of the determined polygon enclosed into the shape defined by the intrados "i" and the extrados "a".

Example 1

A plate-bande, made by sandstone, has two supports without friction and it is loaded by an uniform pressure q.

The experimental results (Fig 5) show us that the structure is an arch outlined in the assigned shape by the fracture pattern. Theorethical solution is obtained in this way: provided that there is a funicular polygon that connects the loads with the assigned directions of the support reactions and entirely contained in the shape of the plate-bande, the solution, or one of the possible solutions, is determined by all the vertical slidings of this polygon entirely contained in the original shape (Fig 6).

Example 2

Another example both theorethical and experimental, concerns the determination of the reactive structure in a beam manifactured with mortar; the fracture pattern are obtained under the imposed deformation; in the initial step we have an elastic response of the entire beam, but when the resistance to tension is going to the limit, we see the apperance of vertical fractures growing up to the pressed areas.

In this situation, we see the bifurcation of the central fracture in two lines, almost parallel to the horizontal side of the beam. At the end, the complete pattern of the fracture results to be as in Figg.7,8,9, [9].

3.3 EXPERIMENTS MADE ON NO-COMPRESSION MATERIALS

The results are purely qualitative; but their simplicity of execution is remarkable with respect to the informations that they supply; above all it is necessary to point out the dependence of the reacting structure on external loads, inside the assigned shape.

The material used is a common domestic paper for cooking or a rubber sheet; the greater difficulties rose in the realization of the constraints. These are made by parallel cuts in the paper for simulating sliding supports: Figg.10,11,12.

4. THE MASONRY PROBLEM

4.1 STATICS

Now, let us consider a two-dimensional problem of stress, defined by a flat disk or wall, subjected to force systems on the edges, and, if necessary, to body forces: it is kept in equilibrium in order to avoid, for the moment, any consideration of a constraint problem; we cannot know *a priori* which parts should be considered as structure and which is not-resistant.

The wall, the assumed plate, is of masonry construction: in our first consideration, we shall assume that there are only external pressures on the edges, and that there are no body forces. The exclusion of the body forces, at the moment, allows us to focus on particular aspects of the problem.

We shall also assume that external self-equilibrated loads act at the boundary; the existence of a solution requires (Fig.13).

(a)   if $p = p(s)$ are surface forces, n is the outward normal and n^T its transposed vector, we must have

$$n^T p < 0$$

(b)   sketching in R an arbitrary cross section from A to B, and assuming a point Q, so that $\xi_A < \xi_Q < \xi_B$, we must have, for moment   equilibrium

$$\int_A^B \left(-p_\xi \eta + p_\eta \xi\right) ds = 0$$

where p_ξ and p_η are the components of $p(s)$ on the axes $\xi, \eta.$,

The stress tensor σ is defined by three stress components, two normal and a tangential; it is obvious that for the particular material assumed, σ must lead, for consistancy with the no-tension material, to two negative principal stress, or a negative principal stress, and a null one, or both null stresses.

Therefore, we have three possible cases at a generic point:

(1) both principal negative stresses;
(2) a negative principal stress and a null one;
(3) both null principal stresses.

These points make up areas and lines, or sub-regions of R, where the material behaves:

(1) in the standard way: sub-regions R_2;
(2) in anomalous, non standard way: sub-regions R_1;
(3) the material does not exist: sub-regions R_0.

The shape of these sub-regions and their arrangements on R are the basic unknown aspect of the problem: later on we will see how they can sometimes be predetermined; for the moment, we will only point out what happens in the various sub-regions R_i:

- in R_2: the behaviour is defined by the classical theory of elasticity, without further specifications, except those regarding eventual anisotropies of the material;
- in R_0: there is simply no stress;
- in R_1: because of the particular kind of assumed material, it is not clear what will happen; therefore by examining it in detail we can obtain much qualitative information of general validity.

It is necessary to underline another aspect of the problem. This concerns the great difference from what generally occurs in structural analysis, namely that the resisting structure here depends on loads and changes whit them. This means that two different load condition produce two different structures; consequently it is not possible, in general, to add up their effects in order to examine their simultaneous actions: for this reason, the main principles of the linear theory of elasticity can no longer be taken as valid

I think I had better begin by describing an image of a physical model, one which clearly represent its characteristic of no resistance to tensile stress. For this purpose, masonry, whit its many variants, mast be set aside. In the image to be proposed the arrangement of bricks, the mortar quality and the disposition of stones are of no importance; but it is significant that they belong to the same class of materials that does not resist tensile stress.

A model that represent very well all of these media already exists; it is made of an assembly of variable size spheres; and as J. Heyman points out, it has been studied for a long time in research on mansonry statics, from the end of the seventeenth century.

Then, the unilaterality of the material is mathematically expressed by inequalities for the stress components; in Mohr's representation of state of stress the circle has to lie within the half-plane of negative normal stress, and thus with its two intersections with the normal stress axis negative, or null and negative.

If we indicate the stress tensor σ by:

$$\sigma = \begin{bmatrix} \sigma_x & \tau_{xy} \\ \tau_{xy} & \sigma_y \end{bmatrix}$$

the conditions are described by

$$\mathrm{tr}\ \sigma \equiv \sigma_x + \sigma_y \le 0$$

$$\det \sigma \equiv \sigma_x \sigma_y - \tau^2_{xy} \ge 0$$

and, consequentely,

in R_0 : tr $\sigma = 0$ det $\sigma = 0$

in R_1 : tr $\sigma < 0$ det $\sigma = 0$

in R_2 : tr $\sigma < 0$ det $\sigma > 0$.

Although it is not possible to give the complete list of the conditons that must be fulfilled, the following example (Fig.14,15) has been set out to explain other singularities of this problem.

Let us consider a rectangular disk; external pressures $p(x)$ and $q(x)$ satisfy the equilibrium equations. Let us begin to demonstrate that the sub-regions a-b-e-f and c-d-h-g are R_0. In fact, let us sketch the cross section i-j and consider the equilibrium of the left part.

For equilibrium to occur in the x direction, there must be

$$\int_i^j \sigma_x \, dy = 0$$

However, given $\sigma_x \leq 0$, we must have, on the entire section i-j, $\sigma_x = 0$
From the condiction $\det \sigma \geq 0$, it follows that $\tau_{xy} = 0$.
It may easily be proved that $\sigma_y = 0$ with the aid of section k-l.
Next, we must discuss the equilibrium of the strip b-c-f-g (Fig.16).
The cross section o-r is close to b-f : for equilibrium in the x direction we have, as in the preceding analysis, $\sigma_x = 0$ and consequently $\tau_{xy} = 0$.
This shows that equilibrium in the y direction can be satisfied if, and only if, the distribution of pressure is such that $p(x) = q(x)$ on the two opposite sides.

This problem suggest further remarks about the regularity of the boundary data. The pressures $p(x)$ and $q(x)$ may be described by functions with concentrated or distributed discontinuities, as Fig.17 shows, provided that equilibrium is verified within each strip of material; the stress are: $\sigma_x = \tau_{xy} = 0$, $\sigma_y = p(x) = q(x)$. In fact, such a solution satisfies the differential equations of equilibrium, the boundary conditions and the conditions $\det \sigma = 0$. Obviously, the compatibility equations in term of stress, $\Delta \left(\sigma_x + \sigma_y \right) = 0$ for isotropic materials, is violated. They are verified only if σ_y -and also $p(x)$, $q(x)$- are linear functions of x: that is the case shown in Fig.18. We shall deal later with such problems, in order to discuss their kinematical aspects.

Similar remarks are valid for load condictions for the problem of a concentrated force acting at the tip of the wedge. If is important to emphasize that in such cases the lines, along which the external thrusts act, are the trajectories of the principal stresses and potential fracture lines.

Such simple remarks can highlight a fundamental aspect of the problem: the failure of De Saint-Venant's postulate and the inability of the medium to spread internally the stresses due to the external actions.

Another question arises when we examine the new problem of a disk which is subjected to a generic self-equilibrating load, a condition which can be generated with an external pressure on the boundary, for example see Fig.19a.

According to the previous results, it is easy to exclude, from this problem, the Ro sub-regions, determined by the lines connecting points A, B...F and the boundary of the disk. It is obvious that there are other sub-regions R_0.

From the equilibrium consideration we deduce that the resultants P of the pression p, S of s and T of t meet at a common point O; let be y the action line of P and x its orthogonal section which divides the disk in two parts the disk.

The internal resultant stresses on this section are σ_y and τ_{xy}. Because of the equilibrium in the x-direction, we must have

$$\int_{x_1}^{x_2} \tau_{xy}\, dx = 0$$

The distribution of τ_{xy} over the section is unknown, but there is a point X: $X_1 < X < X_2$, so that

$$\int_{x_1}^{x} \tau_{xy}\, dx = S_x$$

where S_X represents the component over x of the resultant S.

So, from this point we trace a new section x-$\overline{x}$ parallel to y; the equilibrium in the x-direction of the left side of the disk requires that (Fig.19a,19b):

$$\int_{\overline{x}}^{x} \sigma_x\, dy = 0$$

i.e., from the no-tension material, $\sigma_x = 0$ for every point of this section and, consequently, also $\tau_{xy} = 0$.

The new section F-$\overline{F}$ determines a new part of the disk, having two boundary lines, that is $F\overline{X}$ and $\overline{X}\overline{F}$, both of them free of tensions: all the part $F\overline{F}\overline{X}$ is in a no-tension state. Therefore, our problem becomes the one represented in Fig.20 , in which it is easy to recognize the R_0 and R_1 sub-regions; the nucleus of the entire disk is candidated to be a sub-region R_2. According to a theorethical point of view, this is a very complex problem because in general, the external pressions although self-equilibrating, are incompatible with the Cauchy's condition at the corners U, V, Z, because we have, for each corner, four conditions of equilibrium in three unknown components of stress. This suggest the idea of a new approach, not presenter here, based on Cosserat's hypotheses.

4.2 KINEMATICS

From a kinematical point of view we may develop some preliminary remarks about the strains within the sub-regions of R (for sake of simplicity, a plane problem will be considered again). If a region R is made up by elastic standard material, the constraints preventing rigid-body motions are sufficient to allow a determination of the displacement functions.

In our problem, on the other hand, the non-solid material does not prevent the points of the unconstrained region R from moving, with the only condition that any two particles cannot move to the same position in the final configuration. In other hand, in the natural state all displacements which cause a separation or a slip between particles are possible.

But such displacements, from a physical point of view, cannot exist without an external cause. Therefore, the positions of the material particles are here uniquely defined, in the natural state, by the corresponding points in the space, as if the material were solid. Every possible compatible displacement is assumed to be zero in the natural configuration; without such an assumption it is clearly impossible to define the region R as an object of investigation.

However, we have already stated that R is divided into sub-regions, as a consequence of the stresses generated by the external forces. A so-called 'regular' behaviour is defined in R_2 when, given the stress tensor σ, the corresponding displacements can be uniquely determined without 'zero-stress' strains δ. In a sub-region R_0 with null stress everywhere, the determination of the displacements has little meaning, given the previously stated assumption in the natural configuration.

On the other hand, in the sub-regions R_1, where the stress is determined by the further condition $\det \sigma = 0$, the evaluation of displacements is particularly significant, since the strain compatibility condition

$$\frac{\partial^2 \varepsilon_x}{\partial y^2} + \frac{\partial^2 \varepsilon_y}{\partial x^2} = \frac{\partial^2 \gamma_{xy}}{\partial x \partial y}$$

is generally not satisfied. In the previous examples, where $\sigma_y = f(x)$, $\sigma_x = \tau_{xy} = 0$, the equation is satisfied in R_1 if and only if $f(x)$ is a linear function. That means that there are no given displacements u, v, such that the differential system

$$\frac{\partial u}{\partial x} = \varepsilon_x; \quad \frac{\partial u}{\partial y} + \frac{\partial v}{\partial x} = \gamma_{xy}; \quad \frac{\partial v}{\partial y} = \varepsilon_y$$

can be integrated. We could therefore introduce in the sub-regions R_1 the so-called 'zero-stress' strains δ,

$$\delta = \begin{bmatrix} \delta_x & \delta_{xy} \\ \delta_{xy} & \delta_y \end{bmatrix}$$

and rewrite the preceding system in the form

$$\frac{\partial u}{\partial x} = \varepsilon_x + \delta_x; \quad \frac{\partial u}{\partial y} + \frac{\partial v}{\partial x} = \gamma_{xy} + 2\delta_{xy}; \quad \frac{\partial v}{\partial y} = \varepsilon_y + \delta_y.$$

The components of δ depend on the stress σ, since they must satisfy the fundamental orthogonality condition

$$\sigma\delta = \sigma_x \delta_x + 2\tau_{xy}\delta_{xy} + \sigma_y \delta_y = 0$$

That is, the constrained energy density must be zero.

Nevertheless, such a condition is not sufficient for the evaluation of the displacements u and v, since there are now four equations and five unknowns. Such kinematical indeterminacy, which is an intrinsic property of the assumed material, may be explained as follows: since there is one principal stress in R_1, two different anelastic strains are possible, namely a positive strain along the zero-stress direction and then a dislocation (slip) along the principal directions. Since the class of positive strains has already been investigated ([10], [11], [12]), we shall deal here mainly with the slip strains (dislocation).

4.3 THE TRAJECTORIES OF THE PRINCIPAL STRESSES IN R_1

Other information can be obtained for sub-region R_1; for this sub-region the equilibrium equations must be added to the condition $\det\sigma = 0$:

$$\frac{\partial\sigma_x}{\partial x} + \frac{\partial\tau_{xy}}{\partial y} + \rho_x = 0; \quad \frac{\partial\tau_{xy}}{\partial x} + \frac{\partial\sigma_y}{\partial y} + \rho_y = 0$$

$$\sigma_x\sigma_y - \tau_{xy}^2 = 0$$

Thus, the problem of the stress analysis is statically determinate according to H. Geiringer [13] because we have three equations and three unknowns; but we do not know the boundary of R_1 and the corresponding conditions; However, it is possible to obtain some general information that can justify the previous results. The mechanical characteristics of the material are not necessary; that is the elasto-kinematics equations can be used only if we want to know the displacements.

The fundamental aspect of this problem concerns the trajectories of the principal stresses whose equations can be obtained using the polar representation of stress tensor:

$$\sigma_x = \sigma\,\text{sen}^2\,\varphi; \quad \tau_{xy} = -\sigma\,\text{sen}\,\varphi\cos\varphi; \quad \sigma_y = \sigma\cos^2\varphi$$

in which $\sigma = \sigma_2$ and φ is the angle between the first principal direction and the x axis.

The condition $\det\sigma = 0$ is automatically satisfied; the equilibrium equations become

$$\frac{\partial\sigma}{\partial x}\,\mathrm{sen}^2\,\varphi-\frac{\partial\sigma}{\partial y}\,\mathrm{sen}\,\varphi\cos\varphi+\sigma\left(\mathrm{sen}\,2\varphi\frac{\partial\varphi}{\partial x}-\cos2\varphi\frac{\partial\varphi}{\partial y}\right)+\rho_x=0$$

$$-\frac{\partial\sigma}{\partial x}\,\mathrm{sen}\,\varphi\cos\varphi+\frac{\partial\sigma}{\partial y}\,\cos^2\,\varphi-\sigma\left(\cos2\varphi\frac{\partial\varphi}{\partial x}-\mathrm{sen}\,2\varphi\frac{\partial\varphi}{\partial y}\right)+\rho_y=0$$

By a simple linear combination, we have

$$\sigma\left[\frac{\partial\varphi}{\partial x}\,\mathrm{sen}\,\varphi-\frac{\partial\varphi}{\partial y}\,\cos\varphi\right]=-\rho_x\cos\varphi-\rho_y\,\mathrm{sen}\,\varphi$$

If we introduce ψ, that is the angle between the direction of principal stress σ and the x axis, and $\tan\psi=z$, we will have

$$\sigma=\frac{\left(1+z^2\right)\left[z\rho_x-\rho_y\right]}{\left(\dfrac{\partial z}{\partial x}\right)+z\left(\dfrac{\partial z}{\partial y}\right)}$$

If we suppose $\rho_x=\rho_y=0$, the significative solutions of the equation

$$\frac{\partial z}{\partial x}+z\frac{\partial z}{\partial y}=0$$

are:

$$\begin{cases}z=const\\[4pt]z=\dfrac{y}{x}\end{cases}$$

and justify the elementary results obtained previously, concerning the equilibrium of masonry walls loaded on the boundary and without gravity forces.

The solution $z=const$ represents the case of parallel trajectories of the principal stresses; the solution $z=y/x$ represents the case of radial trajectories.

We can note the analogy between the previous equations and those regarding the statics of granular media [14].

4.4 SUB-REGIONS AND CONSTITUTIVE EQUATIONS

The state of stress in the sub-regions of the domain R have a corresponding state of strain; these are completely elastic in R_2, elastic and anelastic in R_1, anelastic in R_0.

The stress-strain relations are:

$$\varepsilon(u) = K\sigma + \delta$$

in which $\varepsilon(u)$ are strains determined by displacement u; K is the constitutive matrix of the material and the term δ represents a tensor of anelastic deformations necessary for the integrability of the elasto-kinematic equations in sub-regions R_1 and R_0, that is for the determination of the displacement components.

The orthogonality condition $\sigma\delta=0$ may now be stated, in the principal representation of σ, as

$$\sigma = \begin{bmatrix} 0 & 0 \\ 0 & \sigma_2 \end{bmatrix}$$

$$\delta = \begin{bmatrix} \alpha & \gamma \\ \gamma & 0 \end{bmatrix}$$

The condition

$$\alpha = \alpha(x,y) \leq 0$$

assures positive strains along the directions of tensile stresses; the function $\gamma = \gamma(x, y)$ does not have, in general, any conditions on its sign; in the sub-regions R_0 the strain is completely anelastic with the same limitation for the scalar function α.

To summarize, these three cases can be described by the first and second invariants of σ and δ:

$$R_2 : \begin{cases} tr\sigma < 0 \\ \det \sigma > 0 \end{cases} \rightarrow \begin{cases} tr\delta = 0 \\ \det \delta = 0 \end{cases}$$

$$R_1 : \begin{cases} tr\sigma < 0 \\ \det \sigma = 0 \end{cases} \rightarrow \begin{cases} tr\delta > 0 \\ \det \delta < 0 \end{cases}$$

$$R_0 : \begin{cases} tr\sigma = 0 \\ \det \sigma = 0 \end{cases} \rightarrow \begin{cases} tr\delta \geq 0 \\ \det \delta > 0 \end{cases}$$

The orthogonality condition $\sigma\delta = 0$ in the sub-region R_1 may be expressed, in coordinate system $(0,x,y)$, by keeping the separation between the normal and shearing components, through the same components of the σ tensor:

$$\delta = \alpha \begin{bmatrix} \sigma_x & -\tau_{xy} \\ -\tau_{xy} & \sigma_x \end{bmatrix} + \gamma \begin{bmatrix} -\tau_{xy} & \dfrac{\sigma_x - \sigma_y}{2} \\ \dfrac{\sigma_x - \sigma_y}{2} & \tau_{xy} \end{bmatrix} = \alpha\sigma^A + \gamma\sigma^C$$

The tensors σ^A and σ^C are obtained by linear tranformations of σ.

$$\sigma^A = \begin{bmatrix} 0 & 1 \\ -1 & 0 \end{bmatrix}\begin{bmatrix} \sigma_x & \tau_{xy} \\ \tau_{xy} & \sigma_y \end{bmatrix}\begin{bmatrix} 0 & -1 \\ 1 & 0 \end{bmatrix}$$

$$\sigma^C = \frac{1}{2}\begin{bmatrix} 1 & -1 \\ 1 & 1 \end{bmatrix}\sigma\begin{bmatrix} 1 & 1 \\ -1 & 1 \end{bmatrix} - \begin{bmatrix} \dfrac{\sigma_x + \sigma_y}{2} & 0 \\ 0 & \dfrac{\sigma_x + \sigma_y}{2} \end{bmatrix}$$

In some cases it may be convenient to express the tensor $\gamma\sigma^C$ by the angle 2φ through the well-known relations

$$\tan 2\varphi = \frac{2\tau_{xy}}{\sigma_x - \sigma_y}$$

$$\mathrm{sen}\,2\varphi = \frac{\pm 2\tau_{xy}}{\sqrt{\left(\sigma_x - \sigma_y\right)^2 + 4\tau^2}}$$

$$\cos 2\varphi = \frac{\pm\left(\sigma_x - \sigma_y\right)}{\sqrt{\left(\sigma_x - \sigma_y\right)^2 + 4\tau^2}}$$

We will obtain

$$\gamma\sigma^c = \gamma \begin{bmatrix} -\mathrm{sen}\,2\varphi & \cos 2\varphi \\ \cos 2\varphi & \mathrm{sen}\,2\varphi \end{bmatrix}$$

From the previous expression we can find an evident analogy between the statics of masonry and the theory of plastic flow; the difference between the two problems lies in the fact that in our case the shearing (anelastic) strain takes place on the principal directions of stress without an energy varation.

The two tensors σ^A and σ^C are orthogonal to each other and to the stress tensor σ, which lies on the cone of equation $\det\sigma = 0$, in the stress space.

The strain tensor $\alpha\sigma^A$ lies on the normal to the same cone; since σ^C is orthogonal to σ^A, we can state that σ^C must lie in a plane tangential to the cone; its direction is also determined by its own orthogonality to σ.

Finally the stress-strain relations are

$$in R_2 : \varepsilon(u) = K\sigma$$

$$in R_1 : \varepsilon(u) = K\sigma + \alpha\sigma^A + \gamma\sigma^C; \quad \alpha \le 0$$

$$in R_0 : \varepsilon(u) = \delta$$

Let us suppose that, for an assigned problem, we have determined the stress tensor; thus, the elastic-kinematic equations are

$$\begin{bmatrix} u_{,x} \\ u_{,y}+v_{,x} \\ v_{,y} \end{bmatrix} = \left(\begin{bmatrix} k_{11} & k_{12} & k_{13} \\ k_{12} & k_{22} & k_{23} \\ k_{13} & k_{23} & k_{33} \end{bmatrix} + \begin{bmatrix} 0 & 0 & \alpha \\ 0 & -2\alpha & 0 \\ \alpha & 0 & 0 \end{bmatrix} + \begin{bmatrix} 0 & -2\gamma & 0 \\ 2\gamma & 0 & -2\gamma \\ 0 & 2\gamma & 0 \end{bmatrix} \right) \begin{bmatrix} \sigma_x \\ \tau_{xy} \\ \sigma_y \end{bmatrix}$$

in which the scalar functions α and γ allow the integration of the system. The integrability condition is expressed by the following equation (note that the 'comma' is a differential operator), in which $K_i{}^T$ are the rows of the elastic-constitutive matrix and σ is the stress tensor regarded as a vector:

$$\left(k_1^T \sigma + \alpha \sigma_y - 2\gamma \tau_{xy} \right)_{,yy} +$$

$$+\left(k_1^T \sigma + \alpha \sigma_y - 2\gamma \tau_{xy} \right)_{,yy} + \left(k_3^T \sigma + \alpha \sigma_x - 2\gamma \tau_{xy} \right)_{,xx} = \left(k_2^T \sigma + 2\alpha \tau_{xy} + 2\gamma \left(\sigma_x - \sigma_y \right) \right)_{,xy}$$

This equation is necessary but not sufficient to determine the displacement components u and v, because of the intrinsic lability of the assumed material.

We can separate the anelastic strains $\alpha \sigma^A$ from the $\gamma \sigma^C$ ones; it is also obviously possible to transfer the deformation of one to the other without changing the state of stress because of the internal lability of the material; in several problems, in order to obtain a solution in terms of displacements, it is necessary to introduce a principle that allows one to obtain a solution that minimizes, for example, an integral quadratic from of the displacements.

It is necessary to remark that non-regular load conditions at the boundary, i.e. discontinuity on the pressure function, lead to discontinuous displacement function.

4.5 NUMERICAL EXAMPLES

(1) The rectangular masonry wall shown in Fig.21 has constant thickness $s=1$; the material is isotropic with $E=1$ and $\upsilon=0$. The boundary conditions are

$$x = \pm a \begin{cases} \sigma_x = 0 \\ \tau_{xy} = 0 \end{cases}$$

$$y = \pm b \begin{cases} \tau_{xy} = 0 \\ \sigma_y = -\dfrac{p}{a^2}\left(a^2 - x^2 \right) \end{cases}$$

The solution in the whole region R, in this case R_1, is:

$$\sigma_x = \tau_{xy} = 0; \quad \sigma_y = -\frac{p}{a^2}\left(a^2 - x^2\right) \equiv \sigma_2; \quad \sigma_1 = 0$$

The trajectories of the principal stress are parallel to the y direction; the kinematical analysis is defined by the differential equations:

$$\frac{\partial u}{\partial x} = 0; \quad \frac{\partial u}{\partial y} + \frac{\partial v}{\partial x} = 0; \quad \frac{\partial v}{\partial y} = -\frac{p}{a^2}\left(a^2 - x^2\right)$$

These are not integrable; therefore we can take into account the inelastic deformations that are compatible with the state of stress and permitted by the material; here, we consider anelastic deformations that correspond to half-fluid behaviour:

$$\delta = \gamma \begin{bmatrix} -2\tau_{xy} & \sigma_x - \sigma_y \\ \sigma_x - \sigma_y & 2\tau_{xy} \end{bmatrix}$$

The elasto-kinematic equations therefore become

$$\frac{\partial u}{\partial x} = 0; \quad \frac{\partial u}{\partial y} + \frac{\partial v}{\partial x} = 2\gamma\frac{p}{a^2}\left(a^2 - x^2\right)$$

$$\frac{\partial v}{\partial y} = -\frac{p}{a^2}\left(a^2 - x^2\right)$$

The scalar function $\gamma = \gamma(x, y)$ can be determined by the condition of integrability; an elementary solution that eliminates the intrinsic indeterminacy in terms of displacements may be obtained with the condition

$$x = 0; \quad y = 0$$
$$u = 0; \quad v = 0$$

Then, the solution of the problem is

$$u = 0; \quad v = -\frac{p}{a^2}\left(a^2 - x^2\right)y$$

characterized by the anelastic shearing strain

$$2\delta_{xy} = 2p\frac{xy}{a^2}$$

Let us consider again the elastic-kinematic equations, where we set:

$$2\gamma\frac{p}{a^2}\left(a^2 - x^2\right) = \Gamma$$

The integrability conditions gives the equation

$$\Gamma_{xy} - \frac{p}{a^2} = 0$$

whose solution is

$$\Gamma = \frac{p}{a^2}xy + f(x) + g(y)$$

and then

$$\gamma = \frac{a^2}{2p\left(a^2 - x^2\right)}\left[\frac{p}{a^2}xy + f(x) + g(y)\right]$$

The functions $f(x)$ and $g(y)$ cannot be determined without a further condition.

However, we have to notice that in the deformed state, characterized by the elastic and anelastic components, further zero-stress strain are admissible, complying with the assigned displacements on the axes x and y, and satisfying the system

$$\frac{\partial u}{\partial x} = \delta_x \geq 0; \quad \frac{\partial u}{\partial y} + \frac{\partial v}{\partial x} = 2\delta_{xy}; \quad \frac{\partial v}{\partial y} = 0$$

Such strains, and the corresponding displacements, cannot be removed, except by the following assumption: we shall consider only the zero-stress strains which are necessary for the compatibility of the system and we shall neglect all the strains which are already possible in the natural state. Such an assumption may be carried out, e.g., by minimizing the euclidean norm of the displacement vector.

(2) Rectangular panel with dead load, supported at the base (Fig.22). It is easy to prove that the whole region is R_1. Let us denote by ρ the unit weight of the material; the static solution is

$$\sigma_x = \tau_{xy} = 0; \sigma_y = -\rho(h-y)$$

If we assume E=1, then the elastic-kinematic equations are

$$u_{,x} = v\rho(h-y)$$
$$u_{,y} + v_{,x} = 0$$
$$v_{,y} = -\rho(h-y)$$

The last equation gives

$$v = -\rho\left(hy - \frac{y^2}{2}\right) + f(x)$$

Since $v=0$ for $y=0 \rightarrow f(x)=0$ the first equation gives

$$u = v\rho(h-y)x + g(y)$$

Let us set $u=0$ for $x=0 \rightarrow g(y)=0$.
The second equation is not satisfied, because it gives

$$u_{,y} + v_{,x} = -v\rho x \neq 0$$

On the other hand, this is the term to be added-by use of the strain $\gamma\sigma^C$ -in order to integrate the system.

(3)Another example, chosen to show the relation between the fracture lines and the trajectories of the principal stresses, is the following: a rectangular wall, made by a material without tension strength, whose self weight has mass density $\rho=1$, is subjected at the boundary to the conditions (all quantities are adimensionalized):

$$x = \pm 1 : \begin{cases} \sigma_x = -1 \\ \tau_{xy} = x \end{cases}$$

$$\begin{matrix} y = 0 \\ y = 1 \end{matrix} : \begin{cases} \sigma_y = -x^2 \\ \tau_{xy} = x \end{cases}$$

We easily see that the stress tensor

$$\sigma = \begin{bmatrix} -1 & x \\ x & -x^2 \end{bmatrix},$$

verifies all the equations of the problem, together with $\det\sigma=0$ that ensures $\sigma_1 = 0; \sigma_2 = -\left(1+x^2\right)$ in the assigned domain: all the region is R₁.

The principal directions of stress are defined by

$$\tan 2\varphi = \frac{2\tau_{xy}}{\sigma_x - \sigma_y} = \frac{2x}{x^2 - 1} = K$$

and they generate in the plane xy two families of lines $y_1(x)$ and $y_2(x)$

$$y'_{1,2} = \frac{-1 \pm \sqrt{1+K^2}}{K}$$

that, after some manipulations, gives (Fig.23):

$$\sigma_1 = 0; \quad y_1 = \lg|x| + C_1; \quad y_2 = -\frac{x^2}{2} + C_2; \quad \sigma_2 = -(1+x^2)$$

Obviously any elementary arch comprised between two consecutive lines of stress is in equilibrium under the action of the self weight and the external action σ_2 at the terminal sections; note that these lines are parabolic in accordance with the uniformity of the loads that they substain (the distance between two consecutive lines are constant).

The presence of the fracture is evident because the elasto-kinematic equation, written for example in the case of isotropic material with E=1 and v=0, are not integrable:

$$\begin{cases} u,_x = -1 \\ u,_y + v,_x = 2x \\ v,_y = -x^2 \end{cases}$$

$$\Delta\left(\sigma_x + \sigma_y\right) = -2 \neq 0$$

It is evident that the fracture lines may be only along the same direction of the transmission of the stress, that is in the y_2 directions.

5. THE FORMULATION OF THE PROBLEM

5.1 DISTORSIONS AS FRACTURES

The general formulation of the problem of plane equilibrium of masonry structures made by theoretically unilateral material, follows the classical formulation that derives from the Haar-Kármán principle [15].

As is well known, we must minimize the complementary energy under the following conditions:

$$div\sigma + \rho = 0 \ (internal - equilibrium)$$

$$n^T\sigma = p \ (boundary - static - conditions)$$

$$\left.\begin{array}{c} tr\sigma \leq 0 \\ \det\sigma \geq 0 \end{array}\right\} material - conditions$$

The first fundamental problem that arises concerns the existence of the solution since, as is known, in some instances there is no solution; from a mathematical point of view this problem is the minimization of a quadratic functional with non-linear condictions.

I shall propose an alternative procedure founded on a principle stated by G. Colonnetti in 1920 and later applied by Prager and Hodge [16] for problems related to plasticity theory. Briefly, the principle states that, in order to modify the state of stress in a given statically indeterminate structure, as many distortion components may be applied, as there are redundancies in that same structure.

In this paper I shall prove two properties of the applied distorsion, that are of great importance in view of a numerical solution. The matrix formulation of the problem will be:

$$D\sigma + \rho = 0$$
$$D^T s = K\sigma \quad \text{in } R$$
$$N\sigma = p \quad \text{on } B_1$$
$$s = 0 \quad \text{on } B_2$$

where R is the region with regular boundary $B = B_1 + B_2$; there are assigned loads p on B_1 and constraints with displacement $s = 0$ on B_2. The σ tensor is written as a vector, with components σ_x, τ_{xy}, σ_y (Fig.24); the matrix D of differential operators and N of direction cosines acting on σ are:

$$D = \begin{bmatrix} \dfrac{\partial}{\partial x} & \dfrac{\partial}{\partial y} & 0 \\[2ex] 0 & \dfrac{\partial}{\partial x} & \dfrac{\partial}{\partial y} \end{bmatrix}$$

$$N = \begin{bmatrix} \alpha_x & \alpha_y & 0 \\ 0 & \alpha_x & \alpha_y \end{bmatrix}$$

It is well known that the system has a unique solution $\sigma = \sigma^e$; $s = s^e$ (the superscript denotes the elastic standard solution). Let us consider now in R a continuous distribution of distortion δ; such a tensor can be represented as a vector with components δ_x, $2\delta_{xy}$, δ_y.

The problem is now governed by the equations:

$$D\sigma = 0$$
$$D^T s = K\sigma + \delta \quad \text{in } R$$
$$N\sigma = 0 \quad \text{on } B_1$$
$$s = 0 \quad \text{on } B_2$$

Let δ be a distorsion $\delta = \delta^0 = K\sigma^0$ such that $D\sigma^0 = 0$ in R and $N\sigma^0 = 0$ on B_1; the elastic-kinematic equation gives: $\sigma = K^{-1} D^T s - \sigma^0$. Then, by substitution we obtain:

$$DK^{-1}D^T s = 0 \qquad \text{in } R$$
$$NK^{-1}D^T s = 0 \qquad \text{on } B_1$$
$$s = 0 \qquad \text{on } B_2$$

Since K is positive-definite we have thus obtained a system of elliptic differential equations, and the related maximum principle allows us to state that the unique solution is $s = 0$ and, consequently, $\sigma = \sigma^0$. Accordingly, the distortion gives rise to self-balanced stresses and zero displacements.

Now, let $\delta = \delta = D^T(s)$ with $s = 0$ on B_2 ; from the elastic-kinematic equations we have $\sigma = K^{-1} D^T (s-s)$; then by the equilibrium equations we obtain the unique solution $s=s$ and $\sigma = 0$.

Therefore, the distorsion δ.produce displacement s and zerostresses. We shall explicitly remark that σ^0 is the solution of the homogeneous equilibrium equations and may be used to solve the unilateral constraint problem.

This can be accomplished, for example, starting from the solution σ^e (standard material) and then correcting those values by superimposing suitable distorsion.

Let use notice also that the Haar-Kármán solution implicitily uses the distorsion δ^0; the solution, if it exists, is unique and satisfies the equilibrium equations and the stress condition. On the other hand, the kinematical problem -i.e. the determination of strains and fractures- is still undetermined, due to the intrinsic kinematical undetermination of the material. From such point of view, it seems reasonable to eliminate from the set of possible displacements all those which might occur in the natural configuration. The same may be accomplished by minimizing a suitable norm of the displacement vector.

5.2 NUMERICAL EXAMPLES

The first example that we shall present is a rough idealization of the problem of square wall, subject to a concentrated load, as in Fig.25. It will be solved by FEM with only two triangular elements.

The complete discussion of this example is useful for the understanding of the following result, obtained by a refined discretization.

The equilibrium equations of nodes give (for typographical reason we have posed $(\sigma_x)_i = X_i$; $(\sigma_y)_i = Y_i$; $(\tau_{xy})_i = Z_i$):

$$X_1 = Z_2; \quad Y_1 = Z_2 - 2F; \quad Z_1 = -Z_2;$$
$$X_2 = -Z_2; \quad Y_2 = -Z_2; \quad Z_2 = Z_2$$

Hence we obtain the complementary energy with $E = 1$, $\upsilon = 0$;

$$C = \tfrac{1}{2}\left[8Z_2^2 - 4Z_2 F + 4F^2\right]$$

Its minimum is attained by: $Z_2 = 0.25\ F$. From this, in the two elements, the stress tensors resulting for the standard case are

$$\sigma^{(1)} = F\begin{bmatrix} 0.25 & -0.25 \\ -0.25 & -1.75 \end{bmatrix}$$

$$\sigma^{(2)} = F\begin{bmatrix} -0.25 & 0.25 \\ 0.25 & -0.25 \end{bmatrix}$$

This state of stress is not admissible in element 1, since it gives a positive principal stress (note that $\det \sigma < 0$). Therefore, it is necessary to introduce the conditions on the unilaterality of the material; for the first element, we have:

$$\det \sigma^{(1)} = Z_2(Z_2 - 2F) - Z_2^2 \geq 0$$

$$tr\sigma^{(1)} = Z_2 - 2F \leq 0$$

while for the second, all conditions are satisfied if $Z_2 \geq 0$.

But the condition $\det \sigma \geq 0$ in the first element requires $Z_2 \leq 0$; therefore it has to be $Z_2 = 0$, and then we have the stress tensors

$$\sigma^{(1)} = F\begin{bmatrix} 0 & 0 \\ 0 & -2 \end{bmatrix}$$

$$\sigma^{(2)} = F\begin{bmatrix} 0 & 0 \\ 0 & 0 \end{bmatrix}$$

Next, let us determine the displacements. The elasto-kinematic equations, written with the obtained stresses, would be ($E = 1;\ \upsilon = 0;\ F = 1$):

$$u_2 = 0;\ u_3 = 0;\ v_3 = -2$$

$$-u_3 + u_4 = 0;\ -u_2 - v_3 + u_4 + v_4 = 0;\ v_4 = 0$$

It is easy to show that the system is incompatible: note that it has five unknowns and six equations. This indicates that the static conditioned solutions cannot be derived by a displacement field; so, it is necessary to add a system of zero-stress strains.

We will solve the problem by extensional deformations of the class $\alpha\sigma^A$ and by shearing strains of the class $\gamma\sigma^C$.

In the first case, we have:

$$u_2 = -2\lambda; \quad u_3 = 0; \quad v_3 = -2; \quad -u_3 + u_4 = \alpha$$
$$-v_3 - u_2 + u_4 + v_4 = 2\beta; \quad v_4 = \delta$$

The parameters α, β, δ, λ, must verify the so-called Lanczos identity:

$$-2\lambda - \alpha + 2\beta - \delta - 2 = 0$$

with

$$\alpha\delta - \beta^2 \geq 0; \quad \alpha + \delta \geq 0; \quad \lambda \leq 0$$

But this condition cannot uniquely determine the displacements u and v; so we will introduce as a further condition the minimization of the norm $|D|^2$ of the displacement vector:

$$\min|D|^2 = \sum_{i=1}^{n}\left(u_i^2 + v_i^2\right)$$

In this way we will obtain the solution:

$$u_2 = 1; \quad u_3 = 0; \quad v_3 = -2; \quad u_4 = 1; \quad v_4 = 1$$

In the second case of deformation (Fig.26), we will have the system:

$$u_2 = 0; \quad u_3 = 4\gamma; \quad v_3 = -2; \quad -u_3 + v_4 = -\beta;$$
$$-u_2 - v_3 + u_4 + v_4 = 2\delta; \quad v_4 = \beta$$

to which we have to add the Lanczos condition

$$-4\gamma - 2 + \beta + 2\delta - \beta = 9$$

The minimum of $|D|^2$ gives the solution:

$$u_2 = u_3 = u_4 = v_4 = 0; \quad v_3 = -2$$

We can note that in element 2 we have a pure shear strain without tension, while in the first case we had extensional strains in the element 1. These last may be also regarded as continuous fracture distribution.

Here it is not possible to give an exhaustive description of the method of numerical resolution, which must be used for the solution of problems that are characterized by a great number of unknowns.

In general, however, the method of investigation is based on imposed distorsions with a numerical step-by-step procedure. I will present the solution of the problem of a square wall, simply supported at the base and vertically loaded at the upper midpoint (Fig.27), which has been obtained with this procedure.

It is important to note that the solution of the problem is easily understandable and that this solution is more easily obtained by rigorous formulation than by numerical procedure. The last example deals with the Duomo façade at Siena, erected in the fourteenth century. It is interesting to remark that in the elastic solution there are positive stress of a negligible order of magnitude, if compared with the smaller negative stresses. That explains, perhaps, many of those intuitive solutions founded by ancient builders. The results of the FEM computation are shown in Fig.28 for standard material (left) and for masonry material (right).

6. SOME REMARKS ON SEISMIC BEHAVIOUR OF MASONRY STRUCTURES AND THE PROBLEM OF THEIR CONSOLIDATION. SOME QUESTIONS CONCERNING THE AUTHENTICITY IN THE CONSERVATION OF MONUMENTS. TWO REAL CONSERVATION CASES.

The recent analysis of masonry structures has been mainly confined within the limits monotonic loads. However, the most important aspects of masonry analysis are relative to seismic actions and to the need of preserving hystorical buildings and monuments from earthqueke damage.

In such cases, cyclic loads actually occur and dynamic effects ought to be treated with consideration. Unfortunately, masonry behaviour can be described as brittle, as long as fractures or crush occur without a significant phase of energy dissipation.

In other words, the energy dissipation characteristic are typical of standard materials (e.g. mild steel and in some cases also reinforced concrete) and this can lead some designers to provide reinforcements within the masonry mass (e.g. steel bars) as a mean to increase ductility of the overall structure.

Moreover, when a fracture opens, there is a risk of successive interactions between the isolated sub-structures, as long as the fracture width is far is far less than the foreseen relative displacement, during the subsequent seismic loading. In fact, every seismic regulation or structural code explicitily provides sufficiently wide joints, in order to avoid structural inter-actions during a dynamic loading.

We can conclude that every fracture pattern, although only slightly dangerous when only dead loads are applied, can be a great risk whenever cyclic-dynamic loads occur. Such a consideration can justify the method of improving the mechanical characteristic of masonry, with the aim to obtain a fairly homogeneous, ductile, tensile-resistant, quasi-standard material.

This is the philosophy that guided many interventions in the last years. Recently some attempts have been re-proposed as consolidation methods deduced by the "rules of well building"; but there is to observe that these suggestions have not yet receved the prove of their efficacy because no earthquakes has happened in places where these buildings stay, and unfortunately the story of the relation between architecture and earthquake has many critical cases in which the excellent building techniques, the sagacity in the distribution of builded masses and other inventions, already known by Plinio, have not been sufficient to preserve them from ruin [17].

If than we reflect on the debate generated by R.Lemaire's [18], one of the most important prophet in the theory of restauration, we must conclude that the problem of monument conservation in seismic area, must be reconsidered inside of the debate on AUTHENTICITY AND THE CONSERVATION OF MONUMENTS.

* * *

As an example, now two tipical cases are shown, for which the proposed, and realized solutions, are completely different.

The first case is the Flavium Amphitheatre in Pozzuoli.

The exceptionality of the monument is due to its geographical site both for the high archeological density and for the high microseismical events density of the site.

It is a very big construction, probably built during the Nerone's empire; certainly it was already built when, in 79 A.C., the town of Pompei was buried by the eruption of Vesuvium. Another eruption followed by the arising of a little mountain, called Monte Nuovo, probably damaged the amphiteatre in 1538. Since this date the plunder process, that continues in the following centuries and gave us the monument in the present situation, begins.

It is to underline that the whole region is subject to the bradyseismic phenomenon and in 1984, since this situation was becoming very urgent, temporary supporting scaffolds were built all over the concrete structures of the two levels of the building. The cracking scene suggested this kind of consolidation project, most of all because there was not any reference that could certify the building conditions before this event.

Actually it is sure that this construction, a ruin because of the spoliations, shows signs of past earthquakes, even in the imperial roman age, and that today, after twenty years of not-stop microseism stresses, its structural behaviour is not basically changed.

It was then decided, after a very careful analysis, to give up any consolidation system and to remove all the temporary works that prevented to admire the exceptional building. This solution was suggested by the fact that, being the masonry structures of remarkable thickness and disposed along elliptical rings, no displacement among the parts could have created the conditions for total or partial collapses.

So no exterior intervention; just preservation and cleaning works and rearrangement of the expectional ancient water system that probably allowed even naval battles performances in the amphitheatre. Naturally this building is subjected to periodical inspections to check possible evolution of the recorded pathologies.

The second example I propose is rather different from the previous since in this case the foreseen consolidation work is unavoidable to reach the goal of the seismical safeguard of the monument.

St. Peter Church in Modica, a baroque monument, is situated in a great seismical zone, as the Val di Noto in eastern Sicily is.

The project has been worked out in two steps: whereas on the one hand there are the necessary protections to ensure the stability of the building submitted to the working load, on the other hand there are those interventions studied to improve the structural behaviour against seismic actions.

The consolidation works concern the foundations, the nave columns, the internal walls of the church, the chains of the central vault and then the façade. Regard to the foundations, the consolidation works are not indespensable for the stability of the building, but they can prove to be useful in the case of seismic actions.

Probably the intervention on the columns is the most necessary as the data recorded from the strength compressive tests on samples taken from the building show. It is a kind of work that tries to maintain as much as possible exterior aspect of the columns, using materials similar to the stone in composition and in appearance. Anyway the original material shows low mechanical resistance in relation to the stress calculated in the column subjected only to working loads.

The interial walls of the church show two kind of cracks: there are some above the lunettes of the central vault caused by material discontinuity, and there are two parallel cracks along the wall under the windows probably caused by the vault kinematic mechanisms.

For both a kind of intervention based on a system of reinforced beams is foreseen and, as regard to the vault movement, pointed out by the cracks along the keystone, the chaining of the vaults is proposed.

REFERENCES

[1]   Castigliano, C.A., *Théorie des systemes élastiques*, Torino 1879.
[2]   Résal, J., *Ponts en maçonnerie*, Paris 1887.
[3]   Viollet Le Duc, E., *Dictionnaire raisonné de l'Architecture...*, Paris 1854-68.
[4]   Viollet Le Duc, E., *Entretiens sur l'Architecture*, Paris 1863.
[5]   Delanges, P., *Statica e Meccanica de' semifluidi*, Atti Soc., It., Sc., Modena T IV 1786.
[6]   Heyman, J., *Coulomb Mémoires on Static*, Cambridge 1972.
[7]   Heyman, J., *Poleni's problem*, Proc. Inst. Civ. Eng. **84**, 1988.

[8] Moreau, J.J., *Some numerical methods in multibody dynamics: application to granular materials*, E. Jour. of Mech.A/Solids **13**, 1994 pp. 93-114.

[9] Briccoli Bati, S., et al., *Confronto tra esperimento e modello teorico per un problema di strutture in muratura*, in Del Piero, G., Maceri, F. (a cura), "Unilateral problems in structural analysis" **2**, Udine 1987.

[10] Di Pasquale, S., *Statica dei solidi murari. Teoria ed esperienze*, Preprint Dip. Costruzioni, Firenze 1984.

[11] Giaquinta, M. and Giusti, E., *Researches on the equilibrium of masonry structures*, Arch. Rat. Mech. Anal.**88**, 1988.

[12] Del Piero, G., *Constitutive equation and compatibility of the external loads for linear elastic masonry-like materials*, in "Meccanica" **24**(3), 1989.

[13] Freudenthal, A. M. and Geiringer, H., *The mathematical theories of the inelastic continuum*, Handbook of Physics, Vol.4, 1953.

[14] Sokolovskii, V. V., *Statics of Granular Media*, Pergamon Press, Oxford, 1965.

[15] Washizu, K., *Variational methods in elasticity & plasticity*, Oxford 1982.

[16] Prager, W. and Hodge, P., *Theory of Perfectly Plastic Solids*, New York 1961.

[17] Di Pasquale, S., et al., *Architettura e terremoti. Il caso di Parma*, Parma 1983.

[18] Lemaire, R., *Authenticité et patrimoine monumental*, in "Restauro" **129**, 1994.

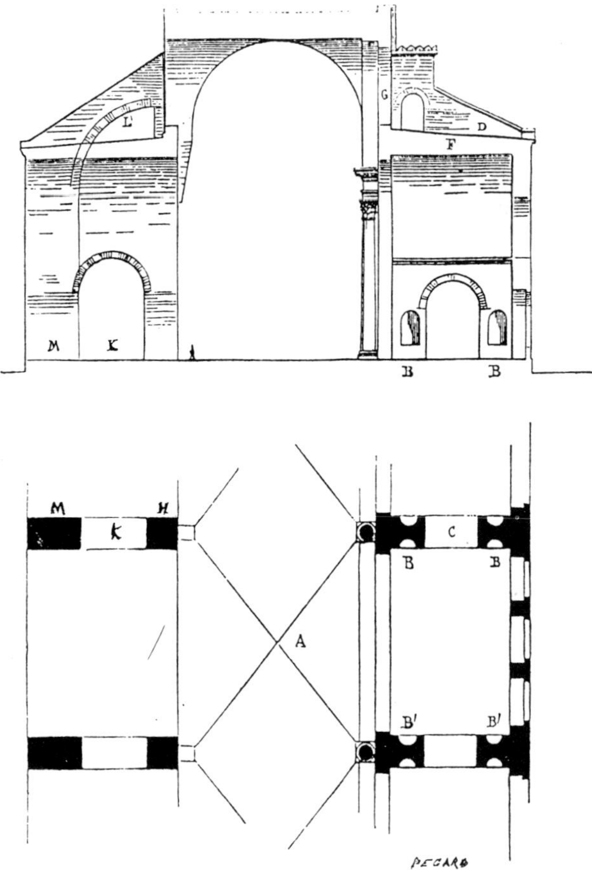

Fig. 1. From E.E. Viollet-le-Duc, "Entretiens".

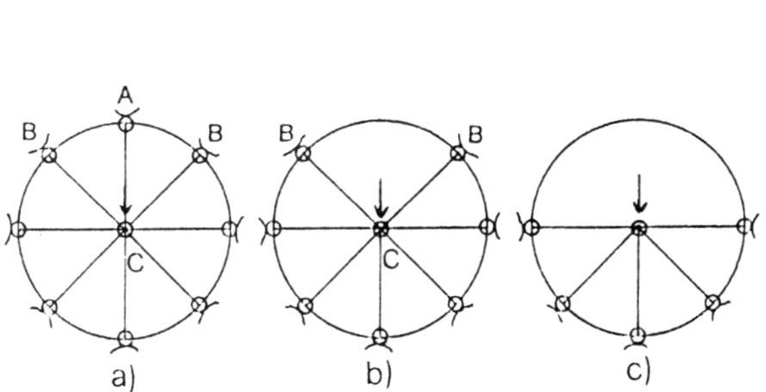

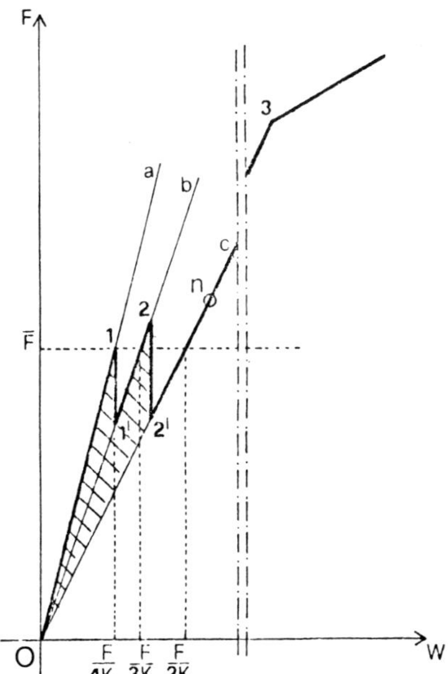

Fig.2.

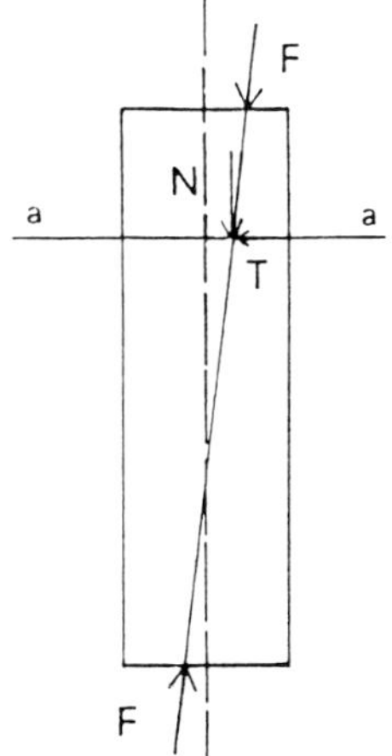

Fig.3.

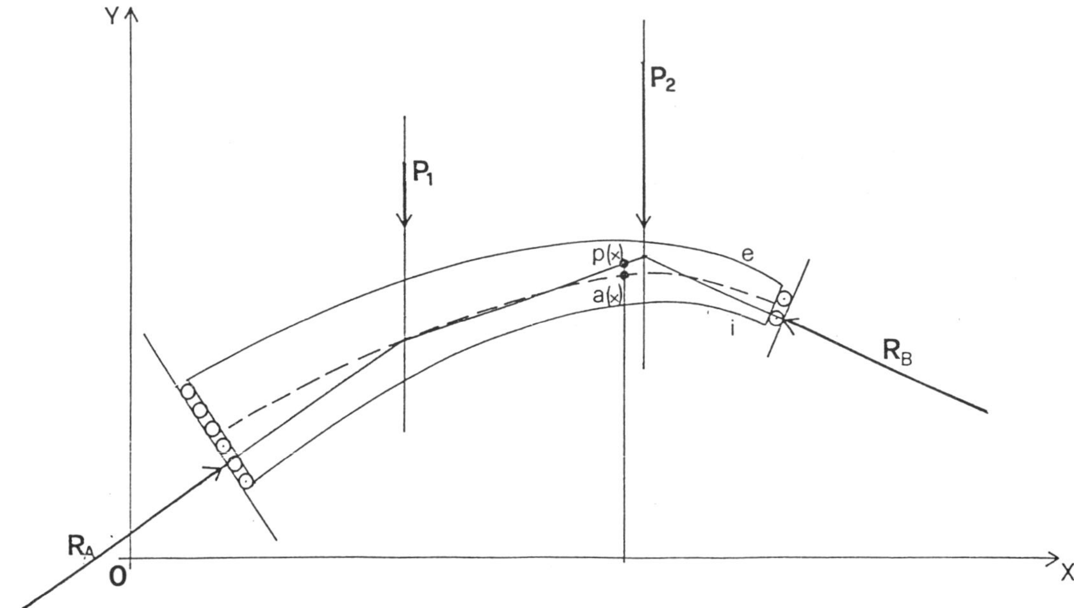

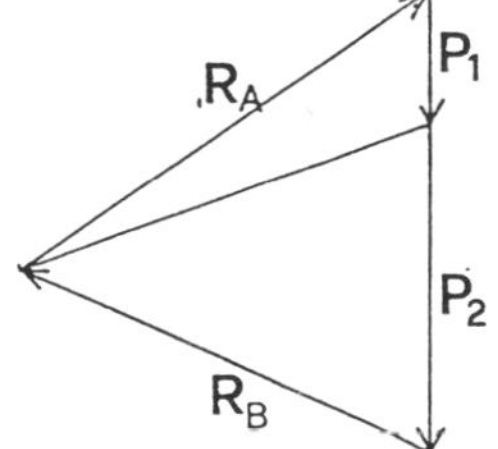

Fig.4.

Fig. 5a

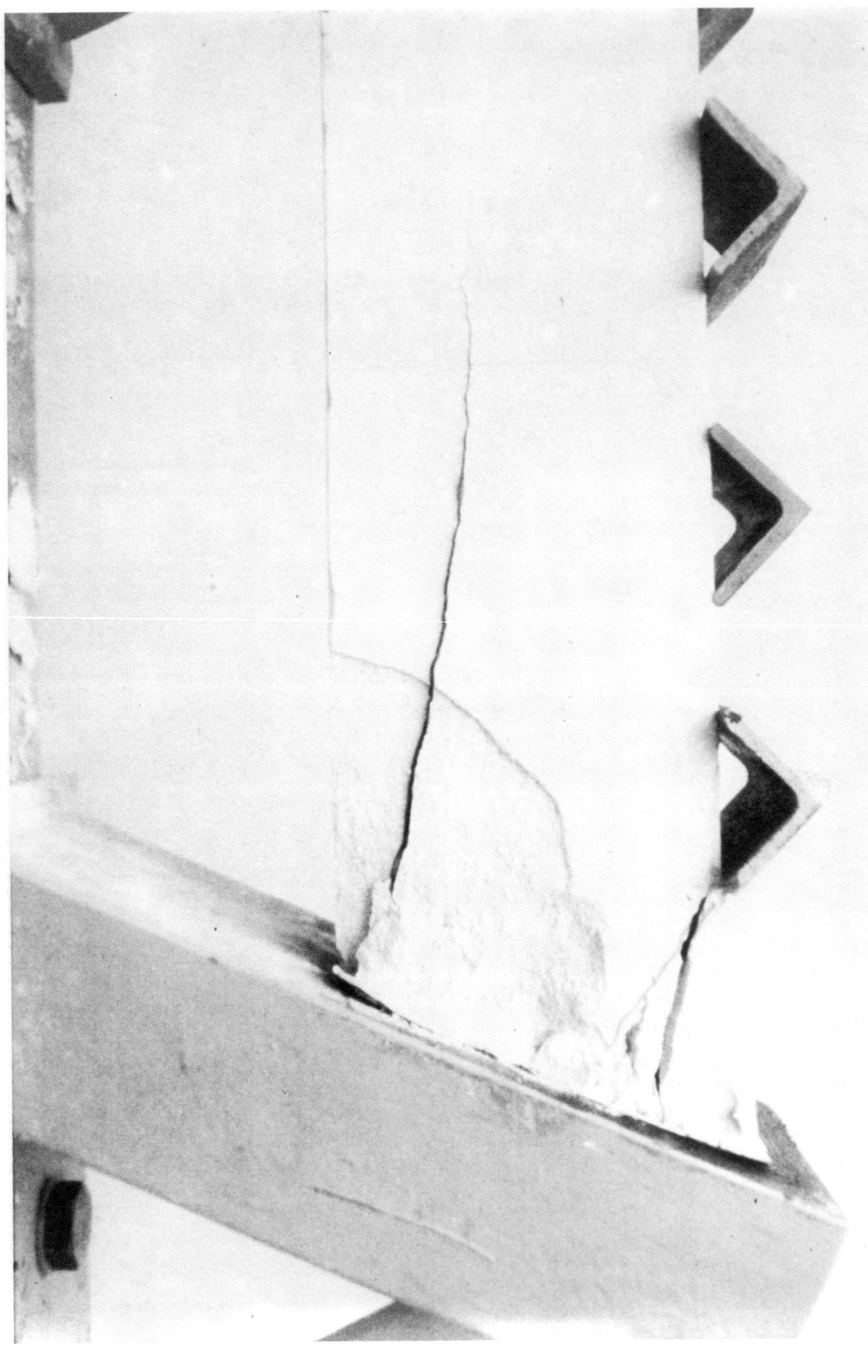

Fig. 5b

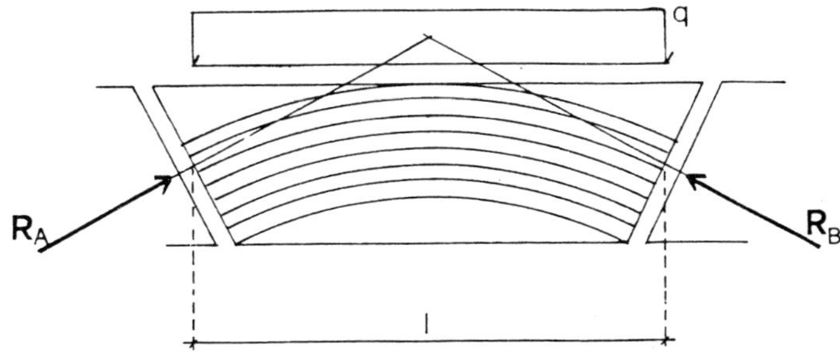
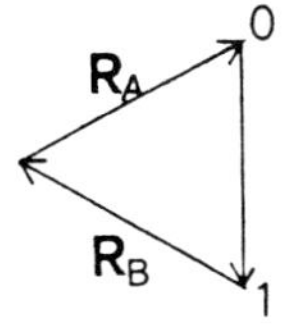

Fig.6.

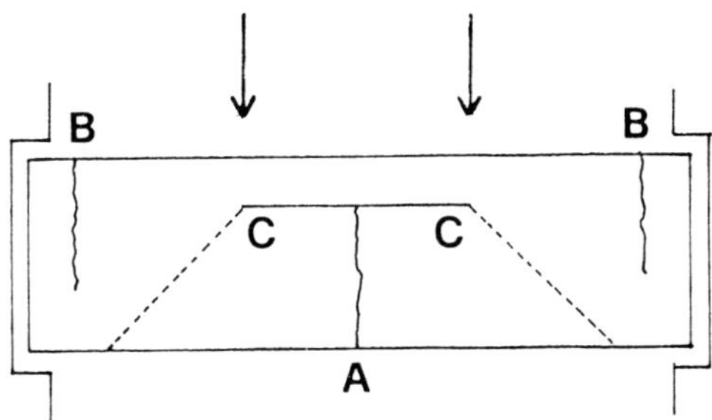

Fig.7.

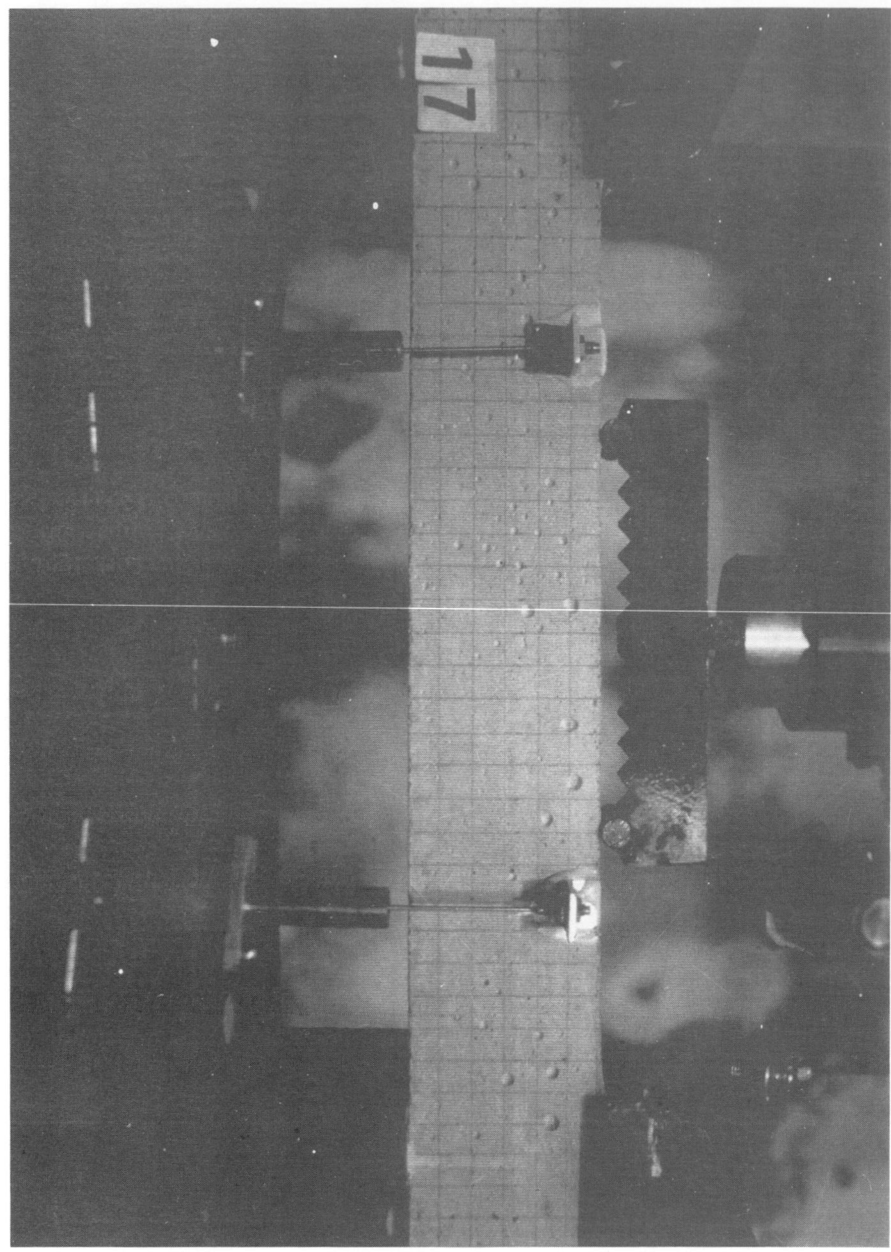

Fig. 8

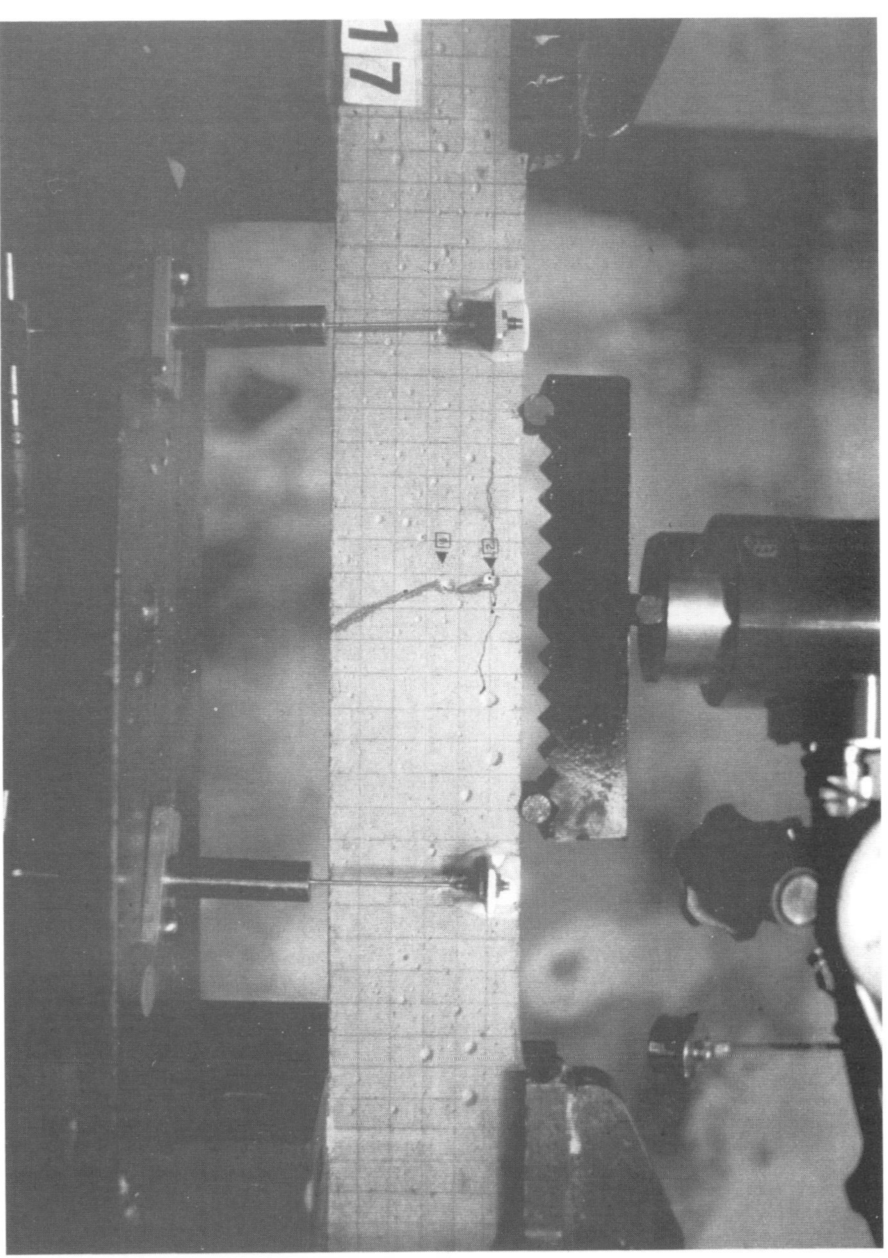

Fig. 9

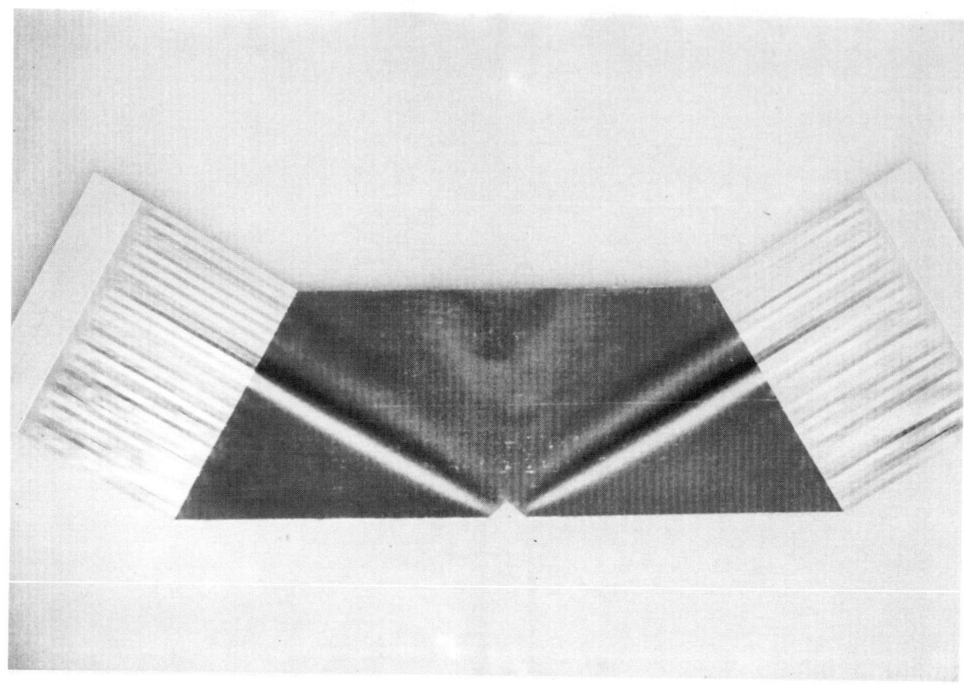

Fig. 10a

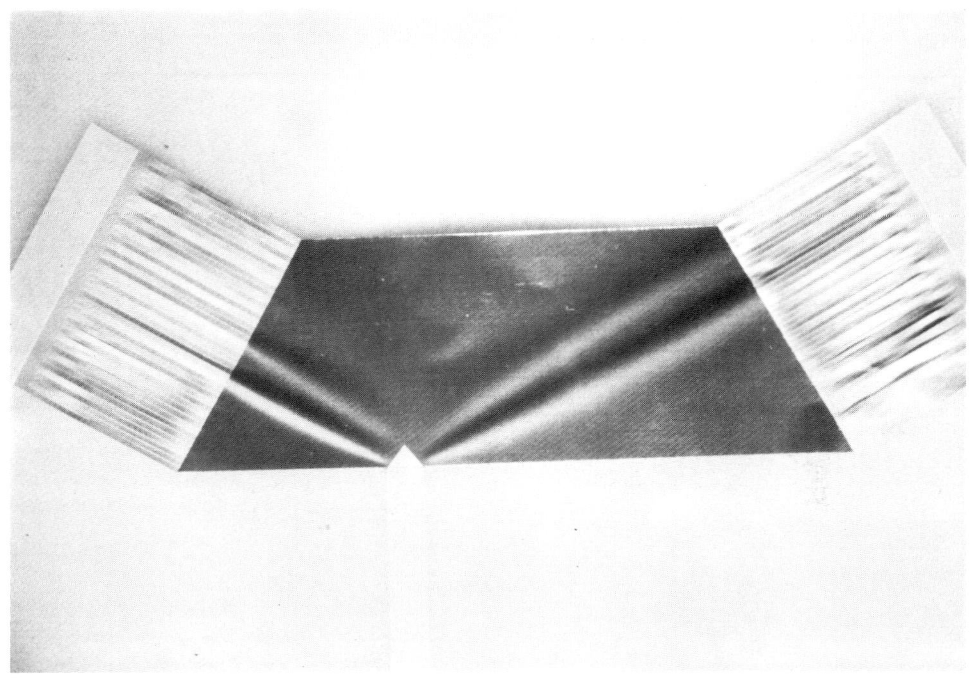

Fig. 10b

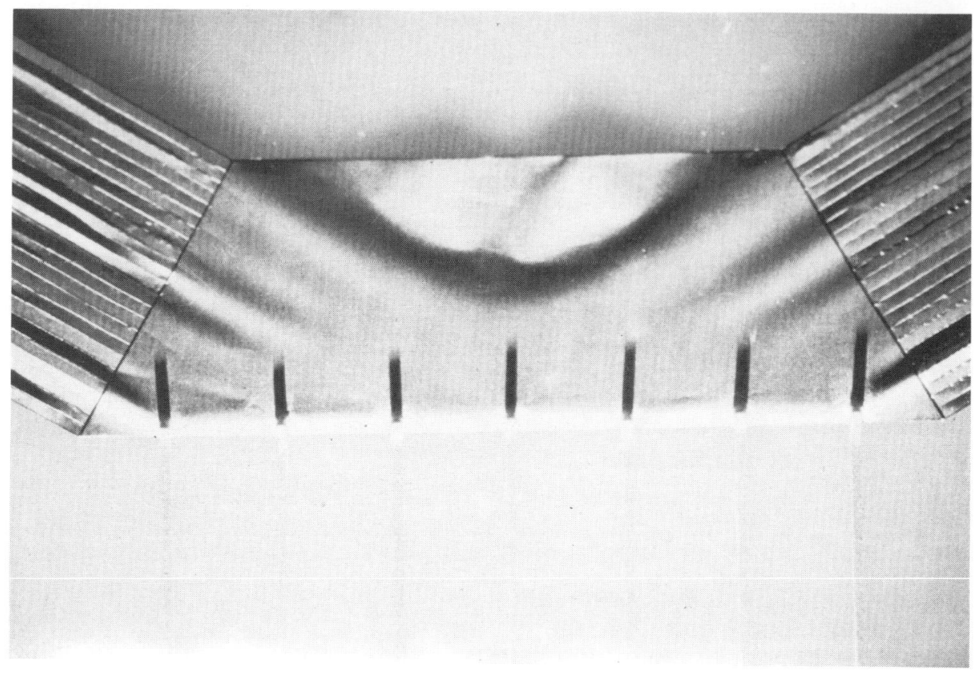

Fig. 10c

Fig. 11a

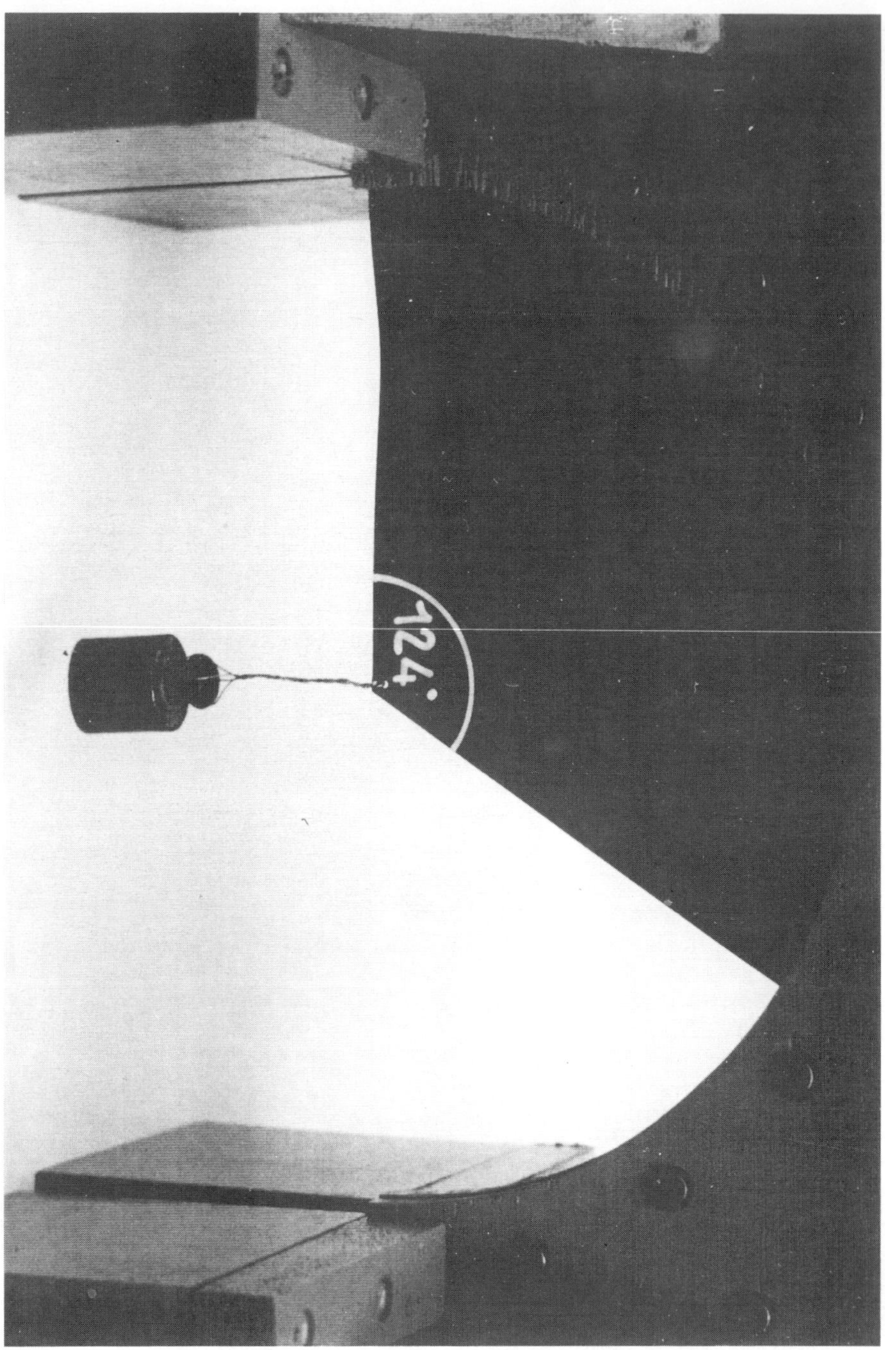

Fig. 11b

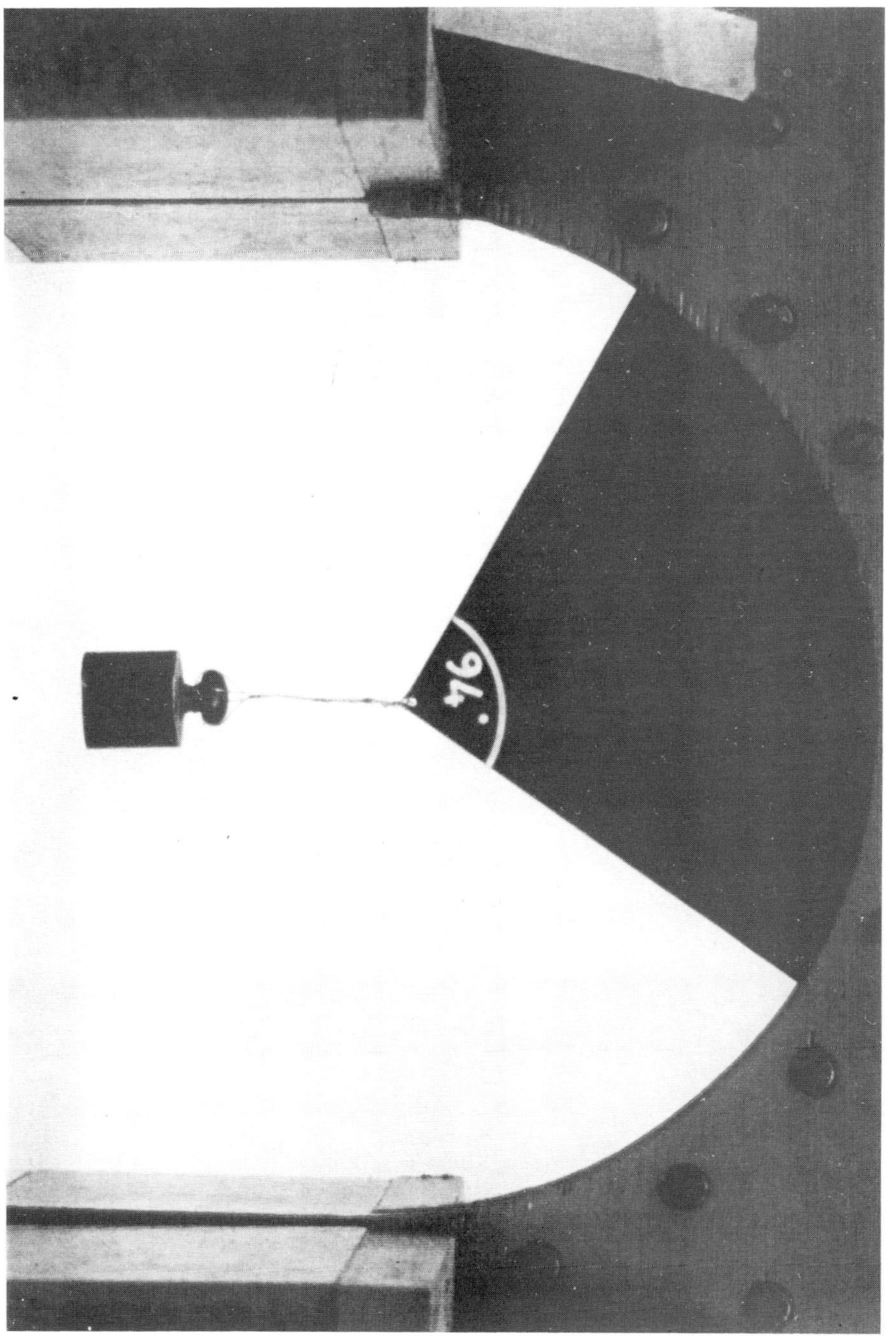

Fig. 11c

                                                    S. Di Pasquale

Fig. 12a

Fig. 12b

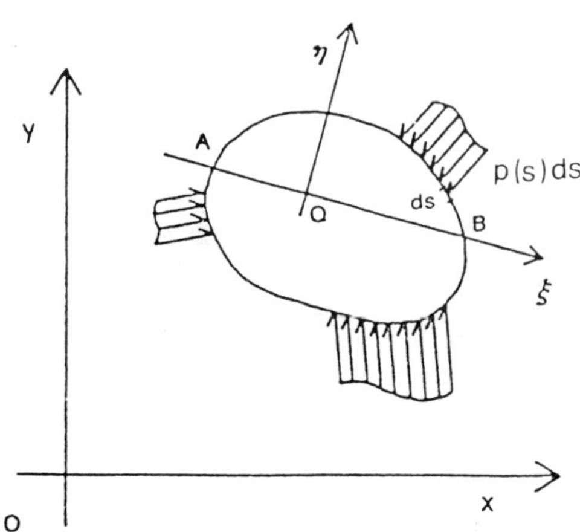

Fig. 13. Plane problem of a disk with self-equilibrated load.

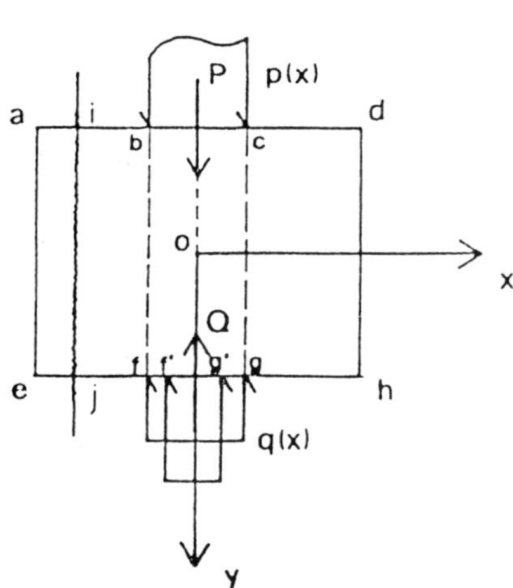

Fig. 14. Equilibrium of the strip a-e-i-j: the stress tensor is null.

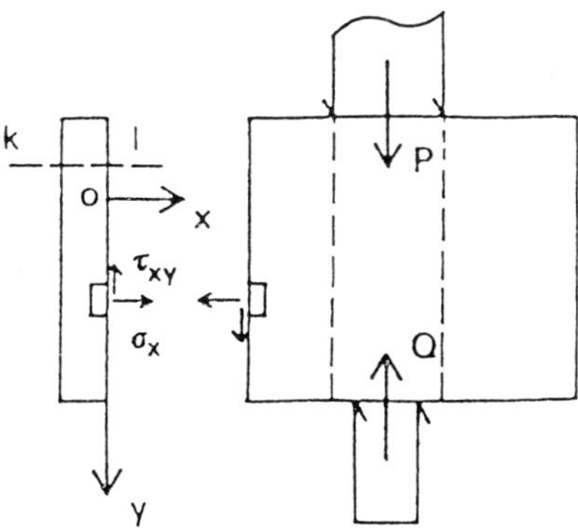

Fig. 15. Equilibrium of the strip *a-e-i-j*: the stress tensor is null.

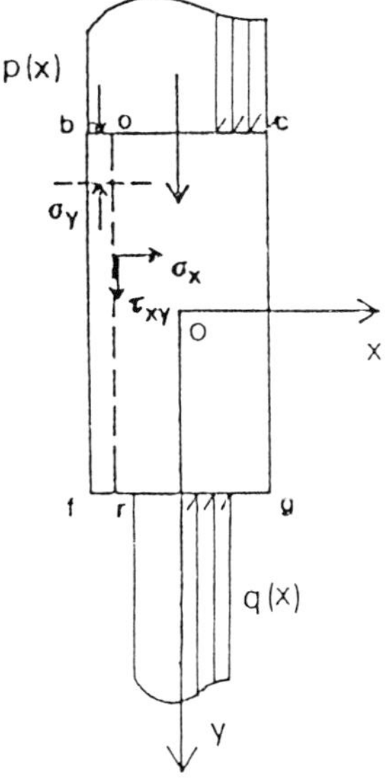

Fig. 16. Equilibrium of the loaded strip.

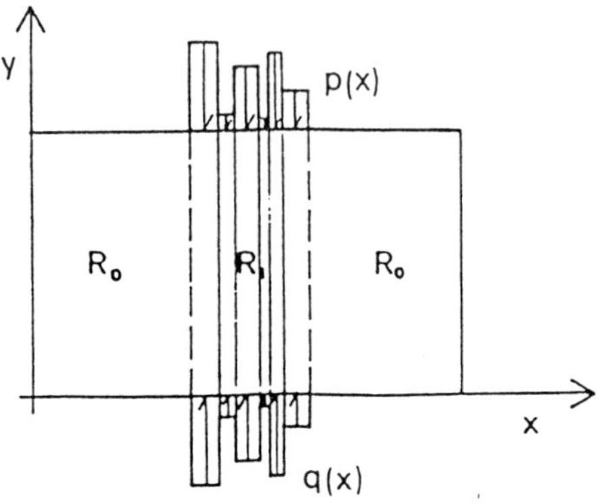

Fig.17. Equilibrium of panel with discontinuous symmetrically distributed loads.

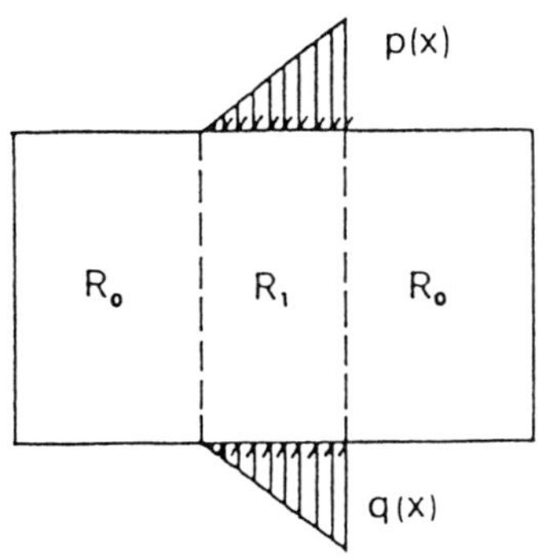

Fig.18. Equilibrium of a panel with discontinuous symmetrically distributed loads

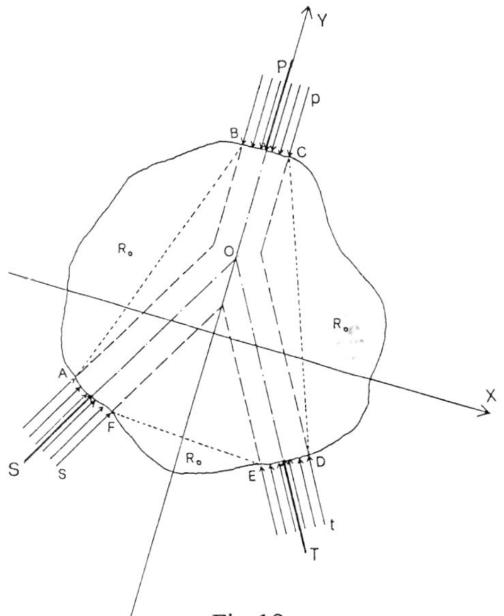

Fig. 19a.

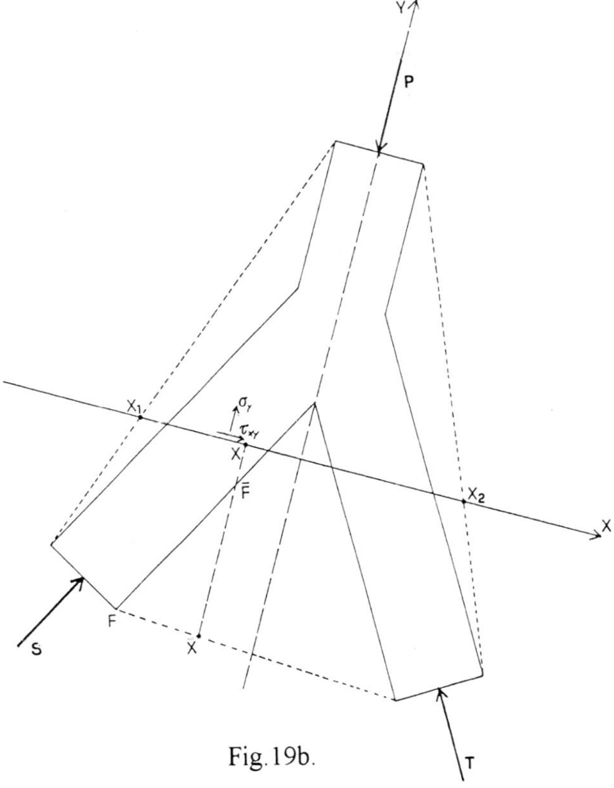

Fig. 19b.

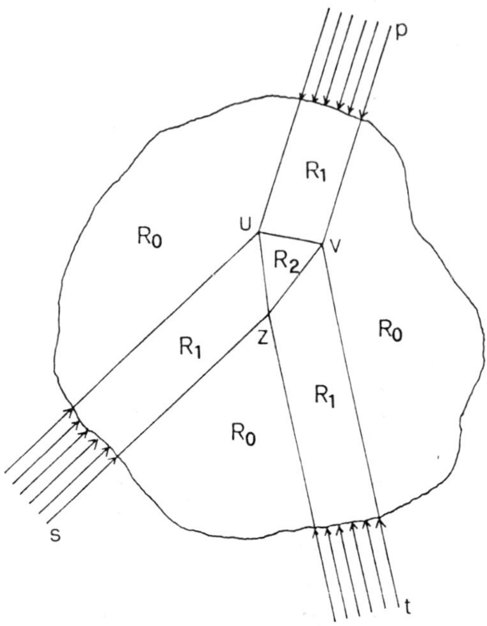

Fig.20.

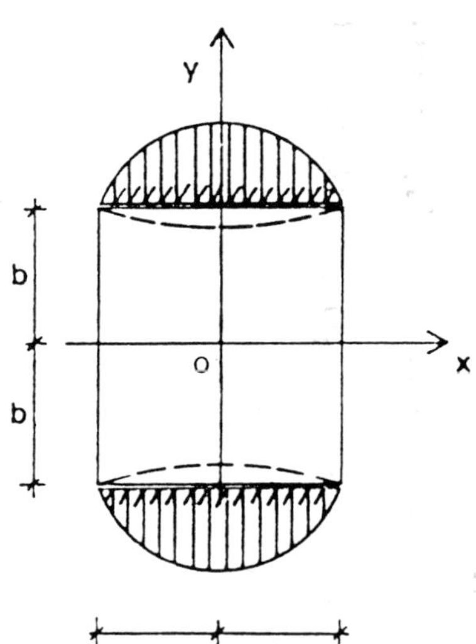

Fig.21 Rectangular panel with self-equilibrated parabolic load; the dashed line represents the deformation of the wall.

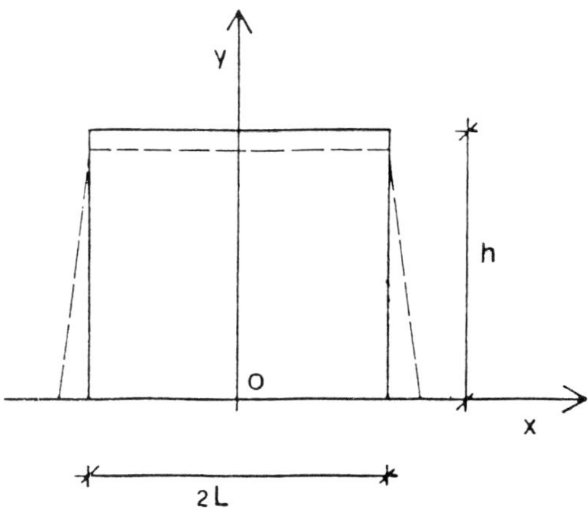

Fig.22. Plane problem of a rectangular wall; the dashed line represents the deformation of
the wall.

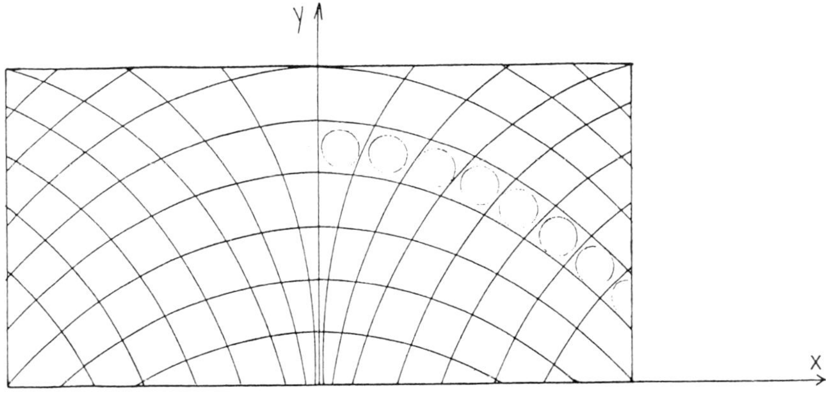

Fig.23.

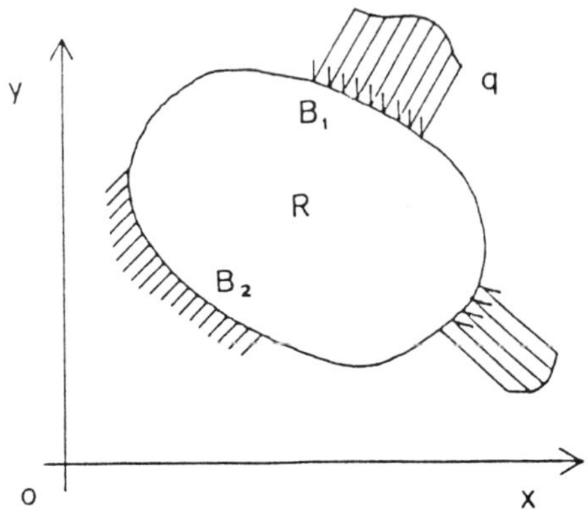

Fig.24. General elastic problem with distorsion for a body occupying the region R.

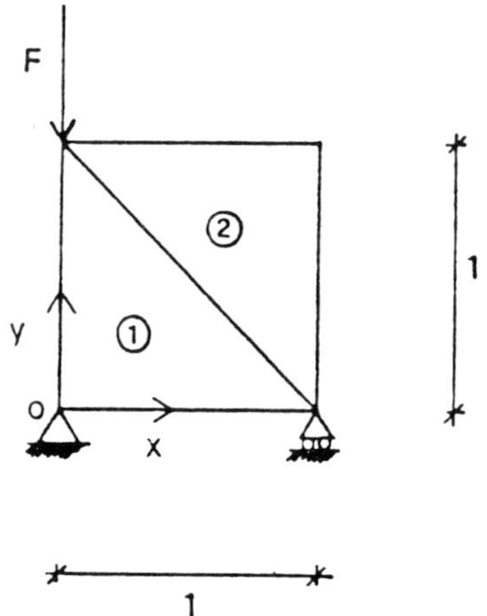

Fig.25. FEM model for a square wall.

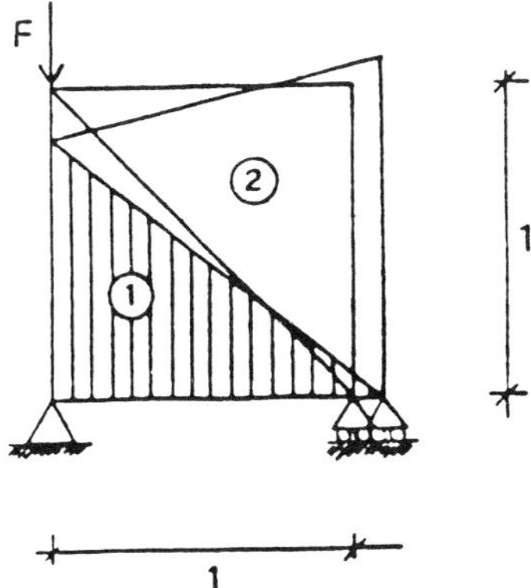

Fig.26. Kinematical solution by extensional deformations and shearing deformations.

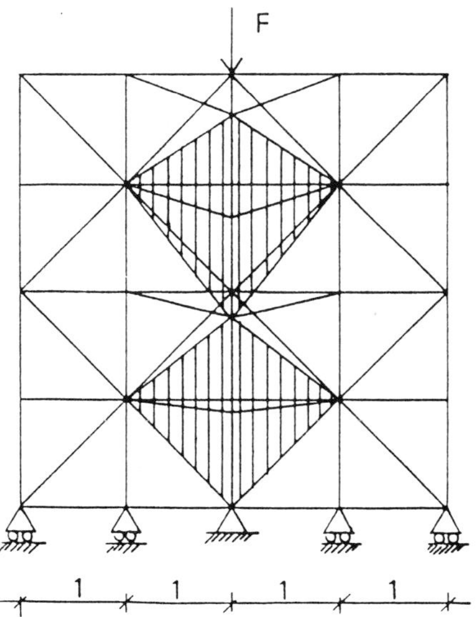

Fig.27. FEM model for a square wall vertically loaded at the upper midpoint.

Fig.28. The Duomo façade at Siena: in the standard solution the maximum tensile stress is 1.1428 N/mm²; in the masonry solution the maximum residual tensile stress is 0.009 N/mm²; for the kinematical indetermination we chose not represent the deformate structure.

The Pozzuoli amphitheatre

 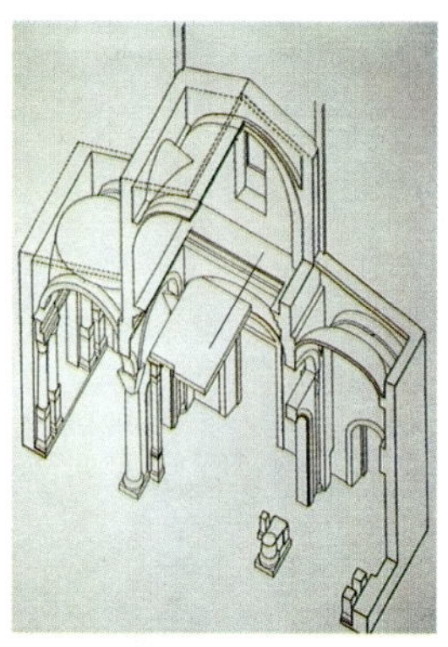

 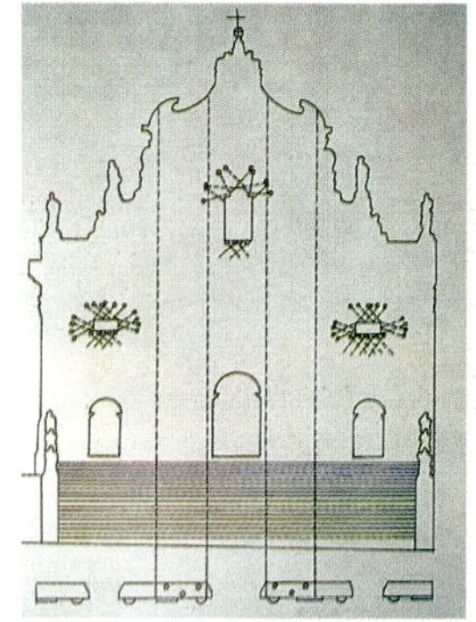

The Modica cathedral

Fig.30a   General view
Fig.30b   An axonometric view
Fig.30c   Preparation of the columns consolidation
Fig.30d   The column consolidation
Fig.30e   The consolidated column
Fig.30f   The facade consolidation

Captions of photos (slides)

Fig.5   Experimental result for a masonry plate-bande.

Fig.10a,b.c   A paper plate bande under various load conditions: notice the different shape of reactive structure.

Figg.11a,b,c   Simulation of Bouninesque problem with a rubber sheet. The fig.11a shows that all the half-circle coincide with the reactive structure; in the fig.11b an sector of the sheet examined presents a buckling of the compressed lower part. In order to avoid buckling, in fig.11c the experimental validity of the theoretical solution is shown.

Figg.12a,b   The fig.12a shows the experimental behaviour of a rectangular masonry wall subjected to a system of uniform pressures on the opposite edge; the pressure action is limited to a region between two vertical lines; fig.12b shows the good agreement with the paper experiment.

SEISMIC DISASTER PREVENTION IN THE HISTORY OF STRUCTURES IN GREECE

P.G. Touliatis

National Technical University of Athens, Athens, Greece

ABSTRACT

Living in a country of frequent and high seismic activity and observing the disastrous influence of that activity to the constructions for several thousands of years, people recognized form the very early stages of the history the necessity of the aseismic behaviour of the buildings.
Many related techniques have been developed for at least 35 centuries and are still developing under the impetus of the modern technology.
This paper is trying to describe some characteristic aseismic traditional techniques in Greece and to formulate their basic principles.

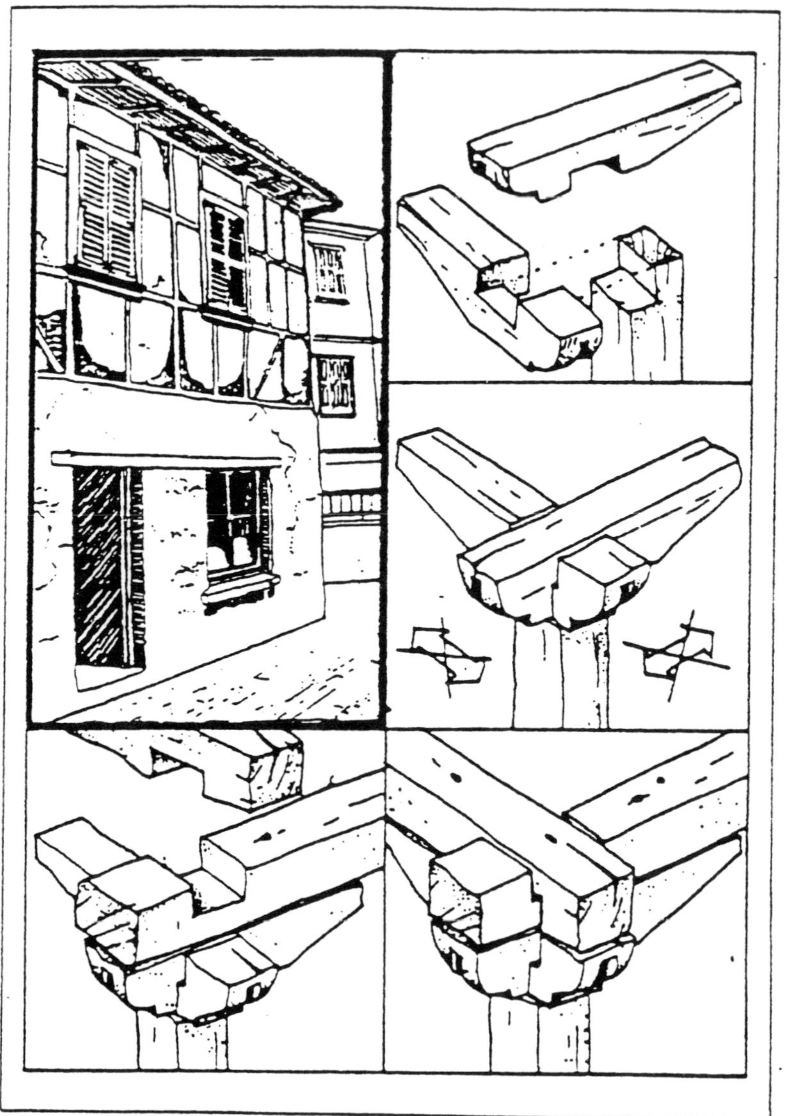

1. INTRODUCTION

All the people living in Eastern Mediterranean have sometimes and in some degree felt the phenomenon of an earthquake and have observed its consequences.

From ancient times, Greek philosophers (as Aristoteles, Pythagoras Hepicouros) have dealt with the earthquake phenomenon and tried to interpret it.

It's estimated that today, a 50% of the annual seismic energy of Europe and a 2% of the annual world seismic energy, is released in Greece.

In this country, people live developing civilizations and constructing their monuments and buildings for many thousands of years. Surviving frequent and disastrous earthquakes they got familiar with the act of observation of the damages on their constructions and so understood, more or less, their behaviour during seismic action. Rebuilding them in better ways, trying to improve their resistance against the dynamic loading, the ancient constructors experimented with different materials, constructional systems, and sometimes, sophisticated detailing. Following long and hard paths of observation, experiments, failures and inventions they created local or more spread around aseismic techniques, concerning basic members of a building (Masonry, roof, e.t.c.), or even a complete building system.

It is a fact that it is impossible to protect completely a construction against the, sometimes out of the human capabilities limits, seismic force. In Greece, monuments, buildings, cities or even whole civilizations have been lost due to seismic or/and volcano activities, since prehistoric times to our days (i.e. Thira volcano eruption 1500 B. C, City of Argostoli complete destruction in 1953, Kalamata severe damages in 1986 e.t.c.).

On the other hand many architectural monuments stand still after more then thousand years (i.e. Parthenon in Athens 438 B.C., Hagia Sofia in Istanbul (CONSTANTINOPLE) 537 a.d., Hosios Loukas Monastery, 955 A.D. e.t.c.) in areas with, some times, high seismic risk. Traditionally constructed buildings and settlements, all over Greece, exist and are used for hundreds of years surviving, repeatedly, seismic action.

2. THE EXPERIENCE OF THE PAST

We should accept today the principle that the design of an aseismic building must be based on the right conception and inspiration from the very beginning. It is wrong to design a building in a seismic area without taking into account the seismic factor and then try to correct the various errors by using complicated calculations and strengthening methods. Today, the correct structural and dynamic analysis and dimensioning are very powerful and valuable weapons in

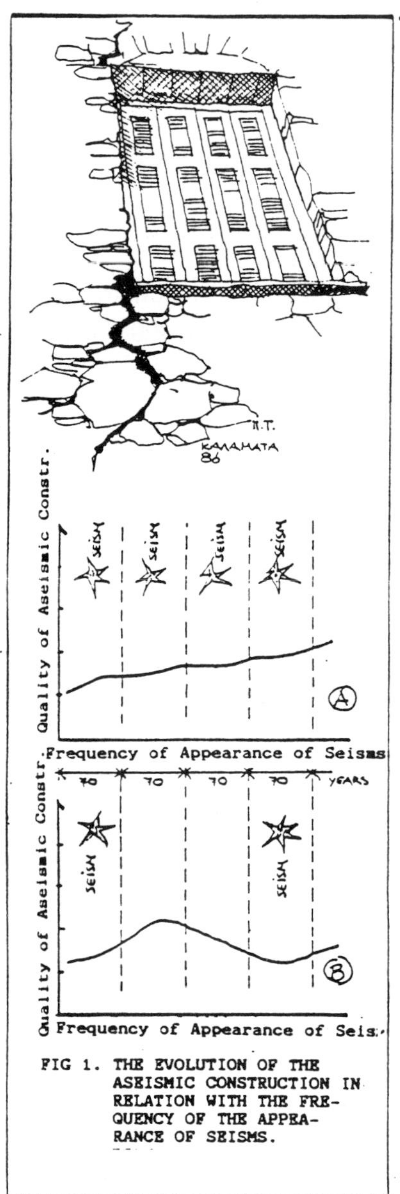

FIG 1. THE EVOLUTION OF THE ASEISMIC CONSTRUCTION IN RELATION WITH THE FREQUENCY OF THE APPEARANCE OF SEISMS.

FIG. 2 KARITENA-GREECE

our hands, in order to design aseismic structures. But, it is well known, that a structure which is based on a faulty conception cannot be totally corrected by any calculation.

On the contrary, when the proper crucial decisions concerning the materials, the load bearing systems, the joints and the forms are taken from the first steps of the design procedure, the correct behaviour of the structure can be guaranteed. It is evident, today, that the designer must develop (through education and praxis) an aseismic perception based on the main principles of the aseismic design.

In older times and despite the fact that the structural and dynamic analysis methods were totally unknown, some very efficient aseismic methods and techniques were developed by local, craftsmen. Nevertheless these skilled workmen had a very deep knowledge of the materials and the building systems of that time, which stayed the same and kept developing for centuries, passing on from one generation to the other. They also had a very good conception of every small detail as well as of the whole of the construction. This deep knowledge accompanied by the observation of the behaviour of structures during earthquakes and the examinations and repairs of the damages led to the invention of very interesting and efficient aseismic construction systems.

Perhaps the following remark can help to detect, study and support such aseismic construction systems :

It is obvious that the evolution of more complete aseismic systems took place in areas where earthquakes were a frequent phenomenon. That is at least one important seismic action during the life period of a generation. Consciousness of danger and personal experience lead traditional constructors not only to the invention of aseismic techniques, but also to their evolution and conservation, as happened in Santorini, Lefkas e.t.c. (Fig. 1,A).

On the contrary, in places where earthquakes are a rare phenomenon and the calm period between two important seismic actions is larger than the average generation lifetime, the attention of constructors tends to diminish during the calm long periods (Fig. 1,B). That happened in Athens where nobody expected the 1981 earthquake and since that time the reinforced concrete aseismic code has been improved repeatedly several times. That also happened in Kairo (October 1992) and, I am very afraid, it can happen any moment in Cyprus. A small, representative example can be taken from the diminishing quality of the construction of the openings in the masonry of the buildings in the village of Karitena in Peloponese. (Fig. 2). After a long calm period from the last significant seismic action the careful, aseismic construction of the perimeters of the openings is forgotten or degenerated into morphocratic repetitions (Fig.2,C).

It would be very interesting to compare maps of the frequency of the repetition of the earthquakes over the country to those of the quality of the aseismic techniques, from place to place. A map of the level and kind of damages can be added.

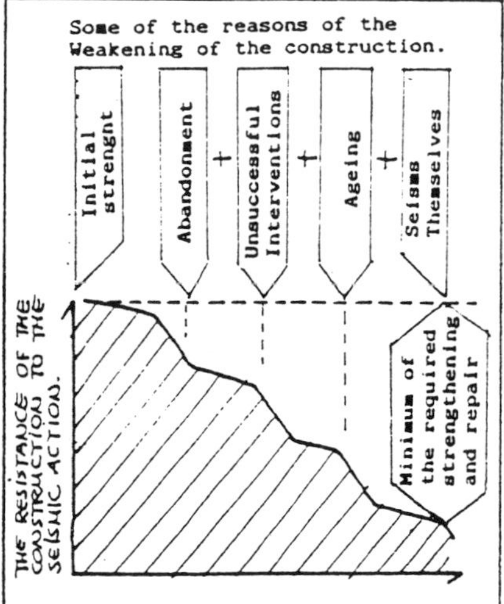

Other points of danger
for an old construction
are:

- The careless use
  of new, not experiened
  materials or construc-
  tion systems which
  are not compatible
  to the structure.

- Careless additions
  over or beside the
  old construction.

- Uncorrect modeling
  and/or difficulties
  in analysis

- Luck of knowledge
  about the traditional
  materials or techni-
  ques.

FIG. 3 THE PROCESS OF
THE WEAKENING OF
AN OLD CONSTRUCTION

Old building in Kalamata
city heavily damaged by the
earhquake of 1986.

3. WEAKENING OF AN OLD STRUCTURE

When, during an earthquake, an old structure is severely damaged, or even destroyed, a quick judgment about improper materials of a wrong building system is not always correct. When we examine an old construction in order to maintain, repair or strengthen it , we should always try to determine its initial condition of strength and ability to resist seismic action. It is easy to observe that the quality of a structures aseismic behaviour usually is weakened through the ages.

The decendance of the resistance to seismic action is due to ageing, abandonment, unsuccessful interventions and the seisms themselves (Fig. 3). We must also add the frequent non compatibility of today's materials and techniques, which we have used carelessly in various interventions and repairs. Also there are great difficulties in modeling and structural analysis of the old structures and a lot of research must be done on this subject (Fig. 3).

So, it is obvious that almost any construction, even with a marvelous aseismic technique applied, reaches a point through-out time when it cannot face successfully the earthquake. Many old, famous monuments, resisting for centuries the dynamic actions at some moment have proved this hypothesis by a local or more general failure. The Parthenon and Hagia Sophia are among them.

In order to intervent at such a moment and bring the construction at least to its initial condition of strength and seismic resistance with the minimum changes of its original constructional conception, it is necessary to detect, study and analyse its construction system and, of course, any existing aseismic method or technique.

4. LOAD BEARING WALLS

Many factors play a crucial role in the determination of a wall's seismic resistance. Basic factors are the materials, which are used, and the construction system.   Many times the old constructors, building in areas with seismic risk, tried to improve both. Accordingly, of course, to the building importance, best materials were chosen for the wall elements (i.e. perfectly curved marble components, high quality bricks e.t.c.). Better cooperation between the wall elements has been achieved by special (i.e. iron embedded in lead) connectors (i.e. classical period monuments). High quality mortars served the same goal (i.e. Byzantine era monumental constructions).

It is very important to mention here that a principle, which is modern today, was known and carefully applied at least 2.500 years ago, during the construction of the Parthenon on the Athens Acropolis. The principle that the connector must be weaker than the connected members.

At least since the Minoan civilization time (about 3.700 years ago) serious attempts have been made to reinforce the masonry, improving its bending

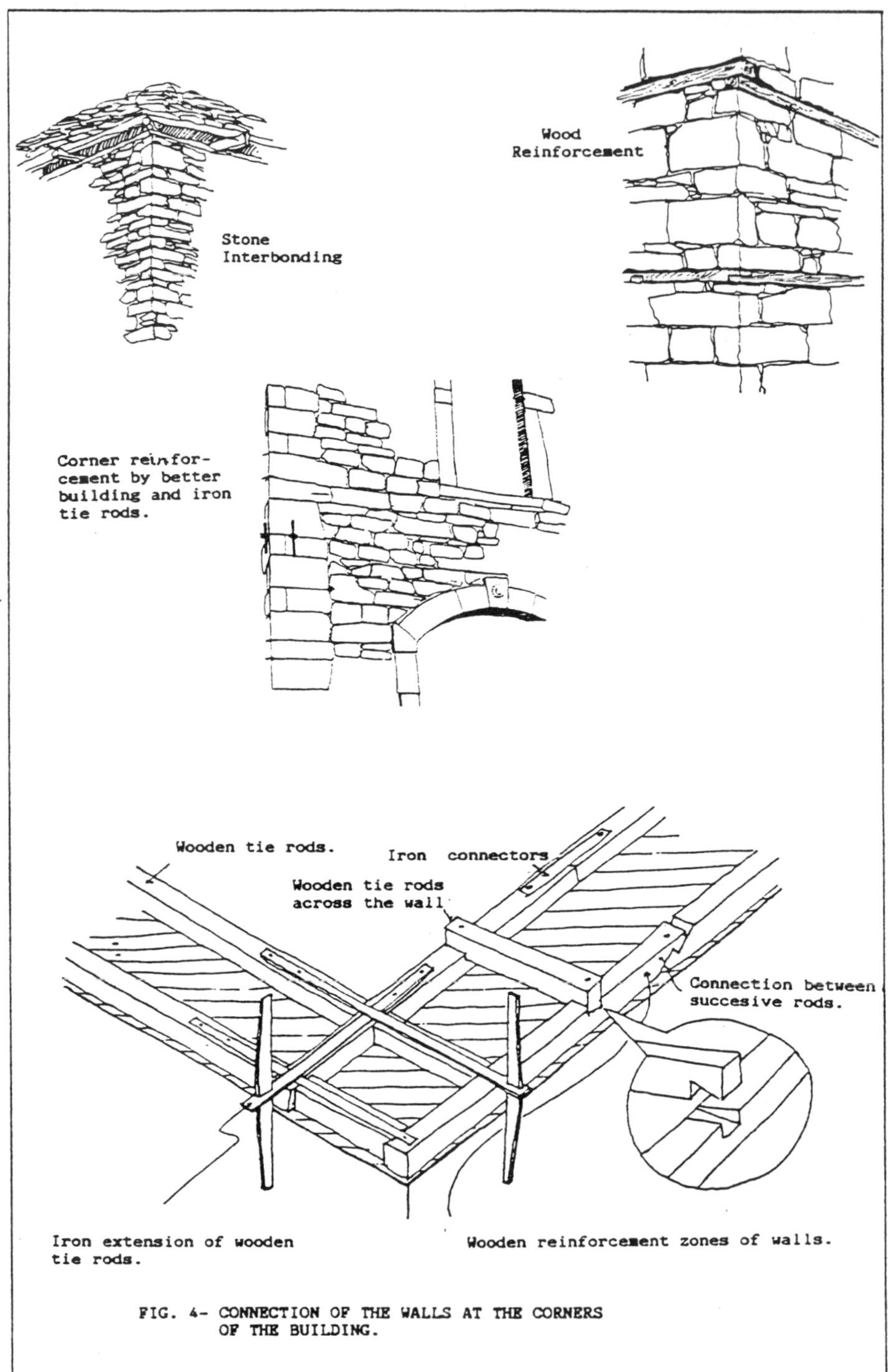

FIG. 4- CONNECTION OF THE WALLS AT THE CORNERS
OF THE BUILDING.

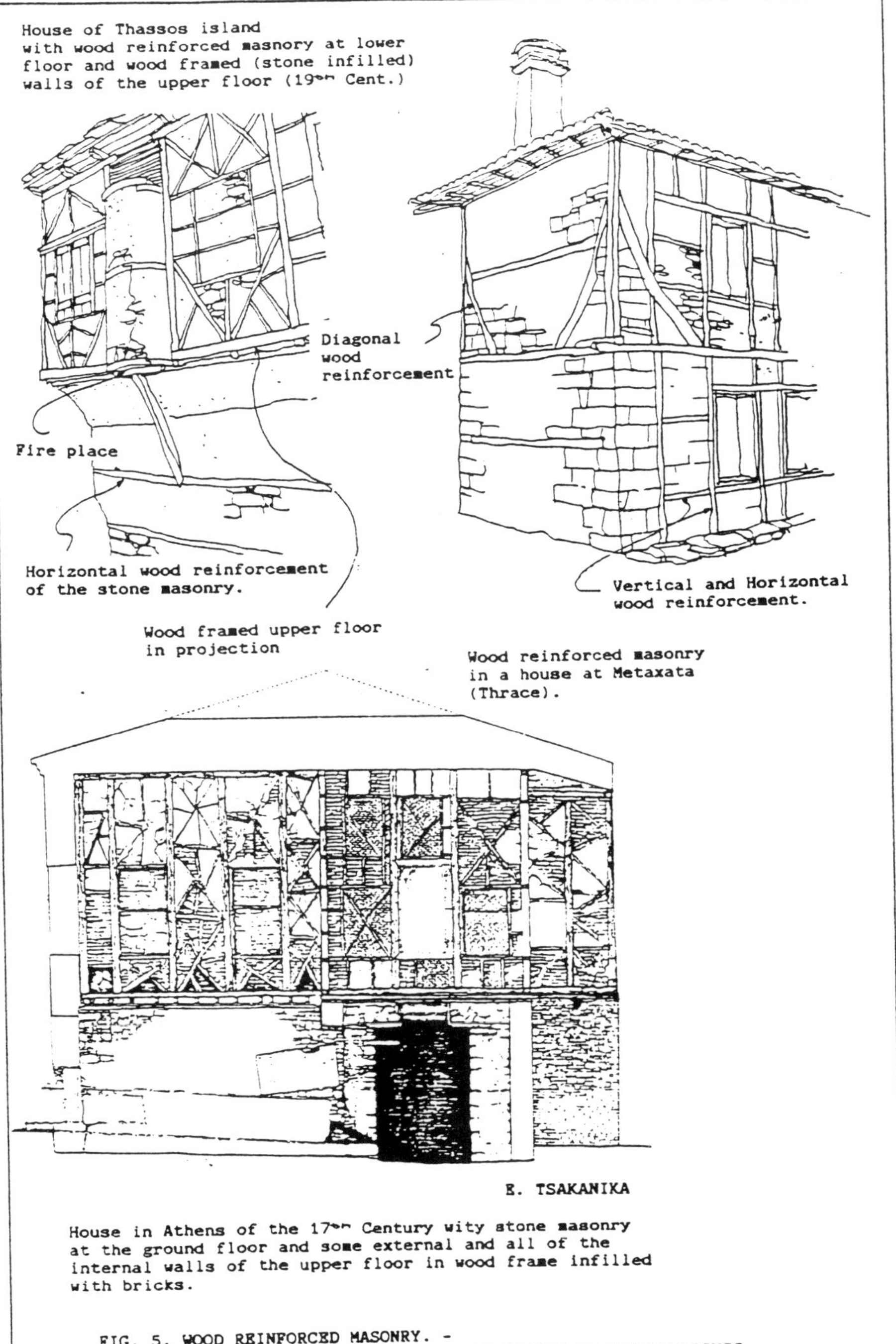

House in Athens of the 17th Century wity stone masonry
at the ground floor and some external and all of the
internal walls of the upper floor in wood frame infilled
with bricks.

FIG. 5. WOOD REINFORCED MASONRY. -
        WOOD FRAMED WALLS AT THE UPPER FLOORS OF THE BUILDINGS

capacity and the ability to undertake tensile forces, by horizontal an, more seldom, vertical reinforcement zones. (Fig. 4). Usually wood was used, some times in a very sophisticated ways, as for an example in the settlement of Akrotiri on the island of Santorini (1.500 B.C.) (Fig. 10).

Trying to minimize the weight (and the mass) at the upper parts of the buildings the external and the internal, load bearing and separating walls of the higher floors were constructed in timer frame (Fig. 5) Those timber framed walls became more stiff by diagonal (and other) bracings, more carefully developed in places with high probability of dynamic action as, for example, the mountains of Pelion in central Greece, with landslides and earthquakes, or the island of Lefkas with very high seismic activity (Fig. 13).

4.1. MORPHOLOGY AND SEISMIC RISK

It is very characteristic that the choice of the same material and the same constructional system didn't produce same architectural and morphological types of buildings. In central and northern Greece the lower floors have stone masonry reinforced by wood and the upper floor has timer framed external and internal walls. The light timber frame, the diagonal timber bracing, the careful interbonding of the timber members e.t.c. permitted more freedom in the design.

So, the upper floor very often projects over the street enlarging or/and correcting the plan and the space of the rooms. On the other hand the upper floor acquired the privilege of many and large windows (Fig. 6).

On the island of Lefkas the same system, in general, was applied. This island is subjected to very strong and frequent earthquakes. Only the ground floor remained in stone. The upper (one or usually two) floors, although built in a very sophisticated timber framed constructional system, keep the strict and conservative morphology of a stone masonry : no projections at all and small openings placed strictly one over the other (Fig. 6,13).

4.2. THE COOPERATION OF THE WALLS

From very early stages of the constructional activity in Greece it was also observed and understood that the load bearing stone or brick walls can resist loads in their plane quite successfully but become very weak in the case of accepting forces perpendicular to their surfase, and suffer more serious damages from bending stresses and in some cases collapse by overturning.

Since prehistoric times many efforts have been done to establish a reliable cooperation between the walls of a building.

Better interbonding of the stones at the corners, continuity of the horizontal reinforcement zones (usually in timber) by careful connections at the corners and

Typical construction of
the houses in Lefkada island
with very poor soil and
very high seismic risk.

The light wood framed upper floor
design, followsthe strict and
conservative forms of the ground
floor masonry:
No projections at all and small
openings placed usually one
over the other

Typical construction of the
houses in Pilion.
(Central Greece with seismic
 Risk and Landslidings).

The light wood framed upper
floors have significant freedom
intheir design:
Free composition in plan.
Daring projections and
high percentage of openings
in the walls.

FIG. 6 THE INFLUENCE OF THE HIGH SEISMIC RISK
ON THE MORPHOLOGY OF THE BUILDINGS.

FIG. 7 THE COOPERATION OF THE WALLS
         IMPROVED BY THE ROOF FRAME
         CONSTRUCTION.

FIG. 8. ROOFS

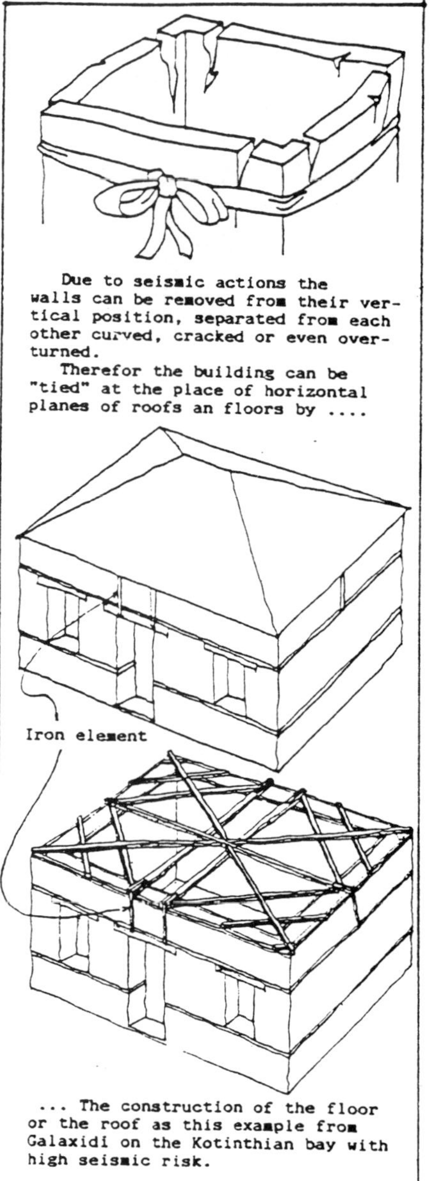

Due to seismic actions the walls can be removed from their vertical position, separated from each other curved, cracked or even overturned.

Therefor the building can be "tied" at the place of horizontal planes of roofs an floors by ....

Iron element

... The construction of the floor or the roof as this example from Galaxidi on the Kotinthian bay with high seismic risk.

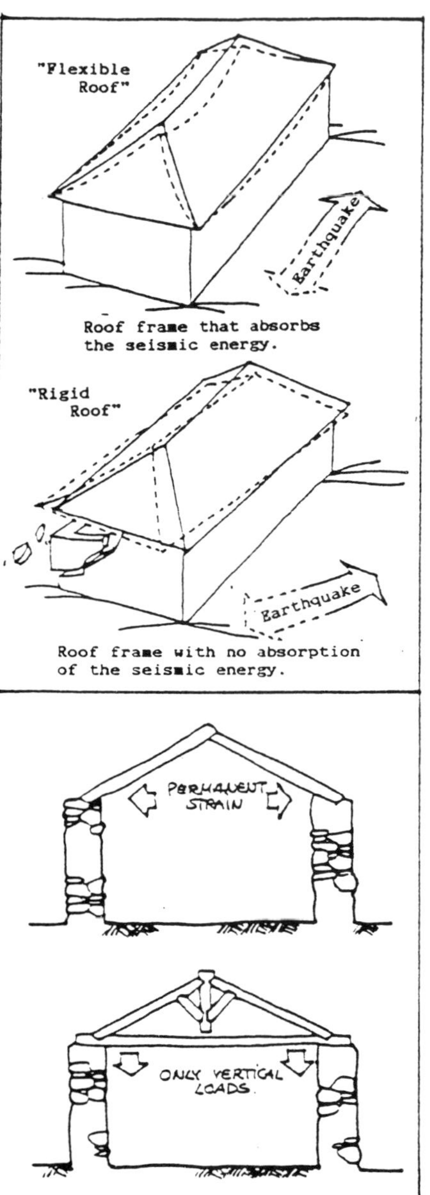

Roof frame that absorbs the seismic energy.

Roof frame with no absorption of the seismic energy.

the use of special tie-rods took place during the centuries (Fig. 4).

In more developed aseismic techniques the builders used the horizontal construction of a floor or a roof (with higher or lower diaphragmatic behaviour) to connect the walls (Fig. 7).

5. HORIZONTAL LOAD BEARING STRUCTURES (FLOORS-ROOFS)

Traditional roofs and floors are usually made of wood. Sometimes we find horizontal load bearing structures of stone or bricks (vaults, domes, arches) metal beams (neoclassical buildings) even reinforced concrete. Especially during seismic action, horizontal load bearing structures, (partially or in the whole) transfer horizontal forces to the vertical load bearing system (Fig. 8).

A general goal is not only to extinguish the transfer of horizontal forces but also to transform these horizontal structures into diaphragms. In this way, with adequate anchoring and joints, the horizontal structures become rigid and connect and strengthen the walls during seismic action (Fig. 7).

In general, there are two large groups of roof construction. The first is a post and beam system. That means that vertical or inclined posts on horizontal beams support other horizontal or inclined beams creating the desired shape of a roof. Usually there is not any special geometric rule of components in that roof system. The advantage of that roof is its significant flexibility and as a result the ability to deform, absorbing energy during seismic action. In that system usually the first, horizontal layer of beams undertake the task of connecting the walls and creating the box-frame action. Also a diaphragm action, if any, is performed by the ceiling construction (Fig. 7,8).

The second group is using more geometrically defined components (as trusses e.t.c.). Here the energy absorbing capacity is much lower, the rigidity higher and the diaphragmatic behaviour of the upper levels of the roof more frequent (Fig. 8).

6. EXAMPLES OF THE ASEISMIC TECHNIQUES (AND CONSTRUCTIONS) IN GREEK HISTORY

6.1. PARTHENON

The temple of Parthenon, on the top of the Acropolis of Athens, built in the unbelievably short time of eight years (447-438 B.C.), has experienced during the last 2.500 years many earthquakes, some quite severe, leaving on the structure unmistakable prints.

Recent works of restoration revealed more details about the perfect work of the architects and craftsmen. Just the description on one such detail, from the

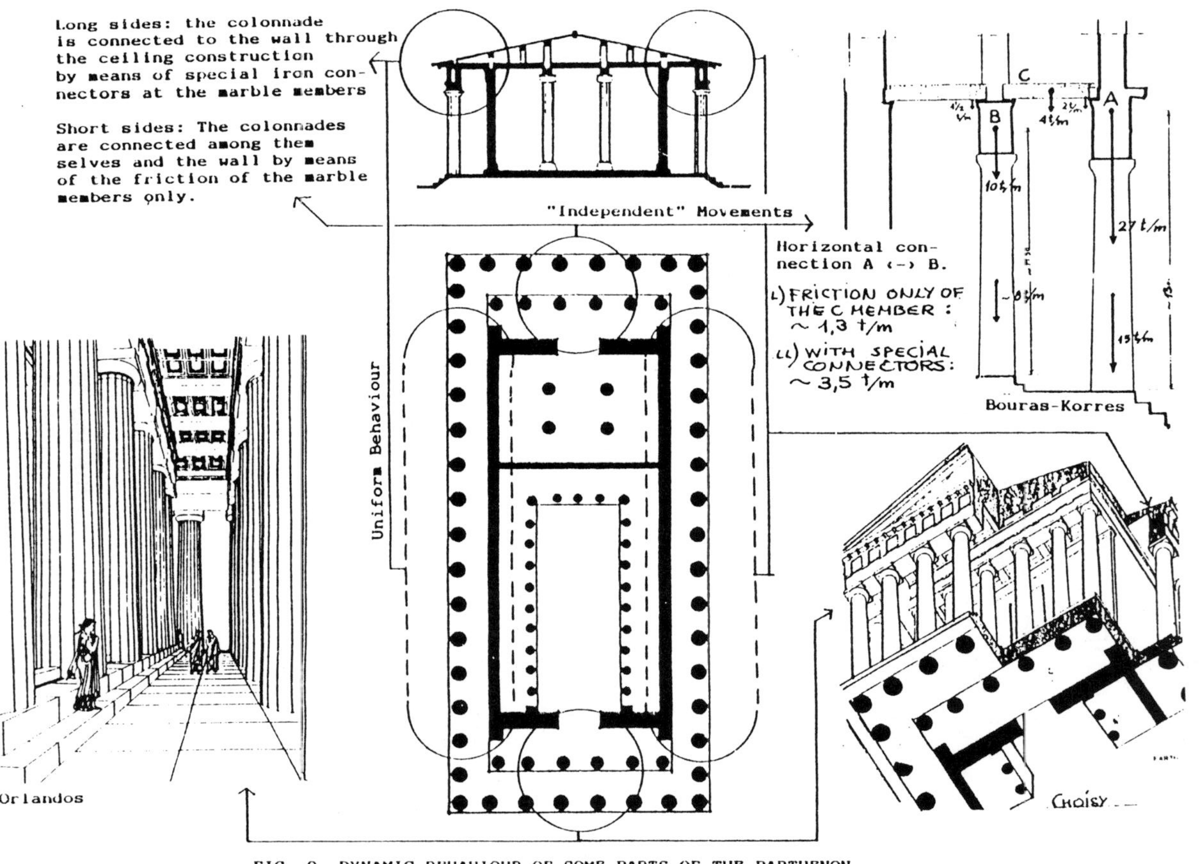

FIG. 9. DYNAMIC BEHAVIOUR OF SOME PARTS OF THE PARTHENON.

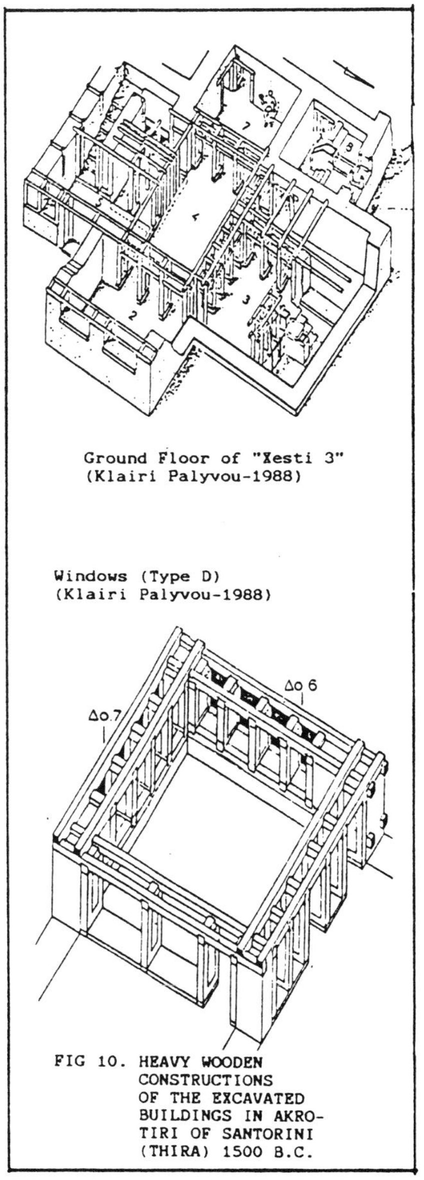

Ground Floor of "Xesti 3"
(Klairi Palyvou-1988)

Windows (Type D)
(Klairi Palyvou-1988)

FIG 10. HEAVY WOODEN
CONSTRUCTIONS
OF THE EXCAVATED
BUILDINGS IN AKRO-
TIRI OF SANTORINI
(THIRA) 1500 B.C.

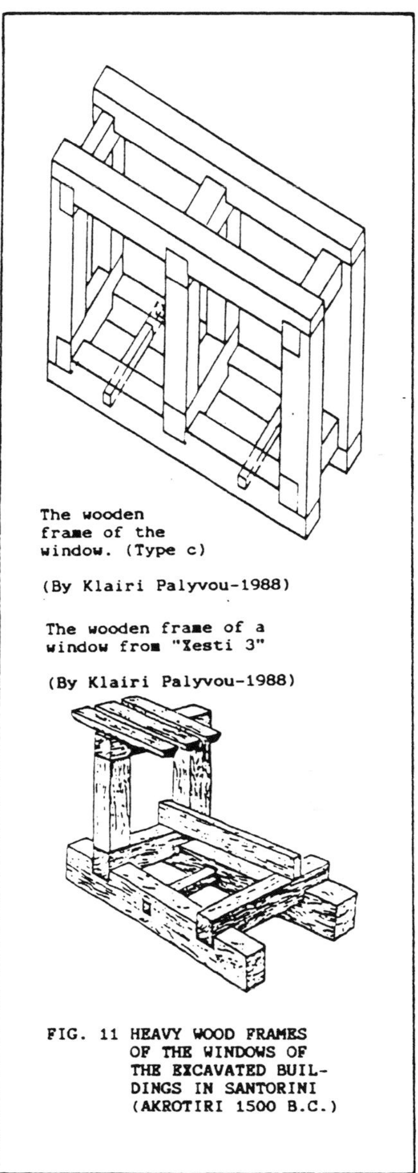

The wooden
frame of the
window. (Type c)

(By Klairi Palyvou-1988)

The wooden frame of a
window from "Xesti 3"

(By Klairi Palyvou-1988)

FIG. 11 HEAVY WOOD FRAMES
OF THE WINDOWS OF
THE EXCAVATED BUIL-
DINGS IN SANTORINI
(AKROTIRI 1500 B.C.)

study of H. Bouras and M. Korres about the parthenon restoration, can determine the level of aseismic technique knowledge 2.500 years ago.

Parthenon has all the features of the Doric Order in their perfection. Designed by Ictinos and Callicrates it was the wider and longer than any other temple of the time with an eight-columned (octastyle) portico rather then the usual six-columned (hexastyle) one.

Along the long sides of the temple the marble components, which cover the space between the colonnade and the wall are also connecting (them by means of special iron joints. In this way the collonnade (with its entablature) has to behave as a whole together with the wall of the nave, during dynamic action. This is performed successfully for 25 centuries because of the similarity of their masses (Fig. 9).

On the contrary, along the other two short sides (the East and the West ones) of the temple, the marble beams, as well as the marble plates, which cover the space between the colonnades and the wall are connecting them at a fixed distance only by means of the friction. That means that the ceiling components of those territories of the temple are simply placed on the entablature of the colonnade and the nave, without any joint element, permitting independent movement. This happens because the mass and the geometry of the colonnade at that point is quite different of those of the wall of the nave. During strong dynamic loading the movements (deflections) of the nave wall have different characteristics from those of the colonnades at the front. The beams would not be able (as the calculation also showed) to keep the distance between the nave wall and the colonnades fixed, even if they were connected by means of the heaviest types of joints, which have been found on the Parthenon.

6.2. "AKROTIRI EXAMPLE (1500 B.C.)

One of the most interesting and early examples of the earthquake resistant building technology is that a prehistoric town-ship at Akrotiri on the island of Thera (Santorini). About 1500 B.C. the volcano of that island erupted, submerging a large part of it and covering all the remaining surface by a thick layer of volcanic ash.

Recent excavations uncovered, after 35 centuries, a neighbourhood with densely built multistory buildings exceptionally decorated with wall-paintings, surrounded by paved streets with a drainage system running underneath.

This very sophisticated building technology, which is now under study, proves a very early experience on aseismic construction in a highly sensitive seismic region.

The main entrance opens into the main street net work and preferably a a point where there is a widening of the public open space.

These entrances are usually non symmetrical in plan, having the one narrow

FIG. 12. THE TRADITIONAL ASEISMIC WOOD FRAMED CONSTRUCTION
ON THE ISLAND OF LEFKAS - GREECE.

The secondary load bearing system of wooden columns:
(A) Activated after the partial failure of the main bearing system of the stone masonry.

(B) During an earthquake.

side wider. The probable reason is that at this side of the main entrance, is constructed next to the door, an "entrance window" which gives sufficient light to the staircase. This is a modern approach for the illumination of this part of the buildings for safety reasons.

It is interesting to underline the existence of a second auxiliary staircase in every building.

The main vertical load bearing system of the building is consisted of stone masonry. The foundation system is still not investigated and very few information is known. The usual height of the buildings is two or three floors.

Among the basic building materials such as : stone, wood, and mud, the role of wood in the defence strategy against seismic risk is surprisingly important.

The wood itself has vanished through out the centuries, but very detailed imprints have been left on the volcanic material or/and on the mud mortar of the construction.

To improve the tension capacity of the stone walls, wood grids are embodied in them with an additional support of vertical studs in some cases.

Unusually reinforced, load bearing frames around door and window openings, are sophisticatedly constructed in three-dimensional systems. Heavy wood profiles are carefully interbonded through mortises and dowels with their ends embedded in the masonry or connected to the wood members which reinforce the masonry.

Similarly constructed wood framed partitions replace in some parts of the building the masonry walls. (Fig. 11).

These wood constructed load-bearing systems, very commonly used in the Akrotiri construction, called :

    a. A pier-and-door partition (polythyron)

    b. A pier-and-window partition (polyparathyron)

    c. A pier-and- closet partition (polyhermarion)

are still a subject under research (Fig.10).

This typology, deriving from the function of each partition is describing a very strong timber framed system of piers and beams in two parallel rows and formed carefully in a three- dimensional load bearing system. The stiffness of this wooden structural system results from :

    a. The frequent regular and well interbonded details of the connections between the beams (longitudinal and transverse) and the columns.

    b. The repetition of the vertical members every 60-80 cm. approximately.

    c. The infill between the wood members by the wall construction.

Wooden construction elements such as wood reinforcement system, lintels e.t.c. were listed according to their place, number and kind, on tablets of linear b (13th century B.C.), which referred to an account for constructional works.

The design of this construction and especially the details of the joints, which are capable for strong tension actions, prove the effort for survival of a people who, living on the slopes of an active volcano, building ships and navigating all

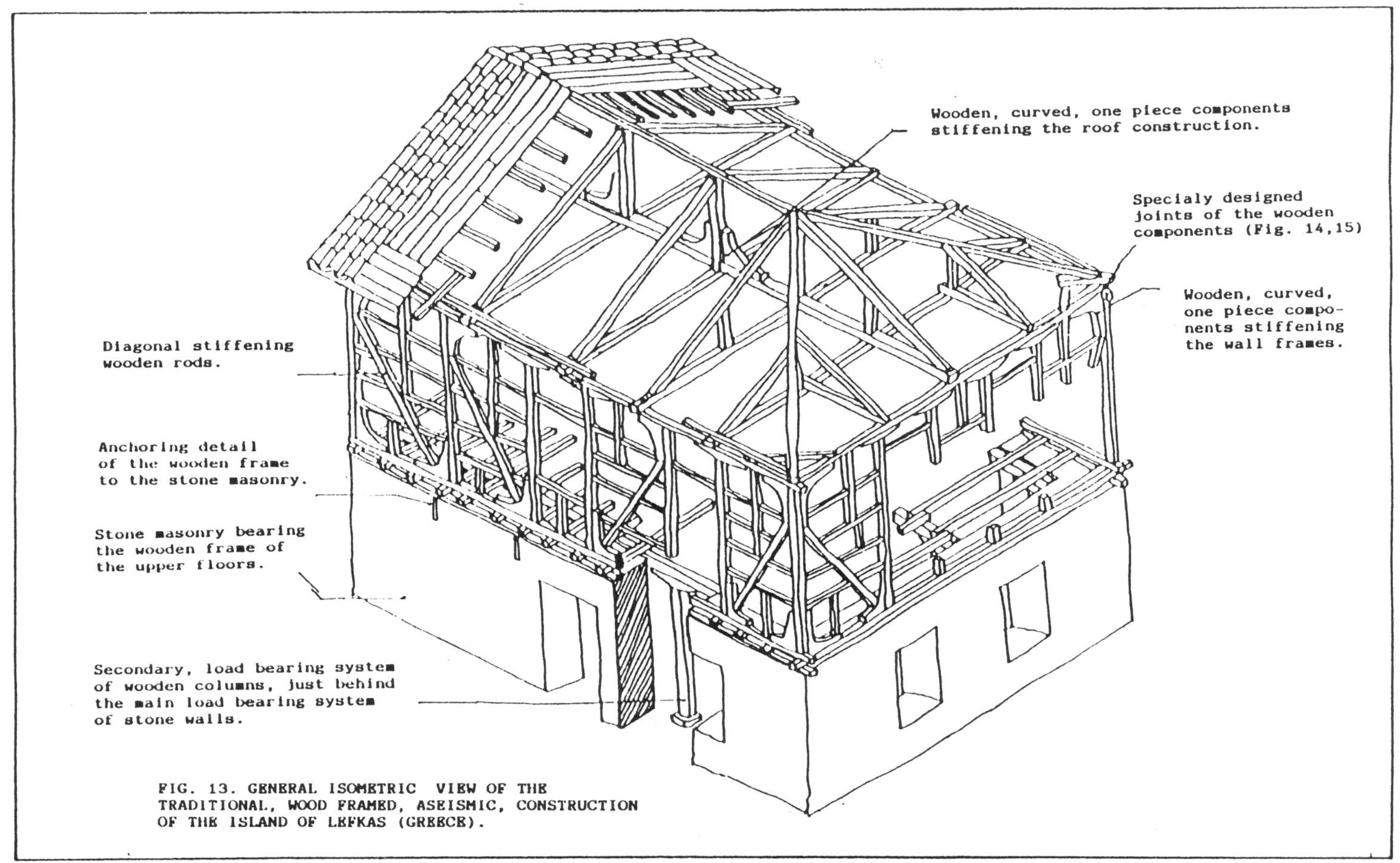

FIG. 13. GENERAL ISOMETRIC VIEW OF THE TRADITIONAL, WOOD FRAMED, ASEISMIC, CONSTRUCTION OF THE ISLAND OF LEFKAS (GREECE).

over the Mediterranean Sea, was familiarized with the constant seismic risk.

6.3. "LEFKAS" EXAMPLE (1825 A.D.)

Lefkas is one of the Greek islands with very high seismic risk. In 1825 the city of Lefkas was destroyed by a severe earthquake. After that, the English who occupied the island (1810-1864) established the first Aseismic Code. In 1827 new regulations about materials and building systems to be used were formulated. Today these systems of wood framed construction are still in common use, responding very statisfactorily to the frequent and strong earthquakes.

Multistory buildings are based upon a foundation which consists of a heavy wood grill covered by sand, stones and puzzolana. The ground floor, is surrounded by stone walls. The timber frame of the upper floors is supported by these walls. A secondary, load bearing system of wooden columns, like a second line of defence, is constructed in parallel line to the internal side of the stone walls, also supporting the same timber frame of the upper building. During severe earthquakes, parts of the stone walls can fall outside leaving the whole wood frame of the multistory building untouched and temporarily supported by the wooden columns until the masonry is repaired. (Fig. 12). While taking advantage of the obvious properties of a stone wall, such as strength, diaphragmatic behaviour, traditional appearance, prestige, security e.t.c. this type of construction system, by means of redundancy does not transfer the severe deformations or possible failures of the weaker parts, which in this case are stone walls, to the light timber structure (Fig. 13).

There it can be argued, that the principle of independently deformed but collaborating parts of a building was established at least two centuries ago, when it was realised that different constructions using various materials present different behaviour under seismic loading. A principle that is very important in the modern aseismic construction as well.

The wood frame, consisting of modulated vertical studs and horizontal beams and girders, is carefully stiffened by slanting wood rods and by wooden corner reinforcements curved out of a whole pieces of wood (usually branches or roots, in right angle, of an olive tree) (Fig. 13).

Sophisticated systems of interbonding of the timber parts are also found. The timber components, properly curved out and using nails, timber dowels and wedges, present resistance to tensile actions and a ductile behaviour always avoiding a too stiff composition (Fig. 14). Thus a second basic principle is recognized : the advantage of using strong timber joints with adequate ductile (or/and energy absorbing) behaviour (Fig. 15). The similarity of the described interbonding of the timer parts in Lefkas with those of the constructions of Akrotiri township in Santorini (13 centuries earlier) is noteworthy.

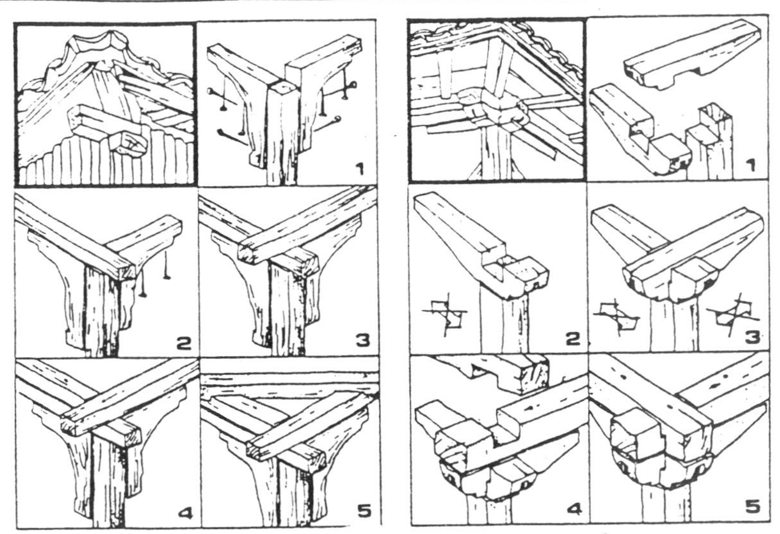

FIG. 14. ISLAND OF LEFKAS IN GREECE.
CHARACTERISTIC EXAMPLES OF THE JOINTS OF THE WOODEN
COMPONENTS OF THE TRADITIONAL WOOD-FRAMED ANTISEISMIC
LOCAL CONSTRUCTION SYSTEM

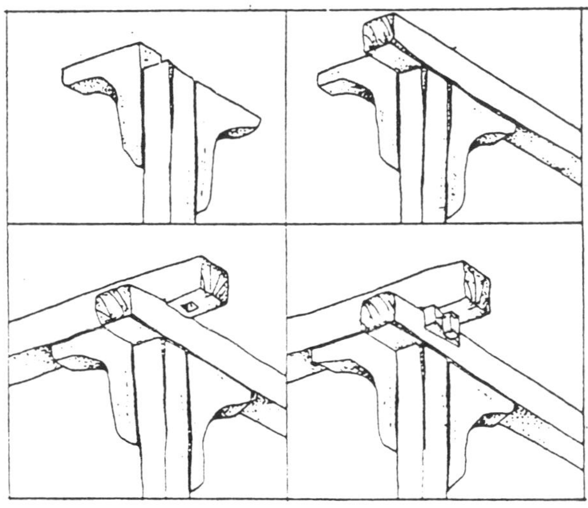

FIG. 15. ISLAND OG LEFKAS IN GREECE
CHARACTERISTIC EXAMPLE OF A JOINT WITH A SIMPLE,
ENERGY ABSORBING MECHANISM. WHEN BROKEN, THE
WOODEN DOWEL, CAN BE EASILY INSPECTED AND REPLACED.

7. EPILOGUE

It is a well established principle that today we must strive to preserve the aseismic behaviour of those old buildings without changing their initial architectural, statical and dynamic conception in any restoration or conservation project because the correctness of the chosen solutions has been proved several times during their existence all those years.

Our goal should be the re-establishment of their initial (at least) strength and resistance to the earthquakes in the most compatible and simple way. In this procedure the best possible knowledge of the relevant aseismic technique which has been used and the aseismic design principles in necessary.

The modern educational, social and administrative systems, world-wide, don't help very much neither the understanding of the traditional local aseismic technique and its design princinples, nor the development of special for each case repair and strengthening methods.

Today the specialist who desing or realize the repair or strengthening projects of a Historical Building frequently have been educated and/or are living in different places, even countries, far away from the subject of their study and its material, loading conditions and constructional originalities. The building regulations, on the other hand, usually don't contribute very much during the procedure of understanding an old structure and deciding the proper interventions. The modern building regulations are, more or less, general, usually have been composed by reinforced concrete specialists, and ignore any specific local environmental or constructional originality. Mostly, they are trying to, protect the historical building morphologically only. And of course the control systems of the proper application even of those non sufficient regulations are, usually, very weak. (Fig. 16).

The only way, in order the specialists, the designers, the constructors, the producers of special materials and the authorities of a territory rich in traditional construction, will be able to understand it and get familiar with its problems, is to develop an organised and analysed, local constructional data bank.

Using the proper information of such a data bank the designers will be able to discover any originality and vulnerability of the traditional structure and to decide correct and compatible intervention methods. The constructors will succeed in organizing the proper specialists for the use of local materials and buildings systems. The authorities will be able to create local regulations, so that the respect, the preservation, the development, and the correct exploitation of the cultural, constructional heritage will be a fact.

Following is a more analytical observation of a traditionally built constructions and their behaviour during seismic action, as well as, some general guidelines for their repair and strengthening.

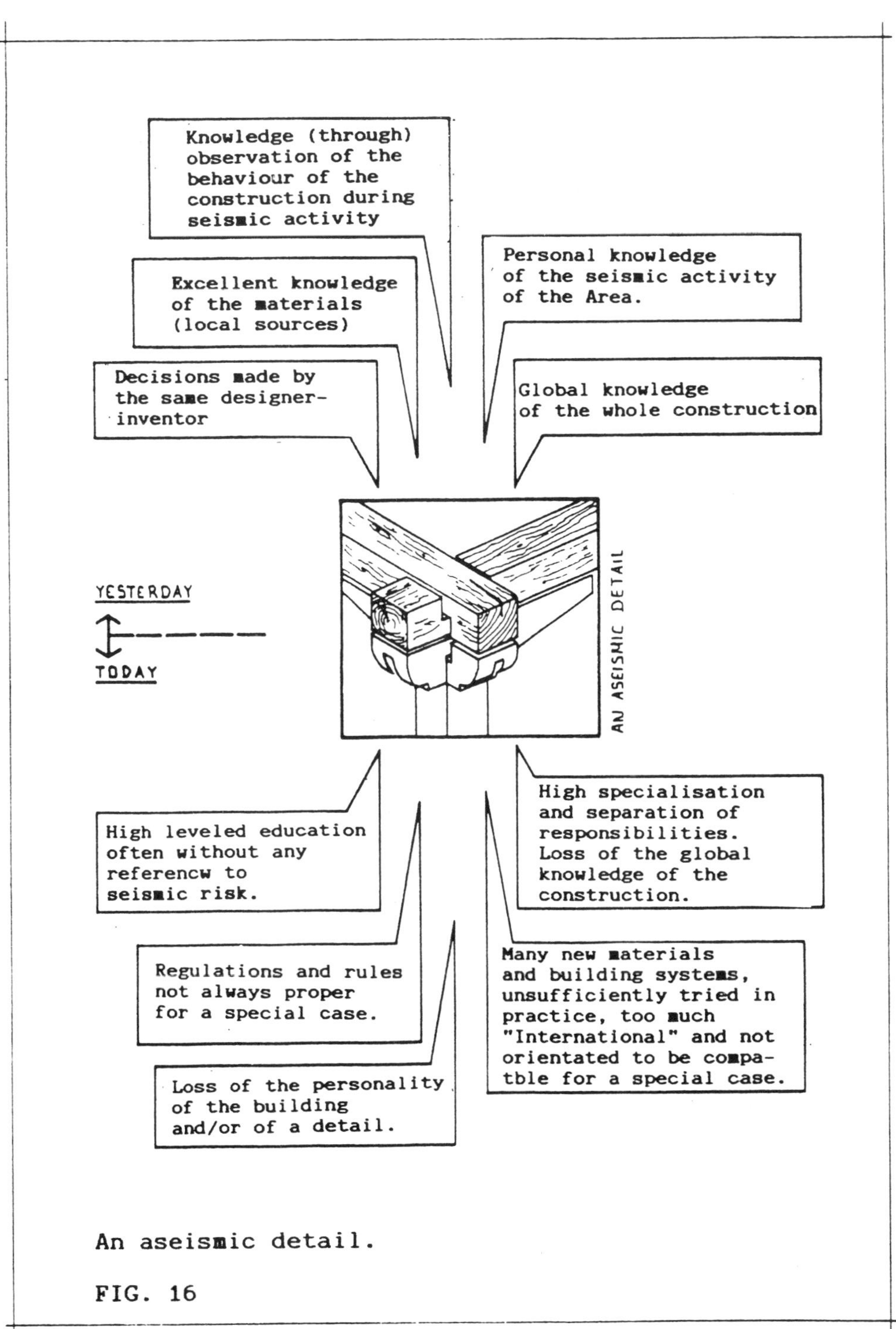

An aseismic detail.

FIG. 16

REFERENCES

1. J. EBERHARD, "ARCHITECTS AND EARTHQUAKES" New York 1975.
2. CHR. ARNOLD, R. REITHERMAN, "BUILDING CONFIGU- RATION AND SEISMIC DESIGN" New York 1982.
3. P. CARYDIS, O. VAGELATOU, I. VLACHOS, M. XINOSTATHI, P. TOULIATOS, K. HOLEVAS, "RESEARCH FOR THE CONSTRUCTIONAL ASEISMIC REGULATION", Athens 1985.
4. DOUMAS C., (AD), "THERA AND THE AEGEAN WORLD I-II", London 1978-80.
5. DOUMAS C., (AD), "THERA AND THE AEGEAN WORLD III", 24-30, London 1990.
6. D. GANTE, M. MITROPOULOS, "LEFKAS EARTHQUAKE - RESISTANT BUILDING TECHNOLOGY OF THE 17th CENTURY. RESEARCH NTUA", Athens 1990.
7. GREEK TRADITIONAL ARCHITECTURE PUBLISHED BY "MELISSA", Athens 1991.
8. M. KORRES, CH. BOURAS, "STUDY FOR THE RESTORATION OF PARTHENON", Athens 1983.
9. C. PALYVOU, "AKROTIRI, THERA : BUILDING TECHNIQUES AND MORPHOLOGY IN LATE CYCLADIC ARCHITECTURE". P.H.F. THESIS N.T.U.A., Athens 1988.
10. ROWLAND MAINSTONE "HAGIA SOPHIA", London 1988.
11. P. TOULIATOS, "INTERVENTIONS ON ARCHITECTURAL MONUMENTS AND HISTORICAL SETTLEMENTS, SEISMIC BEHAVIOUR OF TRADITIONALLY CONSTRUCTED BUILDINGS REPAIRS AND REINFORCEMENTS", Heraklio Crete 1988.
12. P. TOULIATOS, "THE SEISMIC DESIGN OF THE TIMBER CONSTRUCTION OF THE PAST AND TODAY", 1st GREEK CONFERENCE ON ASEISMIC ENGINEERING AND SEISMOLOGY, 2nd VOLUME, Athens 1992.
13. E. TSAKANIKA, S. LASOURAS, "TRADITIONALLY BUILT STRUCTURE OF THE 16th CENTURY". DIPLOMA THESIS N.T.U.A., Athens 1988.

SEISMIC PRECAUTIONARY MEASURES IN GREECE
ARCHANES CASE STUDY

P.G. Touliatos
National Technical University of Athens, Athens, Greece

ABSTRACT

Since very early times of human civilisation, like the prehistoric settlement of 1.500 B.C. at Akrotiri on the island of Santorini in Greece, the habitants of areas with high seismic risk had understood the necessity of a general communal resistance.
Today, all over the world and in Greece, efforts are done to organise such risky communities to resist at all levels of their structure against the seismic risk.

1. INTRODUCTION

Great civilizations had suffered great loses due to the anger of Engelados (the ancient Greek god of earthquakes). The long history of the building techniques in Greece is based upon the knowledge of the necessity of the aseismic construction.

Today, the direct cost of earthquake activity in Greece tolls about 20.000.000.000 DR. (7.000.000 ECU.) annually, but the indirect cost of the goods, the human activity and especially the cultural heritage is invaluable. Monuments, historical buildings and traditional constructions, suffer the most.

As a consequence of these facts, the earthquake problem in Greece is of primer concern since many centuries ago. Aseismic construction principles were implemented since prehistoric (Akrotiri in Santorini, Minoian Crete) and historic (Parthenon, Hagia Sophia during Byzantine times) periods.

Greece contributes in the mitigation of seismic risk through aseismic regulations at least since about two centuries ago. In Lefkas, (Ionian island) in 1825 were established the first aseismic construction regulations.

Since 1958, in Greece, it is obligatory the structural study of every new construction to take also into account the aseismic control of the construction. Therefor, 4% of the total cost of the construction is for aseismic reasons.

2. THE VULNERABILITY

The vulnerability of a populated area concerning the seismic risk, depends on many factors except the resistance of the constructions against dynamic loading. Some of these factors are : the ability of vital public services as hospitals, police stations, fire brigades, e.t.c. to operate successfully during and after the earthquake period, the preparedness of the community to organize the living facilities (tents, food e.t.c.) for the homeless population in prearranged areas of the city, the preparedness of the population to face a disaster without panic and despair, the capability of the community to repair the damages and reactivate the normal flow of life as soon as possible after the catastrophe e.t.c.

According to that philosophy, densely populated old cities could be more vulnerable than other settlements even if they exist in areas with lower seismicity.

All over Greece, especially in territories with high seismicity, parts, or preferably whole cities are preparing programs of global anti seismic protection taking into account the land-planning, the urbanistic, the construction regulations, the geophysical, the historical, the seismic intensity e.t.c. factors.

So, "Preparedness and Earthquake Protection Program" by various local and prefectural governments, are under study and development in Patras, Athens, Heraklion, Egion, Archanes e.t.c.

Seismological Institutions, University Laboratories, National and European Seismology and Earthquake Engineering Centres in Greece are co-operating closely in that effort.

One, such, characteristic example is of the town of Archanes in Crete island.

3. THE TOWN OF ARCHANES

Archanes is a small town (or a "large village") 15 Kilometers from the capital of Crete, the Heraklion city, to the south,

Archanes, are situated at the foot of the "Giouhta" mountain, a sacred mountain, in a very fertile territory, which was systematically cultivated for thousands of years.

The fertility of the earth, the short distance from the city of Heraklion and its harbour and the activity of the citizens brought Archanes to an outstanding and wealthy position.

The region of Archanes was inhabited very early since neolithic periods (6.000-2.800 B.C.).

Under the most important part of the town, a Minoian Palace has been discovered, built around 2.000 B.C. After the earthquake of 1.700 B.C. the Palace was rebuilt. Around the settlement, some very important sanctuaries of that period had been found. The most important is a temple of Anemospilia on the north side of the Giouhta mountain. The earthquake of 1.700 B.C. (and the fire that followed) interrupted a ceremony of human sacrifice, killing the 3 priests around the sacrificed body of the altar.

About 1.600 B.C. another earthquake, probably related to the volcano explosion of the island of Thira (Santorini), damages the constructions which are again rebuilt to live a flourishing period until the 1.450 B.C. The severe earthquake of 1.450 B.C., destroys everything on the island and stops the prosperous and glorious development of Archanes.

As the recent research proves, Archanes continues existing after 1.450 B.C. and is present and active in all the following historical periods.

Perhaps as the Architect - Archaeologist C. Tsompanaki notes : "after the Minoian, the second more significant period for the town is the recent one, after the 1898 until our days".

Two crucial factors make the case of Archanes especially interesting :
- The existence of the Minoian Palace under the central area of the settlement, helped its buildings to survive. The citizens avoided to demolish any old construction out of the fear of the excavations and tried to maintain and preserve them. There are buildings which use parts of the Minoian prehistoric masonry (up to 2.5 meters high, with stone blocks up to 1.5 meters length) in their foundations, or/and even their walls.
- The seismicity of the territory of Archanes (and Heraklion) is one of the highest

in Greece and Europe. Except the already mentioned catastrophic earthquakes of the prehistoric period the following examples can show the intensity of the local seismic activity :
1. In 1810, one third of the buildings in Heraklion city were destroyed as well as an old Monastery.
2. During the severe earthquake of 1856 the city of Heraklion (the capital of Crete island) was almost completely destroyed. From 3.620 buildings only 18 remained standing and functioning.
3. In 1926, an earthquake destroyed in Heraklion 200 buildings and caused serious damages to the Archaeological Museum and its treasures.

4. PREPAREDNESS AND EARTHQUAKE PROTECTION PROGRAM OF ARCHANES

The scope of this project is to organize and co-ordinate a general preparedness and protection program for the specific urban community in case of a natural disaster and especially for an earthquake.

The project started with the second semester of the 1992 year. The first stage ended with the end of 1993. It was financed by the "Integrated Mediterranean Program" (MOP) for Crete (which is estimated to reach 9.000.000 ECU and the 70% comes from the European Community).

Many originalities were realized during this project, such as :
- The European Community contribution came directly to the municipality of Archanes avoiding the usual way through the Central Administration of the Government. The building of these money was realized by the Municipality of Archanes itself.
- The Greek contribution was achieved by the citizens of Archanes themselves and not the central administration. This fact secured the enthusiastic co-operation of the population throughout the duration of the first stage, offered a great flexibility, caused the public sensitisation and permitted the citizens to obtain direct experience of the actions and the direct feeling of the benefits of the results.

The administrative support for the project was realized by the specially developed : "Development Organization of Temenos and Pediada".

But the technical support was organized by the Archanes Municipality itself.

The usual way of organization and decision of the type of the different interventions through the National regulations and instructions was completely avoided. The Municipality in close co-operation with the National Technical University of Athens prepared simultaneously :
a. Special studies for the repair and strengthening of each building.
b. Special studies for the aesthetical, urban and social improvement of the central parts of the town.

c. Special studies for the global aseismic policy of the town.

5. REPAIR AND STRENGTHENING

The Municipality of Archanes in co-operation with various departments of the National Technical University of Athens proceeded to the following activities :
- Research and specification of the local geophysical and seismical characteristics.
- Research and specification of the local Architectural and Constructional particularity. For that purpose the local constructional systems were studied and analysed. Local or more general damages and failures were detected and analysed. The causes of their appearance were explained. In that procedure a technical questionnaire - list was prepared and completed for each building.

So the local constructional system(s) was oriented and every typical weakness or/and defectiveness was detected.

Then by some seminars and publications the local scientific and labour potentialities were organized and properly instructed to carry out the repair and strengthening works, under the supervision of the "Technical Counsellors" (i.e. The Technical Division of the Municipality, N.T.U.A.).

During the first stage (until the end of 1993) 60 buildings were repaired, strengthened and renovated.

There are at the moment 280 new applications of the owners of buildings for the same procedure for the second stage of intervention.

6. URBAN, AESTHETICAL AND SOCIAL IMPROVEMENT

Besides the structural repair and strengthening of the constructions of the town, some very important goals were simultaneously obtained, (by special studies) :
- The restoration of the buildings to a more healthy and viable condition.
- The preservation, encouragement and/or restoration of the traditional and cultural character or the town. The relevant study for the central area of the town is completed during the first stage.
- The resurgence of the social acceptance of these cultural and antiseismic preventive goals.

7. GLOBAL ASEISMIC POLICY

During the second stage (which is expected to start during the 1994) a general seismic disaster confrontation plan for the whole community will be

completed with the following main topics :
- Local aseismic constructional regulations concerning the interventions to the traditional building.
- Detailed geophysical study.
- Classification of the buildings to three (or five) categories of resistance to dynamic action, and the relevant strengthening.
- Organization of places of assembly of the population after the disaster with water, first aids, food, communication and shelter potentials.
- The special education and preparation for the seismic disaster of all the social groups as children, teachers, special services, doctors, technical scientists, e.t.c.

CONCLUSION

The seismic activity is unavoidable in our country. The prediction is a very useful weapon in our hands and strong efforts are realised towards the two more developed earthquake prediction systems which are under study in Greece. But even if we reach the point of practically effective prediction system, still the preparedness shall be our main goal. Still the prevention will be better than cure. Still the necessity to protect our cultural heritage our property and our lives will be urgent.

The Archanes Earthquake Protection Program is ambitious to become the pilot program for the rest of the, for 4.000 years alive and stubbornly fighting the seismic disaster, population of Crete.

REFERENCES

1. I. LIAPIS - A. VOULGARIS-VEZIROGLOU, "GENERAL UPGRADING AND BUILDINGS RESTORATION OF THE CENTRAL PART OF ARCHANES", N.T.U.A., Athens 1992.
2. J. AND E. SAKELLARAKIS, "ARCHANES-CRETE", Athens 1991.

SEA OF CRETE

SEA OF CRETE
ΟΛΠΟΣ ΗΡΑΚΛΕΙΟΥ
IRAKLION
ΗΡΑΚΛΕΙΟ
ΚΟΛΠΟΣ ΜΑΛΙΩΝ
Λιμένας Χερσονήσου
ΚΝΩΣΟΣ
Μάλια
Μοχός
Κράσι
Καρέ
ΚΑΤΩ ΑΡΧΑΝΕΣ
Επισκοπή
ΕΠΑΝΩ ΑΡΧΑΝΕΣ
ARKHANES
Καστέλι
Αγ. Ηλίας
Λουτρά
Θραψανό
Athens
GREECE
Aegean Sea
Cyclades
Thera
Iraklion
KNOSSOS
Crete
ZAKROS
HAGIA TRIADA
ARKHANES
PHAISTOS
ANEMOSPILIA
ANEMOSPILIA
ΚΑΤΩ ΑΡΧΑΝΕΣ
Γιούχτας
ΕΠΑΝΩ ΑΡΧΑΝΕΣ
ARKHANES
Σπήλιος Στραβομύτη
Ξερή Καρά
Βαθύπετρο

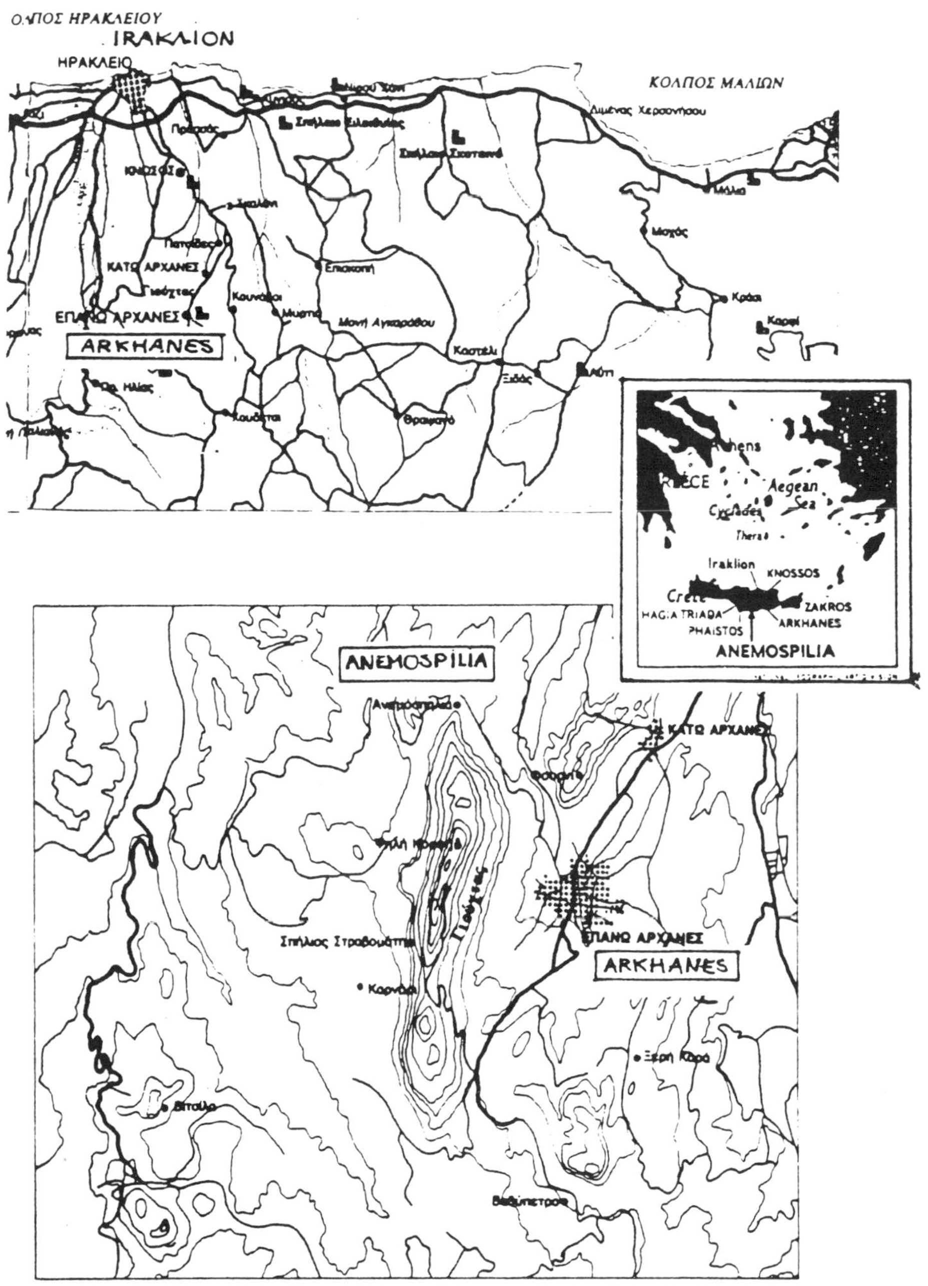

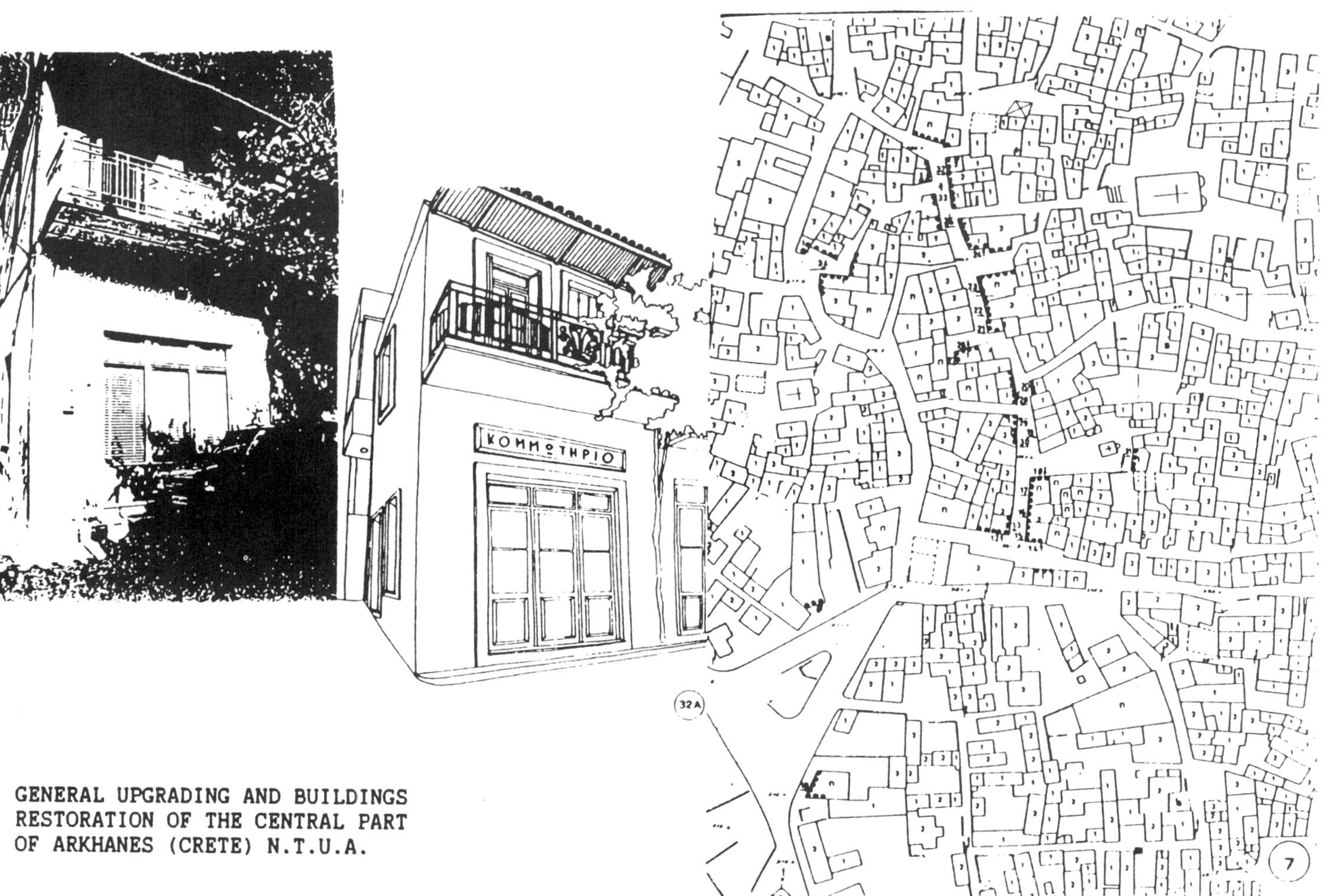

GENERAL UPGRADING AND BUILDINGS RESTORATION OF THE CENTRAL PART OF ARKHANES (CRETE) N.T.U.A.

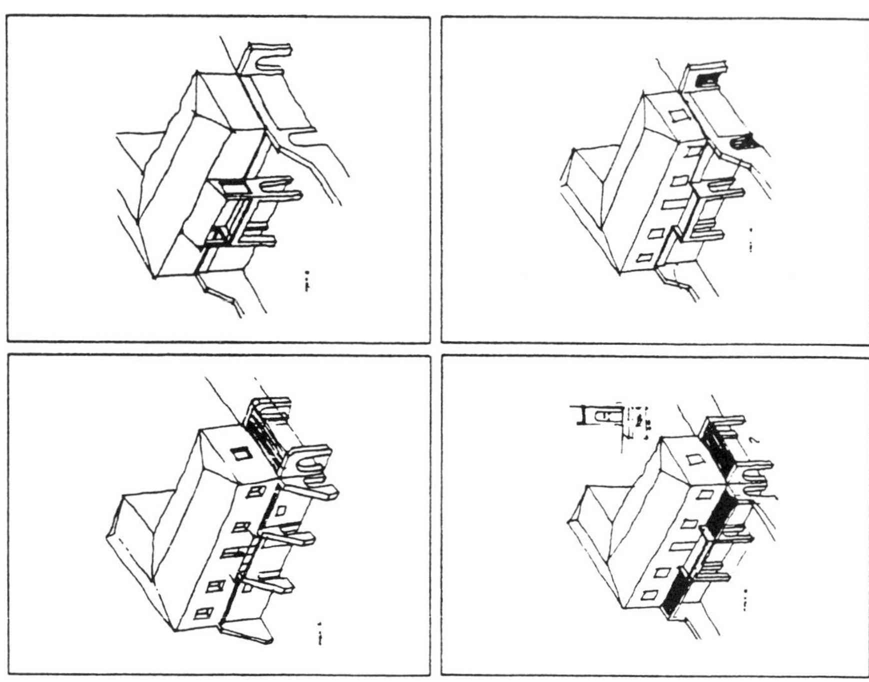

SEVERAL INITIAL PROPOSALS TO SUPPORT A
HEAVILY DAMAGED BUILDING IN ARCHANES

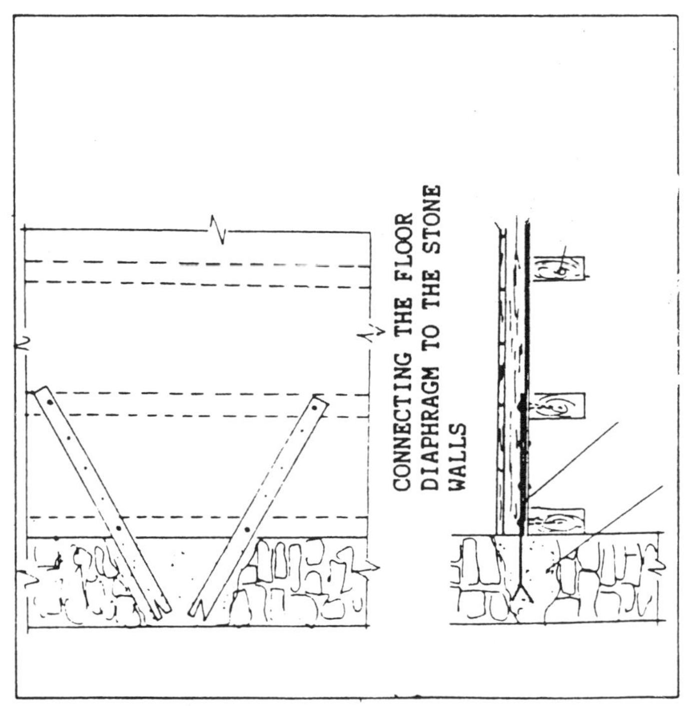

CONNECTING THE FLOOR
DIAPHRAGM TO THE STONE
WALLS

TENTATIVE GUIDELINES FOR PROTECTION AND REHABILITATION OF HISTORICAL BUILDINGS SUBJECTED TO EARTHQUAKES

C.A. Syrmakezis

National Technical University of Athens, Athens, Greece

ABSTRACT:
In this text, tentative guidelines for protection and
rehabilitation of historical buildings subjected to
earthquakes are presented. After an introduction referring
to the task of intervention on a historical building, the
restrictions due to the lack of information and the nature
of the historical building are briefly described. The
intervention process on a historical building is
subsequently presented, and its phases are briefly
described. Descriptions refer to the following points of the
historical building intervention process: the identification
of the basic structural character of the building, the
investigation and documentation of its present condition, of
the analysis of the structure, of the repair and
strengthening methods and technics, and of the special needs
for supervision of the repair work.

1. INTRODUCTION

Speaking from the engineering point of view about architectural heritage, one has to include all historical structures, built by all previous human generations. Historical buildings, to be considered in this text, compared to the other types of structures (e.g. bridges, viaducts, etc), represent the majority of the historical structures. Monuments (seen as historical buildings of extremely high importance), as well as traditional buildings (seen as old buildings of much lower importance), are also represented in most of the statements of this report.

Repairing and strengthening of a historical building is indeed a difficult and challenging task. Its cultural value and the desire to preserve it for future generations, demands a high level protection against any possible future destruction under future actions, in which earthquake is included. At the same time, any intervention to accomplish this task, has not to harm what has survived till today.

To accomplish this task, a precise understanding of the problems facing the structure, the reasons for them, as well as a sound knowledge of the effect any intervention might have on the structure, is needed, so that intervention does not become the cause of future damage to the historical building under consideration.

The task of such an intervention on a historical building is not an easy one. The fact that these structures have survived in the past, does not guarantee that they will survive in the future also. The damage sustained by a historical building throughout its history, whether it was caused by natural phenomena (e. g. earthquakes), or by man's willful destruction or even good intentions, may render the structure incapable of withstanding future shocks.

The assessment of a historical building resistance, so that it may be strengthened and thus preserved, presents special problems, not encountered in everyday structures. These problems stem from the ways in which the structures were built, the types of materials used, the complex past histories of successive changes and their progressive deterioration over the years, as well as its cultural value, sometimes as a work of art.

Special attention has always to be paied on the earthquake action. Although the damage caused to a modern day structure by an earthquake may run in a very high cost, the damage caused to a historical building is much higher. These structures are surviving representatives of man's past achievements and are looked upon today with admiration and respect.

It has to be reminded that for masonry structures in seismic zones, earthquake is the number one "enemy", due to the unpleasant coincidence of zone seismicity and masonry very bad structural response to earthquakes.

The burden of protecting a historical building falls mainly on the shoulders of the designer. Definitely, a successful intervention on a historical monument, prerequisites a good comprehension of its structural response, either under static, or under dynamic (earthquake) loading.

For the designers (architects, civil engineers, joint teams) taking part to the preservation process of a historical building, means mainly being faced with the demanding task of providing the historical building with the ability to withstand future actions with the minimum possible amount of damage, while bearing in mind the characteristics and values which make this building sometimes unique and worthy of special attention.

It is necessary that several steps be taken, in order to understand the structural behaviour of the historical building, especially under earthquake forces, and if necessary, methods of intervention, namely restoration, strengthening, anastylosis, taking into account their priority at national, as well as, sometimes, international level.

Definitely, a successful intervention on a historical building prerequisites a good comprehension of its structural response, either under static, or under dynamic (earthquake) loading.

It is to be noted that, till today, relatively little research has been conducted, and very little has been written to aid in overcoming the many problems associated with the design of the repair and/or strengthening process. Studies are needed on the different varieties of structural elements and structural systems to give a better understanding of the behaviour of the structures under consideration. These might include experimental testing also. Research on the materials of the past versus those available today, as well as on the techniques suitable for intervention, should be conducted.

Guidelines should also be provided, wherever possible. This text represents an initial effort, for such a tool, towards this direction.

Finally objective critiques of the repair and strengthening work already completed should be made. These critiques should deal with specific aspects of the past work on such structures, and should perhaps provide comparison studies of the techniques used to accomplish the same goal.

It is to be noted that, till today, the research that has been conducted on the questions, although significant, can not be considered yet as complete. And much more work is needed, to support reliably the designers efforts in overcoming the problems associated with the analysis of the repair and/or strengthening process of a historical building. Studies are needed on the different varieties of structural elements and structural systems to give a better understanding of the behaviour of the structures under consideration. Due to the complexity of the problem, these studies have to include experimental testing also, for a better check of analytical results. Research on the materials of the past versus those available today, as well as on the techniques suitable for intervention, should be conducted. Finally objective critiques of the repair and strengthening work already completed should be made. These critiques should deal with specific aspects of the past work on such structures, and should perhaps provide comparison studies of the techniques used to accomplish the same goal. Documentation of past work is an important and obvious prerequisite for this task.

Of course, the inherent structural complexity, the variability of building materials as well as the degree of damage and deterioration that the structure has sustained throughout its long existence, makes each case a unique one for study. Nevertheless, any effort lightening the burden of the engineer and helping him to achieve his goal, is highly desirable.

2. RESTRICTIONS DUE TO THE LACK OF INFORMATION AND THE NATURE OF THE HISTORICAL BUILDING

Unfortunately, for historical buildings, there are no at the moment satisfactory codes of practice, design aids or recommended procedures to aid in the analysis and design for repair and strengthening to accomplish the aforementioned process.

Under this circumstances, collection of experience and bibliography, as well as exchange of information, are of primary importance for the near future.

On the other hand, there are quite a few restrictions which provide a further barrier to be overcome. These barriers come in the form of documents on which present day protection practices are based. These documents look mainly at the cultural, historic, and artistic point of view of preserving the building.

Preservation of a historical building demands that the architectural, archaeological, and historical aspects be taken fully into account by the engineer, and as such involves the unified effort of architects, archaeologists, seismologists and other professionals with relevant expertise, to the engineer's task. Under these circumstances, the effects of restoring - repairing - strengthening a historical building, in relation always with enforcing a great number of principles, is a rather difficult but also attractive task for an experienced engineer.

Several examples of such restrictions can be found, for instance for the case of monuments, in the chart of Venice of 1964. This document include general articles which are concerned with restoration from the archaeological - architectural point of view, and not specifically with the need for intervention to provide the resistance of the structure. However, the principles are relevant, and although they are mainly related to the effort of a structural rehabilitation of monuments, they definitely cover cases of historical buildings as well, at least the most important of them.

Here are some examples of this kind, taken from the aforementioned chart, to which, correlation of technical decisions is necessary.

"The process of restoration is a highly specialized operation. Its aim is to preserve and reveal the aesthetic and historical value of the monument and is based on respect for original material and authentic documents. It must stop at the point where conjecture begins, and in this case moreover any extra work which is indispensable must be distinct from the architectural composition and must bear a contemporary stamp. The restoration in any case must be preceded and followed by an archaeological and historical study of the monument. " (Article 9).

"The valid contributions of all periods to the building of a monument must be respected, since unity of style is not the aim of a restoration. When a building includes the superimposed work of different periods, the revealing of the underlying state can only be justified in exceptional circumstances and when what is removed is of little interest and the material which is brought to light is of great historical, archaeological, or aesthetic value, and its state of preservation good enough to justify the action. Evaluation of the importance of the elements involved and the decision as to what may be destroyed cannot rest solely on the individual in charge of the work" (Article 11).

"Replacements of missing parts must integrate harmoniously with the whole, but at the same time must be distinguishable from the original so that restoration does not falsify the artistic or historical evidence" (Article 12).

"Additions cannot be allowed except in so far as they do not detract from the interesting parts of the building, its historical setting, the balance of its composition and its relation with its surroundings" (Article 13).

"Ruins must be maintained and measures necessary for the permanent conservation and protection of architectural features and of objects discovered must be undertaken. Furthermore, every means must be taken to facilitate the understanding of the monument and to reveal it without ever distorting its meaning. All reconstruction work should however be ruled out a priori. Only anastylosis, that is to say, the reassembling of existing but dismembered parts can be permitted. The material used for integration should always be recognizable and its use should be the least that will ensure the conservation of a monument and the reinstatement of its form" (Article 15).

"In all works of preservation and restoration there should always be precise documentation in the form of analytical and critical reports, illustrated with drawings and photographs. Every stage of the work of clearing, consolidation, rearrangement, and integration as well as technical and formal features identified during the course of the work should be included. This record should be placed in the archives of a public institution and made available to research workers. It is recommended that the report should be published" (Article 16).

3. PROCESSING TO AN INTERVENTION ON A HISTORICAL BUILDING

Preserving a historical building, means mainly being faced with the demanding task of providing the historical building with the ability to withstand future earthquakes with the minimum possible amount of damage, while bearing in mind the characteristics and values which make this historical building, sometimes, a rare architectural example, worthy of special attention.

In the case of an historical building, the engineer must have a sound appreciation of its structural characteristics and its expected behaviour, in order that he may effectively assess the present condition of the structure, model it for the purpose of analysis, and finally make the appropriate choice of the methods of repair and strengthening he will use. It is therefore of great importance to understand how the structures of yesterday differ from the various types of structures of today.

Namely, it is necessary that several steps be taken, in order to understand the structural behaviour of the historical building (especially under earthquake forces), and if necessary, the methods of intervention, namely restoration, strengthening, anastylosis, taking into account their priority at national, as well as international level.

The basic steps of this complex procedure, which must be followed in meeting the challenge of preserving historical building(generally, but against earthquakes also), can be considered as follows.

■ Identification of the essential structural character of the historical building (e. g. what are the basic structural elements, how does the structure behave as a whole during an earthquake, etc).

■ Investigation and documentation of the present condition of the structure (including any emergency action such as temporary shoring to safeguard the damaged structure). In this step, careful examination of the structural system, the materials, the degree of deterioration, etc. is needed.

■ Analysis of the structure to determine its strength in relation to future seismic loading.

■ Determination of the methods and techniques to be used for the repair and strengthening of the structure.

■   Supervision of the repair work and introduction of any modification that is deemed necessary in light of new observations and facts.

■   Documentation of all the above steps for future reference.

These steps will be presented briefly in the next paragraphs of this chapter.

4. IDENTIFICATION OF THE BASIC STRUCTURAL CHARACTER

A sound appreciation of a historical building's structural characteristics and its expected behaviour during an earthquake is absolutely necessary step: for the effective assessment of the present condition of the structure, for its modeling for the purpose of analysis, and finally for making the appropriate choice of the methods of repair and strengthening to be used.

It is therefore of great importance to understand how the structures of yesterday differ from the various types (e. g. moment resisting frames, shear walls etc.) of structures of today. However historical buildings are generally complicated structures and there is an unfortunate task of knowledge and information concerning the behaviour of such structures, especially under seismic loads. What can be generally said about these structures is that they are typically more massive than todays structures and that they usually have to carry their loads primarily in compression, even under earthquake loading.

A historical building, is usually defined by a set of more or less independent members. Table 1, gives a list of some typical structural elements which are usually found in a historical buildings or a monument. The selective combination of these typical structural elements may form almost all the various known types of structures of the type considered.

The construction materials of yesterday also present a problem, in that little is known about their properties. The effective strengths of these materials are usually much lower than those used today and these already low strengths may be further reduced by the effects of weathering and other forms of deterioration. The engineer will have to rely on testing to determine the materials properties and use the result cautiously since the properties of a material

TABLE 1: TYPICAL STRUCTURAL ELEMENTS vs. TYPICAL RESPONSE AND PROBLEMS

Typical structural element	Typical response and problems
I. COLUMNS (FREE STANDING) -Monolithic -Superimposed drums	Rocking motion Failure usually by overturning Drum relative displacements
II. WALLS AND PIERS	For lateral loading, failure usually by overturning(slender walls) or by primary internal disruption (thicker walls) For in -plane loading, failure usually by shear
-Unmortared unshaped blocks	Imperfect bearing of blocks resulting excessively high local compression
-Unmortared fully shaped blocks	Separation of blocks
-Facing of unmortared shaped blocks with loose fill	Separation of facing, bowing outwards of facing
-Mortared fully shaped blocks	Originally weak mortar or mortar which has weakened with time
-Timber framed walls	Decay of timber
III. BEAMS AND LINTELS -Timber beams -Stone lintels	 Decay of timber Flexural failure
IV. ARCHES	Outwards thrust at support level Slip of the keystone
V. VAULTS	Outwards thrust at support level Dropping of crown
VI. DOMES	Radial cracking as a result of outward thrusting at the base of dome
VII. TIES -Timber ties	 Decay of timber Inadequate end anchorages
-Iron ties	Rusting of iron Inadequate end anchorages
VIII. BUTRESSES	

may vary from location to location in the same structure.

It should be noted that most of the historical buildings were built with specific consideration given to their geometry and aesthetic quality and not necessarily to their structural integrity. In addition, the majority of them, existing today, have been subjected to many additions, changes in style, devastations from fires, earthquakes, and wars, reconstructions and attempts at restorations and consolidations, all of which result in a structure which may have a passing resemblance only to the original.

5. INVESTIGATION AND DOCUMENTATION OF PRESENT CONDITION

The investigation of the present condition of a structure is always the first step for its repair and strengthening. In the case of structures of cultural or historical significance, where it is necessary to preserve the original character and fabric, this takes on added importance.

The first step in such an investigation involves a visual inspection of the historical building and/or monument, and a precise recording of all damage to the structure. These recordings should include all of the following information.

- cracking of masonry,

- deformations of arches and vaults,

- tilting of walls, piers, or columns,

- differential settlements,

- slips or failures of tie members,

- condition and safety of decorations or other contents of value,

- manner of construction,

- state of deterioration due to weathering, erosion etc. ,

- any other information which might be deemed helpful in the analysis and repair work.

Ideally this information should be recorded before a structure suffers the damaging effects of another earthquake. Unfortunately, this is not usually the case. The engineer is usually called to make an accurate assessment of the condition of the structure and the need for repair after an earthquake has occurred. He is asked to make his assessment without any accurate and systematic architectural or structural plans, with no record of the successive additions and changes to the structure in the past, with no record of the damage, cracks and inclinations of the structure before the earthquake and with no specifications of the strengths and properties of the construction materials used.

Of course, visual inspection is not always enough, as not all defects are visible to the naked eye. A case in point is the Parthenon in Athens. The marble blocks of the inside of the Parthenon, including those of the floor, were subjected to thermal fracture during a fire at the end of the Roman period. The injure sustained by the blocks went unnoticed behind their well preserved surfaces due to the fact that the pieces of marble affected by thermal fracture adhere to the firmer marble from which they have separated but not broken away.

Damage of this kind, as well as damage which has been cosmetically repaired in past interventions, require more sophisticated methods of inspection with personnel trained in their use.

For the calculation of the seismic resistance of the structure, a more systematic assessment of the actual structural condition of the historical building has to be performed, based mainly on the existing information, as well as on the pathology image of the structure.

General deterioration and specific damages to masonry structures are usually due to several coexisting causes (combination of damaging actions). This is valid even for cases where damage seems to be suddenly produced, after an accidental event (e.g. earthquake, landslide, etc). An early investigation of the permanently existing causes, combined with an early intervention against them, should make the structures response more reliable against future causes, regular or unpredictable.

Correct classification, as well as deep comprehension, of all causes, are prerequisites to this task.

Pathological causes can be either i n t e r n a l

causes, related to the structure, or   e x t e r n a l
causes, related to its environment.

Masonry pathological causes are always due to an
internal cause, activated by an external one. In the Tables
2 and 3, a classification of these pathological causes is
presented.

TABLE 2: INTERNAL CAUSES

- ■ RELATED TO MATERIAL BEHAVIOUR:
  - General Quality
  - General Response (unreliable)
  - Special Characteristics (e.g.
    limited tensile strength)
  - etc.
- ■ RELATED TO ELEMENT BEHAVIOUR:
  - Unreliability of Materials quality
  - Use of various incompatible materials
  - Materials degradation
  - Understrength - Overstress
  - Low Construction Quality
  - Bad connections to other members
  - etc.
- ■ RELATED TO STRUCTURE BEHAVIOUR:
  - Bad conception
  - Unsuccessful morphology
  - Understrength - Overstress
  - Foundations inefficiency
  - etc.

On the other hand, one must have a knowledge of the
densities, unit strengths and moduli of elasticity of the
materials used in its construction.

However, as mentioned earlier, there are usually no
specifications available giving the properties of materials
used in historical buildings, and testing is proved always
a reliable methodology. This is of course a problem in
itself. There will be difficulties in extracting samples
suitable for tests. Taking core samples may not be possible
as it is destructive and therefore not permitted (e. g. by
the Venice Charter).

Consideration must also be given to the fact that most
testing laboratories are geared for the needs of modern

structures. Finally because of the lack of homogeneity and uniformity of the masonry construction, testing may give unit strengths for constituent materials which are not representative of the overall strength. Properties of a

TABLE 3: EXTERNAL CAUSES

<table>
<tr><td>

■ ENVIRONMENTAL ACTIONS,
  MAINLY OF LOW INTENSITY AND PERMANENT

- Mechanical Effects (e.g. vibrations due to normal traffic)

- Thermal effects (e.g. dilatation problems)

- Humidity effects (e.g. wetting - drying cyclic action, water penetration, etc)

- Physical - mechanical effects (e.g. erosion by rain)

- Physical - chemical effects (e.g. physicochemical attack mainly for embedded elements: wood, steel,etc)

- Chemical Attack (e.g.sulfate attack due to crystalization of salts)

- Corrosion attack (e.g. on wooden elements, steel elements etc)

- Root action

- etc.

</td></tr>
<tr><td>

■ ACCIDENTAL ACTIONS,
  MAINLY OF HIGH INTENSITY AND RARE

- Fire

- Foundations Settlement or Sliding

- Earthquake

- etc.

</td></tr>
</table>

particular type of material may vary, depending on theworkmanship degree of weathering, distribution of cracks, etc.

Further information necessary for the analysis, such as the fundamental period or the ductility ratio, are desirable but cannot always be determined by testing, as would be necessary, as such a testing can be proved destructive for the structure.

Finally all aspects of the investigation must be scrupulously recorded for future use.

6.   ANALYSIS OF THE STRUCTURE

6.1   General

One of the biggest problems facing the engineer is the analysis of the structure to determine its seismic resistance, thus enabling him to decide on which actions must be taken to strengthen it.

The analysis must consider the structure as it is, which usually involves, taking into account an irregular structure in both plan and elevation, with many damaged or even missing structural elements. The structural system will be a combination of walls, columns, beams, arches, vaults, and/or domes, constructed of various types of materials, subjected to various degrees of deterioration, and almost assuredly cracked. And it is very difficult, if not impossible, for the engineer to take into account the cracks in a material, whose properties and strength he knows little about.

Later, having accomplished the first analysis with due consideration given to all the factors which affect the seismic performance of a structure, the engineer must decide on what interventions will be necessary to strengthen the monument. The structure must then be reanalyzed taking into account all the proposals for strengthening, since any alteration to the present state may alter the behaviour of the structure. Prerequisite to that is of course the additional task of studying and assessing the historical substance before repair and restoration measures can be considered.

At present little is known of the dynamic properties of all types of structures, of a historical building or a monument. Not enough will be known of its materials, past history and present condition to justify a precise non linear analysis. The usual recourse, as past causes have indicated, will be to use static analysis. A dilemma arises here in that static analysis is usually recommended for regular structures, whereas dynamic analysis is used for structures with large torsional eccentricities, unusual mass or stiffness distributions, unusual foundation conditions, etc. , a perfect description of most structures of the type considered.

The analysis of a modern structure is usually based directly on some national code. The underlining philosophy of such codes is the prevention of major failure and loss of life. In a modern code, damage is acceptable as long as it does not lead to collapse. These codes provide the earthquake loading to be used for purposes of analysis.

The loading is usually given in terms of a base shear which is then distributed over the height in accordance with an assumed mode of response. The factors which usually contribute to the determination of the base shear are the well known factors as follows.

■   Acceleration factor - horizontal ground acceleration for a specific seismic zone which has been determined based on a chosen return period (usually 100 years).

■   Seismic response factor - reflects the dependence of the seismic acceleration on the fundamental period of the structure.

■ Structural factor - accounts for the energy - absorption capacity of the structure.

■ Foundation factor - accounts for the influence of the soil conditions at a site.

■ Importance Factor - reflects the importance of a structure.

The criteria of such codes of practice however, are not entirely applicable when it comes to the repair and strengthening of structures of cultural and historic interest. The primary concern is the limitation of future damage. On the other hand, Codes for the repair of old buildings would carry the danger of having more attention paid to the fulfillment of these codes than to the specific conditions of the individual building.

In front of such a demand, one is faced once again

with difficult decisions, this time in trying to adapt e. g. the seismic loading coefficients.

A typical example of problem which can be mentioned here, is the choice of the appropriate return period. This period should be considered in view of the objective to ensure the survival of the monumental structure under consideration for future generations, and considering the fact that the longer the return period, the larger the demands on the structure, and thus additional enhancements which are contrary to the principles of restoration.

Answers have to be given also to the questions how the fundamental period of the structure and its ductility can be determined, as well as what degree of importance is to be considered. This last point is very critical, as historical buildings are usually structures of high degree of importance, whereas higher importance results increased seismic forces.

Vertical acceleration is also another factor to be taken into consideration, as many times, especially for structures including arches, vaults, and domes, are proved very sensitive to this type of seismic movement.

6.2 Analysis of a historical building

For modern structures, cautious analysis following modern principles of design and analysis, up to a generally accepted admissible level of accuracy is always enough. But for the analysis of historical buildings, this is not enough. The question how cautious the analysis has to be, is equally important and has to be answered.

Taking structural decisions about repair and/or strengthening of a historical building, these have to be fully clarified and justified in advance, as there are quite a few peculiar aspects, which have to be taken into account, in comparison to a modern structure.

Historical buildings are complicated structures and there is an unfortunate lack of knowledge and information concerning the behaviour of such structures under seismic loads. What can be said about these structures is that they are typically more massive than todays structures and that they usually carry their actions primarily in compression.

In order to show the difficulties and the complexity of the problem, it should be noted here that most of these historical buildings were built with specific consideration given to their geometry and aesthetic quality and not to their structural integrity. In addition, the majority of these historical buildings existing today have been subjected to many additions, changes in style, devastations from fires, earthquakes, and wars, reconstructions and attempts of restorations and consolidations, all of which result in a structure which may have a passing resemblance only to the original.

The analysis of a historical building must be based on the structure as it is, which usually involves irregularities in both plan and elevation, and many damaged or even missing structural elements. The structural system will be a combination of walls, columns, beams, arches, vaults, and/or domes, constructed of various types of materials, subjected to various degrees of deterioration or damages. And it is very difficult, if not impossible, for the engineer to take into account the cracks and damages in a material, whose properties and strength he knows little about. A historical building is usually built with traditional natural materials collected in general in the area of the structure, with technic based on the experience, and without any special design. Its long time survival, when existing, is a very good probabilistic guarantee for its structural quality, tested already in the time.

Under these circumstances, modern methods of analysis, developed generally on abstract mathematical models, should be very carefully applied on historical (e.g. masonry) structures. For modern structures, with new industrial materials used (reinforced concrete, steel, etc), the development of a reliable mathematical model is usually possible, due to the fact that, materials and members characteristics are uniform and mostly explicitly known. Oppositely, for the case of masonry, and especially for the traditional plain one, it seems that there is a long research way to go, until reaching a similar level of confidence.

The analysis of a historical building, compared to the analysis of a modern structure, has to be taken into account in general with a low general validity, as it has to be considered depending m o r e, on the validity of the models used (material response models, structural response model, etc), and l e s s, on the choice of the proper calculation method. Analysis ought also to be regarded more, as showing a possible response of the structure under the idealized conditions chosen, and less, as the unique solution of the problem.

TABLE 4: ANALYSIS PROCESS

---

1. MODELING

- Structure
- Actions
- Materials

---

2. ANALYSIS

- According to type of external actions:
  - Static Analysis
  - Dynamic Analysis

- According to method of Analysis:
  - Elastic Analysis
  - Elastic-plastic Analysis

- According to type of model used:
  - 3-D (whole structure)
  - 2-D (separate walls)
  - 1-D (separate elements: piers etc)

---

3. RESULTS VERIFICATION

- Critical "translation" of analytical results
- Experimental cross-check (calibration)

---

4. DIMENSIONING
  - Allowable stress methods
  - Ultimate strength methods

Reliability of the analysis process becomes higher, if a parametric study of the problem is performed, related at least to the more uncertain parameters, and if envelop values for the design are used.

Analysis of a historical structure (e.g. masonry structure), has many similarities, as well as many differences, compared to the analysis of a modern structure (e.g. reinforced concrete shear wall structure).

Similarities refer generally to the general assumptions, and to the mathematical models used. Differences, refer to the materials properties, the structural system characteristics (e.g. monolithic connections), the distribution of inertial (e.g. earthquake) loads along the height of the building, etc. For an adequate modeling of a masonry building, proper modification of the mathematical models already used for R.C. shear wall buildings, as well as proper modification of models concerning connections, is required. As an example, the inadequacy of the lumped mass system model for earthquake resistant design is mentioned here, due to the relativelylarge masses of the vertical walls, compared to the masseson the horizontal (usually wooden) diaphragms in ahistorical building.

Speaking about analysis, two similar actions have obviously to be undertaken. The first, refers to the analysis of the existing structure, as it is today. The second, to the future structure, as it will become after its repair and/or strengthening. For both cases of course, analysis itself is the middle stage of the whole three - step process, between the initial step of conception - modeling, and the final step of evaluation of analysis results. The diagram of the whole process is shown on table IV.

The transition from the natural image of the real structure observed, to the mathematical solution of the analysis is a very difficult task. The requirements are always put on the real structure, to be finalized in the future, while the analysis is performed on the mathematical model, existing in advance. The real structure follows the (complicated) physical laws governing its response, but will not necessarily follow the mathematical laws of the model chosen. Oppositely, the model follows its mathematical laws, but not necessarily the physical laws of the real structure.

Under this complicated situation, the conception quality, that means the degree of success for the model selection, determines definitely the "distance" between the response of the model and the response of the real structure. And this distance c a n n o t be altered later, through laborious calculations on the model.

6.3 The earthquake effect

Special attention has always to be paid on the earthquake action, as for the case of a historical building, and especially for the common case of masonry historical buildings, built in an area of high seismicity, the problems to be answered by the engineer become more complex.

For the analysis of a historical building under earthquake loading, static action is generally recommended. Dynamic earthquake action is recommended for tall slender type of structures only (e.g. minarets, columns, etc).

For the analysis, the total seismic force can be taken according to the modern codes, equal to:

$$F = \beta(T) \cdot W$$

where, W is the weight of the total mass of the structure, T is the fundamental vibration period of the linear-elastic structure, and $\beta(T)$ the (normalized to the gravity acceleration g) design spectral value of acceleration. Values of $\beta(T)$ depend on the dynamic characteristics of both, ground motion and building, the ductility of the structure, the ground acceleration, the importance of the building, the soil - foundation conditions, etc. A minimum value, according to the draft of the new Eurocode 8, can be taken equal to 0.20a, where a is the pick ground acceleration.

Natural periods of masonry buildings, are usually very short. This means, that for economic design purposes, use of first linear spectral branch is n e c e s s a r y.

It has to be reminded that for masonry structures (especially for plain masonry structures) in seismic zones, earthquake is always the number one "enemy", due to the unpleasant coincidence of zone high seismicity and masonry very bad structural response to earthquakes.

In front of such demands, the engineer is faced once again with difficult decisions, this time in trying to adapt e.g. the seismic loading coefficients to a historical building. A typical example of problem which can be mention ned here, is the choice of the appropriate return period. This period should be considered in view of the objective to ensure the survival of the monument for future generations and considering the fact that the longer the return period, the larger the demands on the structure, and thus additional enhancements which are contradictory to the principles of restoration.

Answers have to be given also to the questions how the

fundamental period of the structure and its ductility can be determined, as well as what degree of importance is to be considered.

This last point is very critical, as monuments and historical buildings are usually structures of high degree of importance, whereas higher importance results in increased seismic forces.

Vertical acceleration is also another factor to be taken into consideration, as many monuments, especially those with arches, vaults and domes,are proved very sensitive to this seismic movement, a point which may be proved critical.

6.4   Final analysis

Finally, having accomplished the first analysis with due consideration given to all the factors which affect the performance of a structure, the engineer must decide on what interventions will be necessary to restore the monument. These interventions have to be properly designed, so that they do not modify significantly the existing structural system.

Such a modification is much undesirable for the case of inertial (earthquake) actions, as they will be also modified.

The structure must be reanalyzed taking into account all the proposals for repair and / or strengthening, since any alteration to the present state may alter the behaviour of the structure.

The analysis has to be extended to the parts of the structure where no intervention is foreseen. Such a check is necessary, due to the redistribution of actions after interventions.

Foundation of the structure has also to be included in the design.

A prerequisite to all these points, is of course the additional task of studying and assessing the historical substance before repair and restoration measures can be considered.

When deciding on the methods and techniques to be used for the repair and strengthening, the engineer must bear in mind the principles set forth, e. g. by the charter of Venice, even though these principles provide him with severe restrictions in his actions. According to these restrictions, the use of any modern technique is permitted only when its "efficiency has been shown by scientific data and proved by experience".

However, in the field "repair and strengthening of historical buildings and monuments" for earthquake resistance, such experience is not yet available. It is for this reason that the principle of reversibility is so important and must be enforced whenever possible. It is also therefore absolutely necessary that full and precise documentation of all stages of the work be kept, so that any work done may be undone with a minimum of damage to the original fabric.

All methods of intervention involve different degrees of interference with the existing fabric and offer different degrees of protection. Some of the problems associated with various facets of intervention are presented below. No attempt is made to present specific methods of intervention since such a discussion would be very lengthly and is beyond the scope of this text. It should be noted however, that the subject of intervention is perhaps the most sensitive and has the most dire need for further research and investigation.

7.1. Temporary shoring and strutting

A historical building or monument in need for strengthening, usually calls for some form of temporary shoring or strutting to reduce the risk of further damage before the final strengthening can be undertaken.

This final strengthening may be several years in coming and as such, the effect of the strutting and shoring on the dynamic response of the structure to some future shock, should also be taken into consideration.

All possible damage to the structure from the temporary supports must be foreseen and avoided.

7.2. Protection of Mosaics and Frescoes

The presence of frescoes and mosaics on the primary structural elements provides for additional complications which must overcome as their existence may prohibit the use of some forms of non destructive testing of materials as well as exclude the use of some strengthening and repairing techniques.

7.3. Choice of new materials

Besides the restrictions put to the introduction of new materials (e. g. chart of Venice, Article 12), according to which the engineer must match the appearance of the new material to the old, the properties also of strength, modulus of elasticity, permeability, thermal expansion, moisture movement, and chemical reactions have to be considered.

Failure to provide a good match may result in separation of the new material from the old or damage to the adjacent old material if it is weaker or more permeable.

There are numerous examples of past interventions which illustrate the potential dangers of such mismatches. One such example can be found in the Parthenon of Athens case, where the rusting of iron clamps embedded in the marble masonry during a past restoration, resulted in serious

The choice of materials, both nonmetallic and metallic, requires special consideration, and despite the many materials available today, the choice is a difficult one indeed.

7.4. Repair of cracks

The most common type of repair is the repair of cracks in masonry wall, domes, vaults, etc.

The techniques used vary according to the type of masonry, the width of the crack, its location, and the need of reestablishing a structural connection to carry tension. The technique most commonly used for the repair of cracks is grouting. However grouting is an irreversible process and the engineer must consider the fact that modern grouts may be proved stiffer and stronger, and can therefore have an undesirable effect.

The long term influence of various grouts on the

initial structure (mechanical, chemical, etc.), due to the relatively short time of experience on such techniques, is another unknown parameter.

7.5. Reconstruction

Reconstruction of parts of damaged structure and/or anastylosis of course is not only the engineer's decision, but it is his responsibility if it is to be carried out.

The rebuilding of parts of a wall or any other structural element, either because the blocks have fallen, or have eroded, or where previously removed, calls for the bonding in of new masonry and the warning of the choice of material is a factor here.

The engineer must also have a good understanding of how any reconstruction is going to effect the overall behaviour of the structure.

7.6. Overall Strengthening

Besides the repair and strengthening of the individual elements of a structure, the engineer must consider the overall seismic response and the rules of a good earthquake engineering practice, such as reduction of mass, improvement of the distribution of mass and stiffness, and improvement of structural interaction. In dealing with historical monuments, however, such practices, though desirable, may be impossible to implement.

8. SUPERVISION OF THE REPAIR WORK

After the analysis has been completed and the methods of intervention have been chosen, the work site must be organized.

This may involve an in depth analysis which considers the degree of safety and stability of the equipment, as well as the degree to which the appearence of the historical building and/or monument will be obscured by that equipment. In certain cases it may be necessary to permit access to the public while the work is in progress.

Finally, after the methods of intervention have been chosen and the work site has been organized comes the actual work on the structure. Here, new facts, which may upset all the carefully detailed planning, may come to light. In most of the renovations done up till today, especially for the case of a monumental structure, new facts concerning the history of the monument have been discovered.

Such discoveries include structural aspects as well as architectural and archaeological aspects. These discoveries result in delays in the repair schedule as well as the need for constant updating of the analysis of the structure and changes in the choice of techniques used.

This, plus the fact that many contractors are not specifically qualified for work on historical monuments, requires the constant monitoring of all reconstruction work by engineers familiar with all the steps of the work done till now.

9. CONCLUSIONS

The aseismic repair and strengthening of a historical building is indeed a difficult and challenging task.

The priceless value, very often, of such structures, and the desire to preserve them for future generations to enjoy, demands that they be protected against possible destruction in a future earthquake while at the same time requires that any intervention to accomplish this task, not harm what has survived till today.

To accomplish this task, a precise understanding of the problems facing the structure, the reasons for them, as well as a sound knowledge of the effect any intervention might have on the structure, are some of the prerequisites, ensuring that intervention does not become the cause of future damage.

Unfortunately, very little research has been conducted in this area, and very little has been written to aid in overcoming the many problems associated with the strengthening process. Studies are needed on the different varieties of structural elements and structural systems, to give a better understanding of the behaviour.

These might include experimental testing also.

Research on the materials of the past versus those available today, as well as on the techniques suitable for intervention, should be conducted.

Guidelines should also be provided, wherever possible, to aid the engineer in his task.

Finally objective critiques of the repair and strengthening work already completed should be made. These critiques should deal with specific aspects of the design work already done, and should perhaps provide comparison studies of the techniques used to accomplish the same goal.

Of course, the inherent structural complexity, the variability of building materials as well as the degree of damage and deterioration that the structure has sustained throughout its long existence, makes each case a unique case for study.

10. SELECTION OF REFERENCES FOR FURTHER STUDY

1. Giuffre A., "Principles for the Philological Restoration of Historical Buildings in Seismic Areas", Proc. Int. Techn. Conf. on Structural Conservation of Stone Masonry",Athens,1989.
2. Giuffre A., "Mechanics of Historical Masonry and Stre ngthening Criteria", Proc. XV Regional Seminar on Earthquake Engineering, Ravello, Italy, 1989.
3. Giuffre A., "Philological Restoration of Historical Monuments ; the Cathedral of Sant"Angelo dei Lombardi in Irpinia", Proc. Int. Techn. Conf. on Structural Conservation of Masonry Structures,Athens,1989.
4. Second International Meeting for the "Restoration of the Acropolis Monument", Athens, 1983 (Proc. by the Ministry of Culture and Sciences, Committee for the Preservation of the Acropolis Monuments,Athens,1985).
5. Syrmakezis C.A., "Seismic aspects of Preservation of Historical Monuments", General Report, 8th European Conference on Earthquake Engineering, Lisbon, September 1986.
6. Syrmakezis C.A., Giuffre A., Wenzel F., Yorulmaz M., "Seismic aspects of Preservation of Historical Monuments", General Report of the Working Group 7, 9th European Conference on Earthquake Engineering,Moscow, September 1990.
7. Third International Meeting for the "Restoration of the Acropolis Monument", Athens,1989 (Proceedings by the Ministry of Culture, (Committee for the preservation of the Acropolis monuments, Athens, 1990).
8. Tomazevic M., " Fundamentals of Earthquake Resistant Design of Masonry Structures", Proc. XV Regional Seminar

on Earthquake Engineering, Ravello, Italy, 1989.

9. United Development Organization project on Building Construction under Seismic conditions in the Balkan Region, Volume 6, "Repair and Strenghtening of Historical Monuments and Buildings in Urban Nuclei", UNIDO, Vienna, 1984.

10. Wenzel F., "The Civil Engineer in Monument Preservation Continuity and change of tasks", Proc. Int. Techn. Conf. on Structural Conservation of Stone Masonry", Athens,1989.

11. Syrmakezis CA, analysis of masonry buildings", report of the workshop on masonry, Istanbul, March 25 - 26, 1991.

12. Second International Meeting for the "Restoration of the Acropolis Monument, Athens, 1983 (Proc. by the Ministry of Culture and Sciences, Committee for the Preservation of the Acropolis Monuments, Athens, 1985).

13. Third International Meeting for the "Restoration of the Acropolis Monument", Athens, 1989 (Proceedings by the Ministry of Culture, Committee for the preservation of the Acropolis monuments, Athens, 1990).

14. Wenzel F, "The Civil Engineer in Monument Preservation Continuity and change of tasks", Proc. Int. Techn. Conf. on Structural Conservation of Stone Masonry", Athens, 1989.

15. Tassios TP, "Repair and Strengthening of Structures and Monuments against Earthquakes", Proc 7th ECEE, Vol 7, Athens 1982, pp 520-532.

16. Atlas of Isoseismal maps of italian earthquakes, CNR, Progetto Geodinamica, Bologna 1985.

17. Giuffre A., "La meccanica nell' Architettura", cap VI, Roma, 1987.

18. Como M. - Grimaldi A., "An unilateral model for the Limit Analysis of Masonry Walls", Conference on "Unilateral Problem in structural analysis", Ravello, 1985 -ed. Springer, 1985.

19. Feller A, "Engineering (final) report of the Augusta - Victoria Church", Jerusalem, 1987.

20. Wenzel F, Frese B, Vratsanou V, "Structural analysis of the Augusta - Victoria Church of the Ascention in Jerusalem", Proc. the Int. Techn. Conference on "Structural Conservation of Stone Masonry", Athens, 1989.

21. Yorulmaz M., Cili F., Ahunbay Z., "Structural conso-lidation of El Nazar Church, Cappadocia, Turkey", STREMA 89, Florence, 1989.

22. Anicic D., "Earthquake Protection of Monuments and old Urban Nuclei in Seismic Regions", Proceedings of 12th Regional Seminar on Earthquake Engineering, September 1985, Halkidiki, Greece.

23. De Angelis d"Ossat g, "Restoration of Monuments and

Intervention on Old Buildings", Proc IABSE Symposium on Strengthening of Building Structures, Venice, 1983, pp 3-11.
24. Ersoy U, "Diagnosis, Assessment and Emergency Interventions for Historic Masonry Structures", Proc. Int. Techn. Conf. on Structural Conservation of Stone Masonry, Athens, 1989.
25. Tomazevic M, "Redesign of Repaired and Strenghtened Structures, Research data", (General Report), Proc. Int. Techn. Conf. on Structural Conservation of Stone Masonry, Athens, 1989.
26. Penelis G., Papayianni J., Karaveziroglou M., "Pozzolanic Mortars for Repair of masonry Structures", STREMA 89, Florence, 1989.
27. Wenzel F., "On the State of Structural Repair of Masonry", Prcc. Int. Techn. Conf. on Structural Conservation of Masonry Structures", Athens, 1989.
28. Anicic D., "Earthquake Strengthening of Historical Monuments in Dubrovnik, Yugoslavia", Proc. 7th ecee, Athens 1982, pp 313-320.
29. Karaveziroglou M., Papayiani J., Penelis G., "Time dependent deformations of brick Masonry", Proc. Int. Techn. Conf. on Structural Conservation of Masonry Structures", Athens, 1989.

PROTECTION OF THE ARCHITECTURAL HERITAGE AGAINST EARTHQUAKES

CONCLUSIONS

M. Save

Polytechnical University of Mons, Mons, Belgium

ABSTRACT.

Balanced capabilities in rehabilitation and structural mechanics are absolutely necessary. Progress remains strongly needed, along the lines of the present course and also in new promising directions.

Conclusions written independently by a single person of a course given by a team of engineers and architects, on a subject with many aspects which may be non-rational or imperfectly defined, can only be tentative. Anyhow, some very apparent features of the problem can be emphasized :
- good familiarity with the theory and practice of protection and rehabilitation of old buildings is absolutely necessary to avoid that the strengthening process might result in partial or total destruction of the very qualities that should have been protected
- the degree of uncertainty and (or) unaccuracy of the data is particularly high, both concerning the actions (time of occurrence, type and magnitude of the earthquakes likely to happen, including the local influence of soil conditions) and the mechanical properties of the constructions (topology of the structure ; material properties, including degradations due to time action, either chemical or mechanical as uneven settlements, etc..)
- even when the degree of confidence in the data can be regarded as reasonably good, there is a strong lack of realistic and sufficiently easily applicable mechanical models enabling the evaluation of the response of the building to earthquake before and after a planned reinforcing process
- when the interest is on sets of usual, or even poor, old buildings as for instance in old small cities or villages, low-cost reinforcing, possibly made by the inhabitants themselves, may be the only possible solution.

From the remarks here above, it turns out clearly that **sound and simple notions of good <u>qualitative</u> actions are of prominent importance.**

Hence, a **necessary** condition to fulfil by the person responsible for intervention is to possess simultaneously a good training in rehabilitation and good basic knowledge (even without much mathematics) of the mechanics of structures and materials. Teaching such a balanced attitude was the goal of the present course. This "attitude" will generally be sufficient to achieve **improvement** in the strength of the construction to resist earthquake. However, the reinforcing might sometime prove to be quantitatively not large enough, especially when financial constraints are strong and no rule or calculation method exist to sustain the quantitative aspect of the intervention.

Hence, we may not satisfy ourselves with the present situation. Obviously, research along the lines described in the present course must be continued, but we must also be attentive to extract, from modern earthquake engineering and solid mechanics, ideas and techniques that could improve our capabilities in relation with the given problem. Realistic and suitable mechanical models and computing tool are one prominent question. We should be prepared to get acquainted with very general and basic theories as multibody dynamics [1] as well as with anchorage techniques [2] or isolation techniques [3], and judge to what respect they can be useful to our purposes. Obviously, continued and coordinated action is thus necessary.

REFERENCES.

1 : J.J. MOREAU, "Some numerical methods in multibody dynamics : application to granular materials", Invited lecture to the 2nd European Solid Mechanics Conference, Genova, Italy, Sept. 12 - 16, 1994, to appear in European Journal of Mechanics, A : Solids.

2 : P. HABIB and C. ROCH, "Soil dynamics : soil-structure interaction ; protection of existing buildings by anchorage", presented at the International Congress on Earthquake Engineering, Vienna, August 1994.

3 : M. FORNI and A. MARTELLI, "Design guide-lines for seismically isolated power plants", Draft final report, Contract ETNU- 0031 -IT (CCSH), European Atomic Energy Community, June 1994.